Algorithms and Computation in Mathematics • Volume 11

Editors

Manuel Bronstein   Arjeh M. Cohen
Henri Cohen   David Eisenbud
Bernd Sturmfels

Victor V. Prasolov

Polynomials

Translated from the Russian by Dimitry Leites

 Springer

Victor V. Prasolov

Independent University of Moscow
Department Mathematics
Bolshoy Vlasievskij per.11
119002 Moscow, Russia
e-mail: prasolov@mccme.ru

Dimitry Leites *(Translator)*

Stockholm University
Department of Mathematics
106 91 Stockholm, Sweden
e-mail: mleites@math.su.se

Originally published by MCCME
Moscow Center for Continuous Math. Education
in 2001 (Second Edition)

Mathematics Subject Classification (2000): 12-XX, 12E05

Library of Congress Control Number: 2009935697

ISSN 1431-1550
ISBN 978-3-540-40714-0 (hardcover)   e-ISBN 978-3-642-03980-5
ISBN 978-3-642-03979-9 (softcover)
DOI 10.1007/978-3-642-03980-5

Springer is a part of Springer Science+Business Media

springeronline.com

© Springer-Verlag Berlin Heidelberg 2004, First softcover printing 2010
Printed in Germany

Typeset by the translator.
Edited and reformatted by LE-TeX, Leipzig, using a Springer LaTeX macro package.
Cover design: *deblik, Berlin*

Printed on acid-free paper

Preface

The theory of polynomials constitutes an essential part of university courses of algebra and calculus. Nevertheless, there are very few books entirely devoted to this theory.[1] Though, after the first Russian edition of this book was printed, there appeared several books[2] devoted to particular aspects of the polynomial theory, they have almost no intersection with this book.

---

[1] The following classical references (not translated into Russian and therefore not mentioned in the Russian editions of this book) are rare exceptions:

Barbeau E. J., *Polynomials.* Corrected reprint of the 1989 original. Problem Books in Mathematics. Springer-Verlag, New York, 1995. xxii+455 pp.;

Borwein P., Erdélyi T., *Polynomials and polynomial inequalities.* Graduate Texts in Mathematics, 161. Springer-Verlag, New York, 1995. x+480 pp.;

Obreschkoff N., *Verteilung und Berechnung der Nullstellen reeller Polynome.* (German) VEB Deutscher Verlag der Wissenschaften, Berlin 1963. viii+298 pp.

[2] For example, some recent ones: Macdonald I. G., *Affine Hecke algebras and orthogonal polynomials.* Cambridge Tracts in Mathematics, 157. Cambridge University Press, Cambridge, 2003. x+175 pp.;

Phillips G. M., *Interpolation and approximation by polynomials.* CMS Books in Mathematics/Ouvrages de Mathématiques de la SMC, 14. Springer-Verlag, New York, 2003. xiv+312 pp.;

Mason J. C., Handscomb D. C. *Chebyshev polynomials.* Chapman & Hall/CRC, Boca Raton, FL, 2003. xiv+341 pp.;

Rahman Q. I., Schmeisser G., *Analytic theory of polynomials*, London Math. Soc. Monographs (N.S.) 26, 2002;

Sheil-Small T., *Complex polynomials*, Cambridge studies in adv. math. 75, 2002;

Lomont J. S., Brillhart J., *Elliptic polynomials.* Chapman & Hall/CRC, Boca Raton, FL, 2001. xxiv+289 pp.;

Krall A. M., *Hilbert space, boundary value problems and orthogonal polynomials.* Operator Theory: Advances and Applications, 133. Birkhäuser Verlag, Basel, 2002. xiv+352 pp.;

Dunkl Ch. F., Xu Yuan, *Orthogonal polynomials of several variables.* Encyclopedia of Mathematics and its Applications, 81. Cambridge University Press, Cambridge, 2001. xvi+390 pp. (*Hereafter the translator's footnotes.*)

This book contains an exposition of the main results in the theory of polynomials, both classical and modern. Considerable attention is given to Hilbert's 17th problem on the representation of non-negative polynomials by the sums of squares of rational functions and its generalizations. Galois theory is discussed primarily from the point of view of the theory of polynomials, not from that of the general theory of fields and their extensions. More precisely:

In Chapter 1 we discuss, mostly classical, theorems about the distribution of the roots of a polynomial and of its derivative. It is also shown how to determine the number of real roots to a real polynomial, and how to separate them.

Chapter 2 deals with irreducibility criterions for polynomials with integer coefficients, and with algorithms for factorization of such polynomials and for polynomials with coefficients in the integers mod p.

In Chapter 3 we introduce and study some special classes of polynomials: symmetric (polynomials which are invariant when the indeterminates are permuted), integer valued (polynomials which attain integer values at all integer points), cyclotomic (polynomials with all primitive nth roots of unity as roots), and some interesting classes introduced by Chebyshev, and by Bernoulli.

In Chapter 4 we collect a lot of scattered results on properties of polynomials. We discuss, e.g., how to construct polynomials with prescribed values in certain points (interpolation), how to represent a polynomial as a sum of powers of polynomials of degree one, and give a construction of numbers which are not roots of any polynomial with rational coefficients (transcendental numbers).

Chapter 5 is devoted to the classical Galois theory. It is well known that the roots of a polynomial equation of degree at most four in one variable can be expressed in terms of radicals of arithmetic expressions of its coefficients. A main application of Galois theory is that this is not possible in general for equations of degree five or higher.

In Chapter 6 three classical Hilbert's theorems are given: an ideal in a polynomial ring has a finite basis (Hilbert's basis theorem); if a polynomial f vanishes on all common zeros of $f_1, \ldots, f_r$, then some power of f is a linear combination (with polynomial coefficients) of $f_1, \ldots, f_r$ (Hilbert's Nullstellensatz); and if $M = \oplus M_i$ is a finitely generated module over a polynomial ring over K, then $\dim_K M_i$ is a polynomial in i for large i (the Hilbert polynomial of M).

Furthermore, the theory of Gröbner bases is introduced. Gröbner bases are a tool for calculations in polynomial rings. An application is that solving systems of polynomial equations in several variables with finitely many solutions can be reduced to solving polynomial equations in one variable.

In the final Chapter 7 considerable attention is given to Hilbert's 17th problem on the representation of non-negative polynomials as the sum of squares of rational functions, and to its generalizations. The Lenstra-Lenstra-Lovàsz algorithm for factorization of polynomials with integer coefficients is discussed in an appendix.

Two important results of the theory of polynomials whose exposition requires quite a lot of space did not enter the book: how to solve fifth degree equations by means of theta functions, and the classification of commuting polynomials. These results are expounded in detail in two recently published books in which I directly participated: [Pr3] and [Pr4].

During the work on this book I received financial support from the Russian Fund of Basic Research under Project No. 01-01-00660.

Acknowledgement. Together with the translator, I am thankful to Dr. Eastham for meticulous and friendly editing of the English and mathematics, to J. Borcea, R. Fröberg, B. Shapiro and V. Kostov for useful comments.

V. Prasolov

Moscow, May 1999

Notational conventions

As usual, $\mathbb{Z}$ denotes the set of all integers, $\mathbb{N}$ the subset of positive integers, $\mathbb{F}_p = \mathbb{Z}/p\mathbb{Z}$ for p prime.

$(\mathbb{Z}/n\mathbb{Z})^*$ denotes the set of invertible elements of $\mathbb{Z}/n\mathbb{Z}$.

$|S|$ denotes the cardinality of the set S.

$R[x]$ denotes the ring of polynomials in one indeterminate x with coefficients in a commutative ring R.

$[x]$ denotes the integer part of a given real number x, i.e., the greatest integer which is $\leq x$.

Numbering of Theorems, Lemmas and Examples is usually continuous throughout each section, e.g., reference to Lemma 2.3.2 means that the Lemma is to be found in subsection 2.3 inside the same chapter 2.

Subsections are numbered separately, so Theorem 2.3.4 may occure in subsec. 2.3.2.

Certain Lemmas and Examples (considered of local importance) are numbered simply Lemma 1, and so on, and, to find it, the page is indicated in the reference.

Contents

1

Roots of Polynomials

1.1 Inequalities for roots

1.1.1 The Fundamental Theorem of Algebra

In olden times, when algebraic theorems were scanty, the following statement received the title of the *Fundamental Theorem of Algebra*:

"A given polynomial of degree n with complex coefficients has exactly n roots (multiplicities counted)."

The first to formulate this statement was Alber de Girard in 1629, but he did not even try to prove it. The first to realize the necessity of proving the Fundamental Theorem of Algebra was d'Alembert. His proof (1746) was not, however, considered convincing. Euler (1749), Faunsenet (1759) and Lagrange (1771) offered their proofs but these proofs were not without blemishes, either.

The first to give a satisfactory proof of the Fundamental Theorem of Algebra was Gauss. He gave three different versions of the proof (1799, 1815 and 1816) and in 1845 he additionally published a refined version of his first proof.

For a review of the different proofs of the Fundamental Theorem of Algebra, see [Ti]. We confine ourselves to one proof. This proof is based on the following Rouché's theorem, which is of interest by itself.

Theorem 1.1.1 (Rouché). *Let f and g be polynomials, and γ a closed curve without self-intersections in the complex plane[1]. If*

$$|f(z) - g(z)| < |f(z)| + |g(z)| \tag{1}$$

for all $z \in \gamma$, then inside γ there is an equal number of roots of f and g (multiplicities counted).

---

[1] The plane $\mathbb{C}^1$ of complex variable.

V.V. Prasolov, *Polynomials*, Algorithms and Computation in Mathematics 11,
DOI 10.1007/978-3-642-03980-5_1, © Springer-Verlag Berlin Heidelberg 2004

Proof. In the complex plane, consider vector fields $v(z) = f(z)$ and $w(z) = g(z)$. From (1) it follows that at no point of γ are the vectors v and w directed opposite to each other. Recall that the *index* of the curve γ with respect to a vector field v is the number of revolutions of the vector $v(z)$ as it completely circumscribes the curve γ. (For a more detailed acquaintance with the properties of index we recommend Chapter 6 of [Pr2].) Consider the vector field

$$v_t = tv + (1-t)w.$$

Then $v_0 = w$ and $v_1 = v$. It is also clear that at every point $z \in \gamma$ the vector $v_t(z)$ is nonzero. This means that the index $\mathrm{ind}(t)$ of γ with respect to the vector field v_t is well defined. The integer $\mathrm{ind}(t)$ depends continuously on t, and hence $\mathrm{ind}(t) = \mathrm{const}$. In particular, the indices of γ with respect to the vector fields v and w coincide.

Let the *index of the singular point* z_0 be defined as the index of the curve $|z - z_0| = \varepsilon$, where ε is sufficiently small. It is not difficult to show that the index of γ with respect to a vector field v is equal to the sum of indices of *singular* points, i.e., those at which $v(z) = 0$. For the vector field $v(z) = f(z)$, the index of the singular point z_0 is equal to the multiplicity of the root z_0 of f. Therefore the coincidence of the indices of γ with respect to vector fields $v(z) = f(z)$ and $w(z) = g(z)$ implies that, inside γ, the number of roots of f is equal to that of g. $\square$

With the help of Rouché's theorem it is not only possible to prove the Fundamental Theorem of Algebra but also to estimate the absolute value of any root of the polynomial in question.

Theorem 1.1.2. *Let* $f(z) = z^n + a_1 z^{n-1} + \cdots + a_n$, *where* $a_i \in \mathbb{C}$. *Then, inside the circle* $|z| = 1 + \max_i |a_i|$, *there are exactly* n *roots of* f *(multiplicities counted).*

Proof. Let $a = \max_i |a_i|$. Inside the circle considered, the polynomial $g(z) = z^n$ has root 0 of multiplicity n. Therefore it suffices to verify that, if $|z| = 1 + a$, then $|f(z) - g(z)| < |f(z)| + |g(z)|$. We will prove even that $|f(z) - g(z)| < |g(z)|$, i.e.,

$$|a_1 z^{n-1} + \cdots + a_n| < |z|^n.$$

Clearly, if $|z| = 1 + a$, then

$$|a_1 z^{n-1} + \cdots + a_n| \le a\left(|z|^{n-1} + \cdots + 1\right) = a\frac{|z|^n - 1}{|z| - 1} = |z|^n - 1 < |z|^n. \quad \square$$

1.1.2 Cauchy's theorem

Here we discuss Cauchy's theorem on the roots of polynomials as well as its corollaries and generalizations.

Theorem 1.1.3 (Cauchy). *Let* $f(x) = x^n - b_1 x^{n-1} - \cdots - b_n$, *where all the numbers* b_i *are non-negative and at least one of them is nonzero. The polynomial* f *has a unique (simple) positive root* p *and the absolute values of the other roots do not exceed* p.

Proof. Set

$$F(x) = -\frac{f(x)}{x^n} = \frac{b_1}{x} + \cdots + \frac{b_n}{x^n} - 1.$$

If $x \neq 0$, the equation $f(x) = 0$ is equivalent to the equation $F(x) = 0$. As x grows from 0 to $+\infty$ the function $F(x)$ strictly decreases from $+\infty$ to -1. Therefore, for $x > 0$, the function F vanishes at precisely one point, p. We have

$$-\frac{f'(p)}{p^n} = F'(p) = -\frac{b_1}{p^2} - \cdots - \frac{nb_n}{p^{n+1}} < 0.$$

Hence p is a simple root of f.

It remains to prove that if x_0 is a root of f, then $q = |x_0| \leq p$. Suppose that $q > p$. Then, since F is monotonic, it follows that $q > p$, i.e., $f(q) > 0$. On the other hand, the equality $x_0^n = b_1 x_0^{n-1} + \cdots + b_n$ implies that

$$q^n \leq b_1 q^{n-1} + \cdots + b_n,$$

i.e., $f(q) \leq 0$, which is a contradiction. $\square$

Remark. Cauchy's theorem is directly related to the Perron-Frobenius theorem on non-negative matrices (cf. [Wi1]).

The polynomial $x^{2n} - x^n - 1$ has n roots whose absolute values are equal to the value of the positive root of this polynomial. Therefore, in Cauchy's theorem, the estimate

the absolute values of the roots are $\leq p$

cannot, in general, be replaced by the estimate

the absolute values of the roots are $< p$.

Ostrovsky showed, nevertheless, that in a sufficiently general situation such a replacement is possible.

Theorem 1.1.4 (Ostrovsky). *Let* $f(x) = x^n - b_1 x^{n-1} - \cdots - b_n$, *where all the numbers* b_i *are non-negative and at least one of them is nonzero.*

If the greatest common divisor of the indices of the positive coefficients b_i *is equal to 1, then* f *has a unique positive root* p *and the absolute values of the other roots are* $< p$.

Proof. Let only the coefficients $b_{k_1}, b_{k_2}, \ldots, b_{k_m}$, where $k_1 < k_2 < \cdots < k_m$, be positive. Since the greatest common divisor of $k_1, \ldots, k_m$ is equal to 1, there exist integers $s_1, \ldots, s_m$ such that $s_1 k_1 + \cdots + s_m k_m = 1$. Consider again the function

$$F(x) = \frac{b_{k_1}}{x^{k_1}} + \cdots + \frac{b_{k_m}}{x^{k_m}} - 1.$$

The equation $F(x) = 0$ has a unique positive solution p. Let x be any other (nonzero) root of f. Set $q = |x|$. Then

$$1 = \frac{b_{k_1}}{x^{k_1}} + \cdots + \frac{b_{k_m}}{x^{k_m}} \leq \left| \frac{b_{k_1}}{x^{k_1}} \right| + \cdots + \left| \frac{b_{k_m}}{x^{k_m}} \right| = \frac{b_{k_1}}{q^{k_1}} + \cdots + \frac{b_{k_m}}{q^{k_m}},$$

i.e., $F(q) \geq 0$. We see that the equality $F(q) = 0$ is only possible if

$$\frac{b_{k_i}}{x^{k_i}} = \left| \frac{b_{k_i}}{x^{k_i}} \right| > 0 \text{ for all } i.$$

But in this case

$$\frac{b_{k_1}^{s_1} \cdot \cdots \cdot b_{k_m}^{s_m}}{x} = \left(\frac{b_{k_1}}{x^{k_1}} \right)^{s_1} \cdot \cdots \cdot \left(\frac{b_{k_m}}{x^{k_m}} \right)^{s_m} > 0,$$

i.e., $x > 0$. This contradicts the fact that $x \neq p$ and p is the only positive root of the equation $F(x) = 0$. Thus $F(q) > 0$. Therefore, since $F(x)$ is monotonic for positive x, it follows that $q < p$. $\square$

The Cauchy-Ostrovsky theorem implies the following estimate of the absolute value of the roots of polynomials with positive coefficients.

Theorem 1.1.5. a) (Eneström-Kakeya) *If all the coefficients of the polynomial $g(x) = a_0 x^{n-1} + \cdots + a_{n-1}$ are positive, then, for any root ξ of this polynomial, we have*

$$\min_{1 \leq i \leq n-1} \left\{ \frac{a_i}{a_{i-1}} \right\} = \delta \leq |\xi| \leq \gamma = \max_{1 \leq i \leq n-1} \left\{ \frac{a_i}{a_{i-1}} \right\}.$$

b) (Ostrovsky) *Let $\frac{a_k}{a_{k-1}} < \gamma$ for $k = k_1, \ldots, k_m$. If the greatest common divisor of the numbers $n, k_1, \ldots, k_m$ is equal to 1, then $|\xi| < \gamma$.*

Proof. Consider the polynomial

$$(x - \gamma)g(x) = a_0 x^n - (\gamma a_0 - a_1)x^{n-1} - \cdots - (\gamma a_{n-2} - a_{n-1})x - \gamma a_{n-1}.$$

By definition, $\gamma \geq \frac{a_i}{a_{i-1}}$, i.e., $\gamma a_{i-1} - a_i \geq 0$. Therefore, by Cauchy's theorem, γ is the only positive root of the polynomial $(x - \gamma)g(x)$ and the absolute values of the other roots of this polynomial are $\leq \gamma$.

If ξ is a root of g, then $\eta = \dfrac{1}{|\xi|}$ is a root of $a_{n-1}y^{n-1} + \cdots + a_0$. Hence

$$\frac{1}{|\xi|} = |\eta| = \max_{1 \le i \le n-1}\left\{\frac{a_{i-1}}{a_i}\right\} = \frac{1}{\displaystyle\min_{1 \le i \le n-1}\left\{\frac{a_i}{a_{i-1}}\right\}},$$

i.e.,

$$|\xi| \ge \delta = \min_{1 \le i \le n-1}\left\{\frac{a_i}{a_{i-1}}\right\}.$$

If condition b) is satisfied, the root γ of the polynomial $(x - \gamma)g(x)$ is strictly greater than the absolute values of the other roots of this polynomial. □

Remark. The Eneström-Kakeya theorem is also related to the Perron-Frobenius theorem, cf. [An2].

An essential generalization of the Eneström-Kakeya theorem is obtained in [Ga1]. However, the formulation of this generalization is rather cumbersome, and therefore we do not give it here.

1.1.3 Laguerre's theorem

Let $z_1, \ldots, z_n \in \mathbb{C}$ be points of unit mass. The point $\zeta = \frac{1}{n}(z_1 + \cdots + z_n)$ is called the *center of mass* of $z_1, \ldots, z_n$.

This notion can be generalized as follows. Perform a fractional-linear transformation w that sends z_0 to ∞, i.e.,

$$w(z) = \frac{a}{z - z_0} + b.$$

Let us find the center of mass of the images of $z_1, \ldots, z_n$ and then apply the inverse transformation w^{-1}. Simple calculations show that the result does not depend on a and b, namely, we obtain the point

$$\zeta_{z_0} = z_0 + n\frac{1}{\frac{1}{z_1 - z_0} + \cdots + \frac{1}{z_n - z_0}} \tag{1}$$

which is called the *center of mass of $z_1, \ldots, z_n$ with respect to z_0.*

Clearly,

the center of mass of $z_1, \ldots, z_n$ lies inside their convex hull.

This statement easily generalizes to the case of the center of mass with respect to z_0. One only has to replace the lines that connect the points z_i and z_j by circles passing through z_i, z_j and z_0. The point z_0 corresponding to ∞ lies outside the convex hull.

Theorem 1.1.6. *Let* $f(z) = (z - z_1) \cdot \ldots \cdot (z - z_n)$. *Then the center of mass of the roots of* f *with respect to an arbitrary point* z *is given by the formula*

$$\zeta_z = z - n \frac{f(z)}{f'(z)}.$$

Proof. Clearly

$$\frac{f'(z)}{f(z)} = \frac{1}{z - z_1} + \cdots + \frac{1}{z - z_n}.$$

The desired statement follows directly from formula (1). $\square$

Theorem 1.1.7 (Laguerre). *Let* $f(z)$ *be a polynomial of degree* n *and* x *its simple root. Then the center of mass of all the other roots of* $f(z)$ *with respect to* x *is the point*

$$X = x - 2(n - 1) \frac{f'(x)}{f''(x)}.$$

Proof. Let $f(z) = (z - x)F(z)$. Then $f'(z) = F(z) + (z - x)F'(z)$ and $f''(z) = 2F'(z) + (z - x)F''(z)$. Therefore $f'(x) = F(x)$ and $f''(x) = 2F'(x)$. Applying the preceding theorem to the polynomial F of degree $n - 1$, and point $z = x$, we obtain the desired statement. $\square$

Theorem 1.1.8 (Laguerre). *Let* $f(z)$ *be a polynomial of degree* n *and*

$$X(z) = z - 2(n - 1) \frac{f'(z)}{f''(z)}.$$

Let the circle (or line) C *pass through a simple root* z_1 *of* f *and the other roots of* f *belong to one of the two domains into which* C *divides the plane. Then* $X(z_1)$ *also belongs to the same domain.*

Proof. In the case of the "usual" center of mass, the circle C corresponds to the line such that all the roots of $f(z)$, except z_1, lie on one side of it. The center of mass of these roots lies on the same side of this line. $\square$

Corollary. *Let* z_1 *be one of the simple roots of* f *with the maximal absolute value. Then* $|X(z_1)| \le |z_1|$, *i.e.,*

$$\left| z_1 - 2(n - 1) \frac{f'(z_1)}{f''(z_1)} \right| \le |z_1|.$$

Proof. All the roots of f lie in the disk $\{z \in \mathbb{C} \,|\, |z| \le |z_1|\}$, and therefore $X(z_1)$ also belongs to this disk. $\square$

Theorem 1.1.9. *Let* f *be a polynomial with real coefficients and define*

$$\zeta_z = z - n \frac{f(z)}{f'(z)}.$$

All the roots of f *are real if and only if* $\operatorname{Im} z \cdot \operatorname{Im} \zeta_z < 0$ *for any* $z \in \mathbb{C} \setminus \mathbb{R}$.

Proof. Suppose first that all the roots of f are real. Let $\operatorname{Im} z = a > 0$. The line consisting of the points with the imaginary part ε, where $0 < \varepsilon < a$, separates the point z from all the roots of f since they belong to the real axis. Therefore $\operatorname{Im} \zeta_z \leq \varepsilon$. In the limit as $\varepsilon \to 0$, we obtain $\operatorname{Im} \zeta_z \leq 0$.

It is easy to verify that it is impossible to have $\operatorname{Im} \zeta_z = 0$. Indeed, let $\zeta_z \in \mathbb{R}$. Consider a circle passing through z and tangent to the real axis at ζ_z. Slightly jiggling this circle we can construct a circle on one side of which lie the points z and ζ_z, and on the other side lie all the roots of f. If $\operatorname{Im} z = a < 0$, the arguments are similar.

Now suppose that $\operatorname{Im} z \cdot \operatorname{Im} \zeta_z < 0$ for all $z \in \mathbb{C} \setminus \mathbb{R}$. Let z_1 be a root of f such that $\operatorname{Im}(z_1) \neq 0$. Then $\lim\limits_{z \to z_1} \zeta_z = z_1$, and therefore $\operatorname{Im} z_1 \cdot \operatorname{Im} \zeta_{z_1} > 0$. $\square$

Our presentation of Laguerre's theory is based on the paper [Gr], see also [Pol].

1.1.4 Apolar polynomials

Let $f(z)$ be a polynomial of degree n and ζ a fixed number or ∞. The function

$$A_\zeta f(z) = \begin{cases} (\zeta - z)f'(z) + nf(z) & \text{if } \zeta \neq \infty; \\ f'(z) & \text{if } \zeta = \infty \end{cases}$$

is called *the derivative of $f(z)$ with respect to point ζ*. It is easy to verify that, if

$$f(z) = \sum_{k=0}^{n} \binom{n}{k} a_k z^k, \tag{1}$$

then

$$\frac{1}{n} f'(z) = \sum_{k=0}^{n-1} \binom{n-1}{k} a_{k+1} z^k. \tag{*}$$

Therefore

$$\frac{1}{n} A_\zeta f(z) = \sum_{k=0}^{n-1} \binom{n-1}{k} (a_k + a_{k+1}\zeta) z^k. \tag{2}$$

Let $z_1, \ldots, z_n$ be the roots of the polynomial (1), and let $\zeta_1, \ldots, \zeta_n$ be the roots of the polynomial

$$g(z) = \sum_{k=0}^{n} \binom{n}{k} b_k z^k. \tag{3}$$

Formula (2) implies that

$$\frac{1}{n!} A_{\zeta_1} A_{\zeta_2} \cdots A_{\zeta_n} f(z) = a_0 + a_1 \sigma_1 + a_2 \sigma_2 + \cdots + a_n \sigma_n,$$

where

$$\sigma_1 = \zeta_1 + \zeta_2 + \cdots + \zeta_n = -\binom{n}{1}\frac{b_{n-1}}{b_n},$$

$$\sigma_2 = \zeta_1\zeta_2 + \cdots + \zeta_{n-1}\zeta_n = \binom{n}{2}\frac{b_{n-2}}{b_n},$$

$$\cdots\cdots\cdots\cdots\cdots\cdots$$

$$\sigma_n = \zeta_1 \cdot \ldots \cdot \zeta_n = (-1)^n \frac{b_0}{b_n}.$$

Hence the equality $A_{\zeta_1} A_{\zeta_2} \cdots A_{\zeta_n} f(z) = 0$ is equivalent to the equality

$$a_0 b_n - \binom{n}{1}a_1 b_{n-1} + \binom{n}{2}a_2 b_{n-2} + \cdots + (-1)^n a_n b_0 = 0. \tag{4}$$

The polynomials f and g given by (1) and (3) and whose coefficients are related via (4) are said to be *apolar*.

A *circular domain* is either the inner or the exterior part of a disk or the half plane.

Theorem 1.1.10 (J. H. Grace, 1902). *Let f and g be apolar polynomials. If all the roots of f belong to a circular domain K, then at least one of the roots of g also belongs to K.*

Proof. We will need the following auxiliary statement.

Lemma 1.1.11. *Let all the roots $z_1, \ldots, z_n$ of $f(z)$ lie inside the circular domain K and let ζ lie outside K. Then all the roots of $A_\zeta f(z)$ lie inside K.*

Proof. Observe first that, if w_i is a root of the polynomial $A_\zeta f(z)$, then ζ is the center of mass of the roots of $f(z)$ with respect to w_i. Indeed, if $\zeta \neq \infty$, then we can express the equality $A_\zeta f(w_i) = 0$ in the form

$$(\zeta - w_i)f'(w_i) + nf(w_i) = 0, \quad \text{i.e.,} \quad \zeta = w_i - n\frac{f(w_i)}{f'(w_i)}.$$

If $\zeta = \infty$, then $f'(w_i) = A_\zeta f(w_i) = 0$, and hence

$$\sum_{j=1}^{n} \frac{1}{z_j - w_i} = \frac{f'(w_i)}{f(w_i)} = 0.$$

Therefore the center of mass of the points $z_1, \ldots, z_n$ with respect to w_i is situated at

$$w_i + \frac{1}{\sum_j \frac{1}{z_j - w_i}} = \infty.$$

Now it is clear that point w_i cannot lie outside K. Indeed, if w_i were situated outside K, then the center of mass of $z_1, \ldots, z_n$ with respect to w_i would be inside K. However, this contradicts the fact that ζ lies outside K. $\square$

With the help of Lemma 1.1.11, Theorem 1.1.10 is proved as follows. Suppose that all the roots $\zeta_1, \ldots, \zeta_n$ of g lie outside K. Consider the polynomial $A_{\zeta_2} \cdots A_{\zeta_n} f(z)$. Its degree is equal to 1, i.e., it is of the form $c(z - k)$. Lemma 1.1.11 implies that $k \in K$. Since f and g are apolar polynomials, it follows that $A_{\zeta_1}(z - k) = 0$. On the other hand, the direct calculation of the derivative shows that $A_{\zeta_1}(z - k) = \zeta_1 - k$. Therefore $k = \zeta_1 \notin K$ and we have a contradiction. $\square$

Every polynomial f has a whole family of polynomials apolar to it. Having selected a convenient apolar polynomial we can, thanks to Grace's theorem, prove that f possesses a root in a given circular domain. Sometimes for the same goal it is convenient to use Lemma 1.1.11 directly.

Example 1. The polynomial

$$f(z) = 1 - z + cz^n, \quad \text{where} \quad c \in \mathbb{C},$$

possesses a root in the disk $|z - 1| \le 1$.

Proof. The polynomials

$$f(z) = 1 + \binom{n}{1} \frac{-1}{n} z + cz^n \text{ and } g(z) = z^n + \binom{n}{1} b_{n-1} z^{n-1} + \cdots + b_0$$

are apolar if

$$1 - n \left(\frac{-1}{n} \right) b_{n-1} + cb_0 = 0, \quad \text{i.e.,} \quad 1 + b_{n-1} + cb_0 = 0.$$

Now let $\zeta_k = 1 - \exp(2\pi i k/n)$ for $k = 1, \ldots, n$, and take $g(z)$ to be

$$g(z) = \prod (z - \zeta_k) = z^n + \binom{n}{1} b_{n-1} z^{n-1} + \cdots + b_0.$$

Then

$$b_{n-1} = -1 \text{ and } b_0 = \pm \prod \zeta_k = 0.$$

Therefore the polynomials $f(z)$ and $g(z)$ are apolar. Since all the roots of g lie in the disk $|z - 1| \le 1$, at least one of the roots of f lies in this disk. $\square$

Example 2. The polynomial $1 - z + c_1 z^{n_1} + \cdots + c_k z^{n_k}$, where $1 < n_1 < n_2 < \cdots < n_k$, has at least one root in the disk

$$|z| \le \frac{1}{\left(1 - \frac{1}{n_1} \right) \cdot \ldots \cdot \left(1 - \frac{1}{n_k} \right)}.$$

Proof. Let us start with the polynomial $f(z) = 1 - z + c_1 z^{n_1}$. Suppose on the contrary that all its roots lie in the domain $|z| > \frac{n_1}{n_1-1}$. Then by Lemma 1.1.11 the roots of the polynomial

$$A_0 f(z) = n_1 - (n_1 - 1)z$$

also lie in the domain $|z| > \frac{n_1}{n_1-1}$. But the root of $A_0 f(z)$ is equal to $\frac{n_1}{n_1-1}$ and we have a contradiction.

For the polynomial $f(z) = 1 - z + c_1 z^{n_1} + \cdots + c_k z^{n_k}$, we use induction on k. Consider the polynomial

$$A_0 f(z) = n_k - (n_k - 1)z + c_1(n_k - n_1)z^{n_1} + \cdots + c_{k-1}(n_k - n_{k-1})z^{n_{k-1}}.$$

In this polynomial, replace z by $\frac{n_k}{n_k-1}w$. By the induction hypothesis, the roots of the polynomial obtained lie in the disk

$$|w| \leq \frac{n_1}{n_1 - 1} \cdot \frac{n_2}{n_2 - 1} \cdot \ldots \cdot \frac{n_{k-1}}{n_{k-1} - 1},$$

and hence the roots of $A_0 f(z)$ lie in the disk

$$|z| \leq \frac{n_1}{n_1 - 1} \cdot \frac{n_2}{n_2 - 1} \cdot \ldots \cdot \frac{n_k}{n_k - 1}.$$

Therefore the hypothesis that all the roots of $f(z)$ lie outside the disk leads to a contradiction. $\square$

Let $f(z) = \sum_{i=1}^{n} \binom{n}{i} a_i z^i$ and $g(z) = \sum_{i=1}^{n} \binom{n}{i} b_i z^i$. The polynomial

$$h(z) = \sum_{i=1}^{n} \binom{n}{i} a_i b_i z^i$$

is called the *composition* of f and g.

Theorem 1.1.12 (Szegö). *Let f and g be polynomials of degree n, and let all the roots of f lie in a circular domain K. Then every root of the composition h of f and g is of the form $-\zeta_i k$, where ζ_i is a root of g and $k \in K$.*

Proof. Let γ be a root of h, i.e., $\sum_{i=1}^{n} \binom{n}{i} a_i b_i \gamma^i = 0$. Then the polynomials $f(z)$ and $G(z) = z^n g(-\gamma z^{-1})$ are apolar. Therefore, by Grace's theorem, one of the roots of $G(z)$ lies in K. Let, for example, $g(-\gamma k^{-1}) = 0$, where $k \in K$. Then $-\gamma k^{-1} = \zeta_i$, where ζ_i is a root of g. $\square$

For polynomials whose degrees are not necessarily equal, there is the following analogue of Grace's theorem.

Theorem 1.1.13 ([Az]). *Let* $f(z) = \sum_{i=1}^{n} \binom{n}{i} a_i z^i$ *and* $g(z) = \sum_{i=1}^{m} \binom{m}{i} b_i z^i$ *be polynomials with* $m \leq n$. *Let the coefficients of* f *and* g *be related as follows:*

$$\binom{m}{0} a_0 b_m - \binom{m}{1} a_1 b_{m-1} + \cdots + (-1)^m \binom{m}{m} a_m b_0 = 0. \tag{5}$$

Then the following statements hold:

a) *If all the roots of* $g(z)$ *belong to the disk* $|z| \leq r$, *then at least one of the roots of* $f(z)$ *also belongs to this disk;*

b) *If all the roots of* $f(z)$ *lie outside the disk* $|z| \leq r$, *then at least one of the roots of* $g(z)$ *also lies outside this disk.*

Proof. [Ru] a) Relation (5) is invariant with respect to the change of z to rz in f and g, and therefore we may assume that $r = 1$. Suppose on the contrary that all the roots of $f(z)$ lie in the domain $|z| > 1$. Then all the roots of the polynomial $z^n f(\frac{1}{z})$ lie in the domain $|z| < 1$. Therefore, from the Gauss-Lucas theorem (Theorem 1.2.1 on p. 13), it follows that all the roots of the polynomial

$$f_1(z) = D^{(n-m)} \left(z^n f\left(\frac{1}{z}\right) \right) = n(n-1) \cdot \ldots \cdot (m+1) \sum_{i=0}^{m} \binom{m}{i} a_i z^{m-i}$$

lie in the domain $|z| < 1$. Therefore all the roots of the polynomial

$$f_2(z) = z^m \sum_{i=0}^{m} \binom{m}{i} a_i \left(\frac{1}{z}\right)^{m-i} = \sum_{i=0}^{m} \binom{m}{i} a_i z^i$$

lie in the domain $|z| > 1$.

Relation (5) means that the polynomials f_2 and g are apolar. Since all the roots of f_2 lie in the circular domain $|z| > 1$, it follows from Grace's theorem that at least one of the roots of g also lies in this domain, and we have a contradiction.

b) All the roots of f_2 lie in the domain $|z| \geq 1$, hence, it follows from Grace's theorem that at least one of the roots of g also lies in this domain. □

1.1.5 The Routh-Hurwitz problem

In various problems on stability one has to investigate whether all the roots of a given polynomial belong to the left half-plane (i.e., whether the real parts of the roots are negative). The polynomials with this property are said to be *stable*. The Routh-Hurwitz problem is

how to find out directly by looking at the coefficients of the polynomial whether it is stable or not.

Several different solutions of the problem are known (see, e.g., [Po2]). We will confine ourselves with one simple criterion given in [St3].

First, we observe that it suffices to consider the case of polynomials with real coefficients. Indeed, if $p(z) = \sum a_n z^n$ is a polynomial with complex coefficients we can consider the polynomial

$$p^*(z) = p(z)\overline{p(\overline{z})} = \left(\sum a_n z^n\right)\left(\sum \overline{a_n} z^n\right).$$

Clearly, the real parts of the roots of $\overline{p(\overline{z})}$ are the same as those of $p(z)$. Moreover, the coefficients of $p^*(z)$ are symmetric with respect to a_n and $\overline{a_n}$. This means that the coefficients of p^* are invariant under conjugation, that is, they are real.

Theorem 1.1.14. *Let* $p(z) = z^n + a_1 z^{n-1} + \cdots + a_n$ *be a polynomial with real coefficients; let* $q(z) = z^m + b_1 z^{m-1} + \cdots + b_m$, *where* $m = \frac{1}{2}n(n-1)$, *be the polynomial whose roots are all the sums of pairs of the roots of* p. *The polynomial* p *is stable if and only if all the coefficients of the polynomials* p *and* q *are positive.*

Proof. Suppose that p is stable. To a negative root α of p there corresponds the factor $z - \alpha$ with positive coefficients. To a pair of conjugate roots with the negative real part there corresponds the factor

$$(z - \alpha - i\beta)(z - \alpha + i\beta) = z^2 - 2\alpha z + \alpha^2 + \beta^2$$

with positive coefficients. Thus all the coefficients of p are positive.

The complex roots of q fall into the pairs of conjugate roots because the coefficients of q are real. Further, the real parts of all the roots of q are negative. The same arguments as for p show that all the coefficients of q are positive.

Next, let all the coefficients of p and q be positive. In this case, all the real roots of p and q are negative. Therefore, if α is a real root of p, then $\alpha < 0$, and, if $\alpha \pm i\beta$ is a pair of complex conjugate roots of p, then $2\alpha = (\alpha + i\beta) + (\alpha - i\beta)$ is a root of q; hence $2\alpha < 0$. $\square$

1.2 The roots of a given polynomial and of its derivative

1.2.1 The Gauss-Lucas theorem

In 1836, Gauss showed that all the roots of P', distinct from the multiple roots of the polynomial P itself, serve as the points of equilibrium for the field of forces created by identical particles placed at the roots of P (provided that r particles are located at the root of multiplicity r) if each particle creates an attractive force inversely proportional to the distance to this particle. From this theorem of Gauss it is easy to deduce Theorem 1.2.1 given below. Gauss himself did not mention this. The first to formulate and prove Theorem 1.2.1 was a French engineer F. Lucas in 1874. Therefore Theorem 1.2.1 is often referred to as the *Gauss-Lucas theorem*.

Theorem 1.2.1 (Gauss-Lucas). *The roots of P' belong to the convex hull of the roots of the polynomial P itself.*

Proof. Let $P(z) = (z - z_1) \cdot \ldots \cdot (z - z_n)$. It is easy to verify that

$$\frac{P'(z)}{P(z)} = \frac{1}{z - z_1} + \cdots + \frac{1}{z - z_n}. \tag{1}$$

Suppose that $P'(w) = 0$, $P(w) \neq 0$ and suppose on the contrary that w does not belong to the convex hull of the points $z_1, \ldots, z_n$. Then one can draw a line through w that does not intersect the convex hull of $z_1, \ldots, z_n$. Therefore the vectors $w - z_1, \ldots, w - z_n$ lie in one half-plane determined by this line. Hence the vectors

$$\frac{1}{w - z_1}, \ldots, \frac{1}{w - z_n}$$

also lie in one half-plane, since $\dfrac{1}{z} = \dfrac{\bar{z}}{|z|^2}$. Hence,

$$\frac{P'(w)}{P(w)} = \frac{1}{w - z_1} + \cdots + \frac{1}{w - z_n} \neq 0.$$

This is a contradiction, and hence w belongs to the convex hull of the roots of P. $\square$

Relation (1) allows one to prove the following properties of the roots of P' for any polynomial P with real roots.

Theorem 1.2.2 ([An1]). *Let*

$$P(z) = (z - x_1) \cdot \ldots \cdot (z - x_n), \text{ where } x_1 < \cdots < x_n.$$

If some root x_i is replaced by $x_i' \in (x_i, x_{i+1})$, then all the roots of P' increase their value.

Proof. Let $z_1 < z_2 < \cdots < z_{n-1}$ be the roots of P', and let $x_1, \ldots, x_n$ be the roots of P. Let $z_1' < z_2' < \cdots < z_{n-1}'$ be the roots of Q' and let $x_1' = x_1, \ldots, x_{i-1}' = x_{i-1}, x_i', x_{i+1}' = x_{i+1}, \ldots, x_n' = x_n$ be the roots of Q. For the roots z_k and z_k', the relation (1) takes the form

$$\sum_{i=1}^{n} \frac{1}{z_k - x_i} = 0, \quad \sum_{i=1}^{n} \frac{1}{z_k' - x_i'} = 0. \tag{2}$$

Suppose that the statement of the theorem is false, i.e., $z_k' < z_k$ for some k. Then $z_k' - x_i' < z_k - x_i$. Observe that the differences $z_k' - x_i'$ and $z_k - x_i$ are of the same sign. Indeed,

$$z_j < x_i, \ z_j' < x_i' \text{ for } j \le i - 1 \text{ and } z_j > x_i, \ z_j' > x_i' \text{ for } j \ge i.$$

Hence, $\dfrac{1}{z_k - x_i} < \dfrac{1}{z_k' - x_i'}$ for all $i = 1, \ldots, n$. But in this case relations (2) cannot hold simultaneously. $\square$

1.2.2 The roots of the derivative and the focal points of an ellipse

The roots of the derivative of a cubic polynomial have the following interesting geometric interpretation.

Theorem 1.2.3 (van der Berg, [Be2]). *Let the roots of a cubic polynomial P form the vertices of a triangle ABC in the complex plane. Then the roots of P' are at the focal points of the ellipse tangent to the sides of $\triangle ABC$ at their midpoints.*

First proof. Observe first of all that if $Q(z) = P(z - z_0)$, then $Q'(z) = P'(z - z_0)$. Therefore we can take any point for the origin.

We can represent any affine transformation of the plane as a composition of an isometry, a homothety, and a transformation of the form $(x, y) \mapsto (x, y \cos \alpha)$ in a Cartesian coordinate system. Therefore we may assume that the triangle ABC is obtained from the equilateral triangle with vertices w, εw and $\varepsilon^2 w$, where $|w| = 1$ and $\varepsilon = \exp\left(\dfrac{2\pi i}{3}\right)$, under the transformation

$$z \mapsto \frac{z + \bar z}{2} + \frac{z - \bar z}{2} \cos \alpha = z \cos^2 \frac{\alpha}{2} + \bar z \sin^2 \frac{\alpha}{2}. \tag{1}$$

Then the semi-axes a and b of the ellipse considered are equal to $\frac{1}{2}$ and $\frac{1}{2} \cos \alpha$; the distance between its focal points F_1 and F_2 is equal to $\sqrt{a^2 - b^2} = \frac{1}{2} \sin \alpha$. Under the dilation with coefficient

$$\left(\frac{1}{2} \sin \alpha\right)^{-1} = \left(\sin \frac{\alpha}{2} \cos \frac{\alpha}{2}\right)^{-1}.$$

points F_1 and F_2 transform into $(\pm 1, 0)$. The composition of transformation (1) and this dilation amounts to the transformation

$$z \mapsto z \cot \frac{\alpha}{2} + \bar z \tan \frac{\alpha}{2}$$

Set $a = w \cot \frac{\alpha}{2}$. Then the polynomial with roots A, B, and C is of the form

$$P(x) = \left(x - a - \frac{1}{a}\right)\left(x - a\varepsilon - \frac{1}{a\varepsilon}\right)\left(x - a\varepsilon^2 - \frac{1}{a\varepsilon^2}\right)$$

It is easy to verify that $P'(x) = 3x^2 + 3\varepsilon + 3\bar\varepsilon = 3x^2 - 3$, and therefore the roots of P' are ± 1. $\square$

Second proof. [Sc5] Let $\varepsilon = \exp\left(\dfrac{2\pi i}{3}\right)$ and let z_0, z_1, z_2 be the roots of the polynomial P considered. Select numbers $\zeta_0, \zeta_1, \zeta_2$ so that

$$z_0 = \zeta_0 + \zeta_1 + \zeta_2, \quad z_1 = \zeta_0 + \zeta_1 \varepsilon + \zeta_2 \varepsilon^2, \quad z_2 = \zeta_0 + \zeta_1 \varepsilon^2 + \zeta_2 \varepsilon, \tag{2}$$

i.e.,

$$3\zeta_0 = z_0 + z_1 + z_2, \quad 3\zeta_1 = z_0 + z_1\varepsilon^2 + z_2\varepsilon, \quad 3\zeta_2 = z_0 + z_1\varepsilon + z_2\varepsilon^2.$$

In what follows we assume that $z_0 + z_1 + z_2 = 0$, i.e., $\zeta_0 = 0$.

It is easy to verify that the curve $\zeta_1 e^{i\varphi} + \zeta_2 e^{-i\varphi}$, where $0 \le \varphi \le 2\pi$, is an ellipse whose semi-axes are directed along the bisectors of the exterior and interior angles of the angle $\angle \zeta_1 O \zeta_2$, where O is the origin, and the lengths of the semi-axes are equal to $|\zeta_1| + |\zeta_2|$ and $||\zeta_1| - |\zeta_2||$. Indeed, the curve considered is the image of the unit circle under the map $z \mapsto \zeta_1 z + \zeta_2 \bar{z}$. Further, if $\zeta_1 = |\zeta_1| e^{i\alpha}$ and $\zeta_2 = |\zeta_2| e^{i\beta}$, then

$$\zeta_1 e^{i\varphi} + \zeta_2 e^{-i\varphi} = |\zeta_1| e^{i(\varphi+\alpha)} + |\zeta_2| e^{i(\beta-\varphi)}.$$

The absolute value of this expression attains its maximum at $\varphi = \frac{\alpha+\beta}{2} + k\pi$ and its minimum at $\varphi = \frac{\alpha+\beta}{2} + \frac{\pi}{2} + k\pi$. These values of φ correspond precisely to the directions of the bisectors indicated.

The focal points f_1 and f_2 of the ellipse $\zeta_1 e^{i\varphi} + \zeta_2 e^{-i\varphi}$ lie on the line corresponding to the angle $\varphi = \frac{\alpha+\beta}{2}$, i.e., $\frac{f_1 f_2}{\zeta_1 \zeta_2}$ is a positive number. Further, the square of the distance of the focal point to the center of the ellipse is equal to the difference of the squares of the semi-axes, i.e., it is equal to

$$\left(|\zeta_1| + |\zeta_2|\right)^2 - \left(|\zeta_1| - |\zeta_2|\right)^2 = 4|\zeta_1 \zeta_2|.$$

Hence $f_1 f_2 = 4\zeta_1 \zeta_2$.

Relations (2) for $\zeta_0 = 0$ show that the vertices z_0, z_1, z_2 of the triangle considered lie on the ellipse $\zeta_1 e^{i\varphi} + \zeta_2 e^{-i\varphi}$ and the mid-points of its sides lie on the ellipse $\frac{1}{2}\left(\zeta_1 e^{i\varphi} + \zeta_2 e^{-i\varphi}\right)$. The mid-point of a chord of the first ellipse lies on the second ellipse only if this chord is tangent to the second ellipse. Therefore we have to prove that the focal points of the ellipse $\frac{1}{2}\left(\zeta_1 e^{i\varphi} + \zeta_2 e^{-i\varphi}\right)$ coincide with the roots of the derivative of the polynomial $P = (z - z_0)(z - z_1)(z - z_2)$. The focal points of the ellipse satisfy the equation $z^2 - \zeta_1 \zeta_2 = 0$, and the roots of P' satisfy

$$3z^2 + z_0 z_1 + z_0 z_2 + z_1 z_2 = 0, \quad \text{i.e.,} \quad 3(z^2 - \zeta_1 \zeta_2) = 0. \quad \square$$

1.2.3 Localization of the roots of the derivative

Jensen's disks

Let f be a polynomial with real coefficients. For every pair of conjugate roots z and $\bar{z}$ of f, the disk with diameter[1] $z\bar{z}$ is called a *Jensen's disk* of f.

Theorem 1.2.4 (Jensen). *Any non-real root of f' lies inside or on the boundary of one of the Jensen's disks of f.*

---

[1] We mean that z and $\bar{z}$ are the endpoints of a diameter of this disk.

Proof. Let $z_1, \ldots, z_n$ be the roots of f. Then

$$\frac{f'(z)}{f(z)} = \sum_{j=1}^{n} \frac{1}{z - z_j}. \tag{1}$$

Let us show first of all that if z lies outside Jensen's disk with diameter $z_p z_q$, then

$$\operatorname{sgn} \operatorname{Im} \left(\frac{1}{z - z_p} + \frac{1}{z - z_q} \right) = - \operatorname{sgn} \operatorname{Im} z. \tag{2}$$

Indeed,

$$\frac{1}{z - a - bi} + \frac{1}{z - a + bi} = \frac{2(z - a)\left((\bar{z} - a)^2 + b^2\right)}{|(z - a)^2 + b^2|^2}$$

and

$$\operatorname{Im}\left((\bar{z} - a)|z - a|^2 + (z - a)b^2\right) = (b^2 - |z - a|^2) \operatorname{Im} z.$$

Let us show now that if $z \notin \mathbb{R}$ and $z_j = a \in \mathbb{R}$, then

$$\operatorname{sgn} \operatorname{Im} \left(\frac{1}{z - z_j} \right) = - \operatorname{sgn} \operatorname{Im} z. \tag{3}$$

Indeed,

$$\frac{1}{z - a} - \frac{1}{\bar{z} - a} = \frac{\bar{z} - z}{|z - a|^2} = \frac{-2 \operatorname{Im} z}{|z - a|^2}.$$

Formulas (1), (2), (3) imply that if point $z \notin \mathbb{R}$ lies outside all the Jensen's disks, then

$$\operatorname{sgn} \operatorname{Im} \frac{f'(z)}{f(z)} = - \operatorname{sgn} \operatorname{Im} z \neq 0.$$

Hence $f'(z) \neq 0$, i.e., z is not a root of f'. $\square$

As a refinement of Jensen's theorem, we prove the following estimate for the number of the roots of the derivative whose real parts belongs to a given segment.

Theorem 1.2.5 (Walsh). *Let $I = [\alpha, \beta]$, and let K be the union of I and Jensen's disks intersecting I. If K contains k roots of a polynomial $f(z)$, then the number of the roots of $f'(z)$ that lie in K is between $k - 1$ and $k + 1$.*

Proof. Let C be the boundary of the smallest rectangle whose sides are parallel to the coordinate axes and which contain K. Consider the restriction to C of the map $z \mapsto e^{i\varphi}$, where $\varphi = \arg \frac{f'(z)}{f(z)}$. Formulas (1), (2) and (3) imply that the image of the part of C that lies in the upper half-plane lies on the half-circle $|z| = 1$, $\operatorname{Im} z \leq 0$, whereas the image of the part of C that lies in the lower half-plane lies on the half-circle $|z| = 1$, $\operatorname{Im} z \geq 0$. Therefore the number of revolutions of the image of C around the origin is equal to either 0 or ± 1. This means that the indices of C with respect to the vector fields $f(z)$ and $f'(z)$ either coincide or differ by ± 1, i.e., the total numbers of the zeros of functions f and f' lying inside C either coincide or differ by ± 1. $\square$

Walsh's theorem

Theorem 1.2.6 (Walsh). *Let the roots of the polynomials f_1 and f_2 lie in the disks K_1 and K_2 with radii r_1 and r_2 and centers at points c_1 and c_2, respectively. Then every root of the derivative of $f = f_1 f_2$ lie either in K_1, or in K_2, or in the disk of radius $\dfrac{n_2 r_1 + n_1 r_2}{n_1 + n_2}$ centered at $\dfrac{n_2 c_1 + n_1 c_2}{n_1 + n_2}$, where $n_1 = \deg f_1$ and $n_2 = \deg f_2$.*

Proof. Let z be the root of f lying outside K_1 and K_2. Then

$$f_1'(z) f_2(z) + f_1(z) f_2'(z) = 0;$$

moreover, $f_1(z), f_2(z), f_1'(z), f_2'(z)$ are nonzero.

Consider ζ_1 and ζ_2, the centers of mass of the roots of f_1 and f_2 with respect to z, respectively. By Theorem 1.1.6

$$\zeta_1 = z - n_1 \frac{f_1(z)}{f_1'(z)}, \quad \zeta_2 = z - n_2 \frac{f_2(z)}{f_2'(z)}.$$

Hence

$$n_2 \zeta_1 + n_1 \zeta_2 = (n_1 + n_2) z - n_1 n_2 \left(\frac{f_1(z)}{f_1'(z)} + \frac{f_2(z)}{f_2'(z)} \right) = (n_1 + n_2) z,$$

i.e., $z = \dfrac{n_2 \zeta_1 + n_1 \zeta_2}{n_1 + n_2}$. Since all the roots of f_i lie in K_i, it follows that $\zeta_i \in K_i$. It remains to observe that if points ζ_1 and ζ_2 of mass n_2 and n_1 lie in disks K_1 and K_2, respectively, then their center of mass z lies in the disk K. $\square$

The Grace-Heawood theorem

Theorem 1.2.7 (J. H. Grace, 1902; P. J. Heawood, 1907). *If z_1 and z_2 are distinct roots of a polynomial f of degree n, then the disk $|z - c| \le r$, where $c = \frac{1}{2}(z_1 + z_2)$ and $r = \dfrac{|z_1 - z_2|}{2} \cot \left(\dfrac{\pi}{n} \right)$, contains at least one root of f'.*

Proof. Let[1] $f'(z) = \sum\limits_{k=0}^{n-1} \binom{n-1}{k} a_k z^k$. Then

$$0 = f(z_2) - f(z_1) = \int_{z_1}^{z_2} f'(z) \, dz = \sum_{k=0}^{n-1} (-1)^k \binom{n-1}{k} a_k b_{n-1-k},$$

where the coefficients $b_0, \ldots, b_{n-1}$ depend only on z_1 and z_2 and not on the coefficients $a_0, \ldots, a_{n-1}$. Therefore, given z_1 and z_2, we can construct a polynomial $g(z) = \sum_{k=0}^{n-1} \binom{n-1}{k} b_k z^k$ apolar to $f'(z)$.

---

[1] This expression of f' differs by a factor of $\frac{1}{n}$ from formula (∗) in sec. 1.1.4.

To obtain an explicit formula for g, set $a_k = (-1)^k x^{n-1-k}$, i.e., consider $h(z) = (x - z)^{n-1}$. In this case

$$g(x) = \sum_{k=0}^{n-1} \binom{n-1}{k} x^{n-1-k} b_{n-1-k} =$$

$$\int_{z_1}^{z_2} (x - 1)^{n-1} dz = \frac{(x - z_1)^n - (x - z_2)^n}{n}.$$

The roots of g are of the form

$$\zeta_k = \frac{z_1 + z_2}{2} + i \frac{z_1 - z_2}{2} \cot \frac{k\pi}{n} \quad \text{for } k = 1, 2, \ldots, n - 1$$

and all of them lie on the boundary of the disk considered. Therefore, by Theorem 1.1.10 (see p. 8), the disk $|z - c| \leq r$ contains at least one root of f'. $\square$

In [Ma7], there are several other theorems on localization of the roots of the derivative.

1.2.4 The Sendov-Ilieff conjecture

In 1962, the Bulgarian mathematician B. Sendov made the following conjecture often ascribed to another Bulgarians mathematician, L. Ilieff:

"*Let $P(z)$ be a polynomial (deg $P \geq 2$) all of whose roots lie in the disk $|z| \leq 1$. If z_0 is one of the roots of $P(z)$, then the disk $|z - z_0| \leq 1$ contains at least one root of $P'(z)$*".

This conjecture is proved for all polynomials of degree ≤ 5 and several particular polynomials (see, e.g., [Sc4]).

We confine ourselves to the proof of the conjecture for polynomials of the form

$$P(z) = (z - z_0)^{n_0} (z - z_1)^{n_1} (z - z_2)^{n_2}.$$

This proof is given in [Co2].

The case when $n = n_0 + n_1 + n_2 \geq 4$ is much the simplest. In this case we have to prove that if $|z_i| \leq 1$ for $i = 0, 1, 2$, then the polynomial

$$P'(z) = n(z - z_0)^{n_0-1}(z - z_1)^{n_1-1}(z - z_2)^{n_2-1}(z - w_1)(z - w_2) \quad (1)$$

has a root lying in the disk $|z - z_0| \leq 1$. If $n_0 > 1$, then z_0 is such a root. We assume therefore that $n_0 = 1$. Let us express $P(z)$ in the form $P(z) = (z - z_0)Q(z)$. It is clear that

$$P'(z_0) = Q(z_0) = (z_0 - z_1)^{n_1} (z_0 - z_2)^{n_2}. \quad (2)$$

It follows from (1) and (2) that

$$n(z_0 - w_1)(z_0 - w_2) = (z_0 - z_1)(z_0 - z_2). \tag{3}$$

Taking into account that $|z_0 - z_1| \leq |z_0| + |z_1| = 2$ and $|z_0 - z_2| \leq 2$, we obtain

$$|z_0 - w_1| \cdot |z_0 - w_2| \leq \frac{4}{n} \leq 1,$$

and hence either $|z_0 - w_1| \leq 1$ or $|z_0 - w_2| \leq 1$.

It remains to consider the case when $n_0 = n_1 = n_2 = 1$. For this we need the following auxiliary statement which we will formulate more generally than is needed for this proof.

Lemma. *Let $P(z)$ be a polynomial of degree n, where $n \geq 2$. If*

$$|P''(z_0)| \geq (n - 1)\,|P'(z_0)|,$$

then at least one of the roots of P' lies inside the disk $|z - z_0| \leq 1$.

Proof. Let $w_1, w_2, \ldots, w_{n-1}$ be the roots of P'. We may assume that the highest coefficient of P is equal to 1. In this case $P'(z) = n \prod_{j=1}^{n-1} (z - w_j)$. If $P'(z) \neq 0$ we may take the logarithm of both sides and differentiate. This gives

$$\frac{P''(z)}{P'(z)} = \sum_{j=1}^{n-1} \frac{1}{z - w_j}.$$

By the hypothesis z_0 is a simple root of P, i.e., $P'(z_0) \neq 0$. Suppose that $|z_0 - w_j| > 1$ for $j = 1, \ldots, n-1$. Then the inequality $|P''(z_0)| \geq (n-1)\,|P'(z_0)|$ implies that

$$n - 1 \leq \left| \frac{P''(z_0)}{P'(z_0)} \right| \leq \sum_{j=1}^{n-1} \frac{1}{|z_0 - w_j|} < n - 1,$$

and we have a contradiction. $\square$

Now let us consider directly the polynomial

$$P(z) = (z - z_0)(z - z_1)(z - z_2) = (z - z_0)Q(z).$$

Clearly

$$\frac{P''(z)}{P'(z)} = 2\frac{Q'(z)}{Q(z)} = 2\left(\frac{1}{z_0 - z_1} + \frac{1}{z_0 - z_2} \right) = \frac{2(2z_0 - z_1 - z_2)}{(z_0 - z_1)(z_0 - z_2)}.$$

Now consider the triangle ABC with vertices $A = z_0$, $B = z_1$, $C = z_2$. Obviously $|z_0 - z_1| = c$, $|z_0 - z_2| = b$ and $|2z_0 - z_1 - z_2| = 2m_a$, where m_a is the length of the median drawn from A. By Lemma the Sendov-Ilieff conjecture holds if $4m_a \geq 2bc$.

By the hypothesis, $b \leq 2$ and $c \leq 2$, and hence $2m_a \geq bc$ holds both for $m_a \geq b$ and for $m_a \geq c$. It remains to consider the case when $m_a < b$ and $m_a < c$.

Relation (3) shows that the Sendov-Ilieff conjecture holds if $bc \leq 3$. Therefore we may assume that $bc > 3$. In this case

$$b^2 + c^2 = (b - c)^2 + 2bc > 6,$$

and therefore $b^2 + c^2 - a^2 > 6 - 4 > 0$, i.e., $\angle A < 90°$. The inequalities $b > m_a$ and $c > m_a$ imply that $\angle C < 90°$ and $\angle B < 90°$, and hence the triangle ABC is acute.

Let R be the radius of its circumscribed circle, h_a the length of the altitude from A. Then $\dfrac{c}{h_a} = \sin B = \dfrac{2R}{b}$, i.e., $bc = 2Rh_a \leq 2Rm_a$. To obtain the inequality desired, $bc \leq 2m_a$, it remains therefore to prove that $R \leq 1$. The acute triangle ABC lies inside the unit circle $|z| = 1$. If the circumscribed circle S of the triangle ABC lies inside the unit circle, the inequality $R \leq 1$ is obvious. Let now S and the unit circle have a common chord. Since ABC is acute, this chord subtends an acute angle φ whose vertex coincides with one of the vertices of the triangle ABC. The same chord subtends the angles ψ and $180° - \psi$, where $\psi \leq 90°$, whose vertices lie on the unit circle. Moreover, $\psi \leq \varphi$. The inequalities $\psi \leq \varphi < 90° < 180° - \psi$ imply that $R \leq 1$.

1.2.5 Polynomials whose roots coincide with the roots of their derivatives

In the paper [Ya] it was stated that if P and Q are *monic* polynomials (i.e., their highest coefficients are equal to 1) and the sets of roots of P and Q coincide, and the sets of roots of the polynomials P' and Q' also coincide, then $P^m = Q^n$ for certain positive integers m and n. Later certain gaps were discovered in the proof of this statement and soon a counterexample was constructed in [Ro2]. The construction of this counterexample is rather complicated. We advise the interested reader to turn directly to [Ro2].

Concerning properties of polynomials whose roots coincide with the roots of the derivatives see also [Do1].

1.3 The resultant and the discriminant

1.3.1 The resultant

Consider polynomials $f(x) = \sum\limits_{i=0}^{n} a_i x^{n-i}$ and $g(x) = \sum\limits_{i=0}^{m} b_i x^{m-i}$, where $a_0 \neq 0$ and $b_0 \neq 0$. Over an algebraically closed field, f and g have a common divisor if and only if they have a common root. If the field is not algebraically closed, then the common divisor could be a polynomial without roots.

The existence of a common divisor of f and g is equivalent, as one can show, to the existence of polynomials p and q such that $fq = gp$, where $\deg p \leq n - 1$ and $\deg q \leq m - 1$. Indeed, let $f = hp$ and $g = hq$. Then $fq = hpq = gp$. Suppose now that $fq = gp$, where $\deg q \leq \deg g - 1$. If f and g do not have a common divisor, then q divides g: a contradiction.

Let $q = u_0 x^{m-1} + \cdots + u_{m-1}$ and $p = v_0 x^{n-1} + \cdots + v_{n-1}$. The equality $fq = gp$ can be expressed as a system of equations:

$$a_0 u_0 = b_0 v_0,$$
$$a_1 u_0 + a_0 u_1 = b_1 v_0 + b_0 v_1,$$
$$a_2 u_0 + a_1 u_1 + a_0 u_2 = b_2 v_0 + b_1 v_1 + b_0 v_2,$$

$$\cdots\cdots\cdots\cdots\cdots\cdots\cdots\cdots\cdots\cdots\cdots\cdots\cdots\cdots$$

The polynomials f and g have a common root if and only if this system has a nonzero solution $(u_0, u_1, \ldots, v_0, v_1, \ldots)$. If, for example, $m = 3$ and $n = 2$, the determinant of this system of equations is of the form

$$\begin{vmatrix} a_0 & 0 & 0 & -b_0 & 0 \\ a_1 & a_0 & 0 & -b_1 & -b_0 \\ a_2 & a_1 & a_0 & -b_2 & -b_1 \\ 0 & a_2 & a_1 & -b_3 & -b_2 \\ 0 & 0 & a_2 & 0 & -b_3 \end{vmatrix} = \pm \begin{vmatrix} a_0 & a_1 & a_2 & 0 & 0 \\ 0 & a_0 & a_1 & a_2 & 0 \\ 0 & 0 & a_0 & a_1 & a_2 \\ b_0 & b_1 & b_2 & b_3 & 0 \\ 0 & b_0 & b_1 & b_2 & b_3 \end{vmatrix} = \pm \det S(f, g).$$

The matrix

$$S(f, g) = \begin{pmatrix} a_0 & a_1 & a_2 & 0 & 0 \\ 0 & a_0 & a_1 & a_2 & 0 \\ 0 & 0 & a_0 & a_1 & a_2 \\ b_0 & b_1 & b_2 & b_3 & 0 \\ 0 & b_0 & b_1 & b_2 & b_3 \end{pmatrix}$$

is called the *Sylvester matrix* of the polynomials f and g. The determinant of $S(f, g)$ is called the *resultant* of f and g and is denoted by $R(f, g)$. Clearly, $R(f, g)$ is a homogeneous polynomial of degree m with respect to indeterminates a_i and of degree n with respect to indeterminates b_j. The polynomials f and g have a common divisor if and only if the determinant of the system considered vanishes, i.e., $R(f, g) = 0$.

The resultant has many different applications. For example, given polynomial relations $P(x, z) = 0$ and $Q(y, z) = 0$ we can, with the help of the resultant, obtain a polynomial relation of the form $R(x, y) = 0$, i.e., eliminate z. Indeed, consider the given polynomials $P(x, z)$ and $Q(y, z)$ as polynomials in z regarding x and y as constants. Then the vanishing of the resultant of these polynomials is exactly the relation desired $R(x, y) = 0$.

The resultant also allows one to reduce the solution of the system of algebraic equations to the search for roots of polynomials. Indeed, let $P(x_0, y_0) = 0$ and $Q(x_0, y_0) = 0$. Consider $P(x, y)$ and $Q(x, y)$ as polynomials in y. For $x = x_0$, they have a common root y_0. Therefore their resultant $R(x)$ vanishes at $x = x_0$.

Theorem 1.3.1. *Let x_i be the roots of f, and y_j the roots of g. Then*

$$R(f,g) = a_0^m b_0^n \prod (x_i - y_j) = a_0^m \prod g(x_i) = b_0^n \prod f(y_j).$$

Proof. Since $f(x) = a_0(x - x_1) \cdot \ldots \cdot (x - x_n)$, it follows that

$$a_k = \pm a_0 \sigma_k(x_1, \ldots, x_n),$$

where σ_k is an elementary symmetric function. Similarly,

$$b_k = \pm b_0 \sigma_k(y_1, \ldots, y_m).$$

The resultant is a homogeneous polynomial of degree m with respect to indeterminates a_i and of degree n with respect to the b_j. Hence

$$R(f,g) = a_0^m b_0^n P(x_1, \ldots, x_n, y_1, \ldots, y_m),$$

where P is a symmetric polynomial in $x_1, \ldots, x_n$ and $y_1, \ldots, y_m$ vanishing at $x_i = y_j$. The formula

$$x_i^k = (x_i - y_j)x_i^{k-1} + y_j x_i^{k-1}$$

shows that

$$P(x_1, \ldots, y_m) = (x_i - y_j)Q(x_1, \ldots, y_m) + U(x_1, \ldots, \widehat{x}_i, \ldots, y_m).$$

Substituting $x_i = y_j$ into this formula we see that U is the zero polynomial. Similar arguments show that P is divisible by $S = a_0^m b_0^n \prod (x_i - y_j)$.
Since $g(x) = b_0 \prod\limits_{j=1}^{m} (x - y_j)$, we have $\prod\limits_{i=1}^{n} g(x_i) = b_0^n \prod\limits_{i,j}(x_i - y_j)$, and therefore

$$S = a_0^m \prod_{i=1}^{n} g(x_i) = a_0^m \prod_{i=1}^{n}(b_0 x_i^m + b_1 x_i^{m-1} + \cdots + b_m)$$

is a homogeneous polynomial of degree n in indeterminates $b_0, \ldots, b_m$. For indeterminates $a_0, \ldots, a_n$, the arguments are similar. It is also clear that the symmetric polynomial $a_0^m \prod_{i=1}^{n}(b_0 x_i^m + b_1 x_i^{m-1} + \cdots + b_m)$ is a polynomial in $a_0, \ldots, a_n, b_0, \ldots, b_m$. Hence $R(f,g) = R(a_0, \ldots, b_m) = \lambda S$, where λ is a number which does not depend on the a_i and b_i.

On the other hand, the coefficient of $\prod x_i^m$ in $a_0^m b_0^n P(x_1, \ldots, y_m)$ and S is equal to $a_0^m b_0^n$, hence, $\lambda = 1$. $\square$

Corollary 1. $R(g,f) = (-1)^{\deg f \deg g} R(f,g)$.

Corollary 2. *If $f = gq + r$, then*

$$R(f,g) = b_0^{\deg f - \deg r} R(r,g),$$

where b_0 is the leading coefficient of g.

Proof. Let y_j be the roots of g. Then $f(y_j) = r(y_j)$. It remains use that $R(f, g) = b_0^{\deg f} \prod f(y_j)$ and $R(r, g) = b_0^{\deg r} \prod f(y_j)$. $\square$

Corollary 3. $R(f, gh) = R(f, g)R(f, h)$

Proof. Let x_i be the roots of f and a_0 its leading coefficient. Then

$$R(f, gh) = a_0^{\deg g + \deg h} \prod g(x_i)h(x_i),$$
$$R(f, g) = a_0^{\deg g} \prod g(x_i),$$
$$R(f, h) = a_0^{\deg h} \prod h(x_i). \ \square$$

Theorem 1.3.2. *Let* $f(x) = \sum_{i=0}^{n} a_i x^{n-i}$ *and* $g(x) = \sum_{i=0}^{m} b_i x^{m-i}$. *Then there exist polynomials* φ *and* ψ *with integer coefficients in indeterminates* $a_0, \ldots, a_n$, $b_0, \ldots, b_m$ *and* x *for which the identity*

$$\varphi(x, a, b)f(x) + \psi(x, a, b)g(x) = R(f, g)$$

holds.

Proof. Let $c_0, \ldots, c_{n+m-1}$ be the columns of the Sylvester matrix $S(f, g)$ and $y_k = x^{m+n-k-1}$. Then

$$y_0 c_0 + \cdots + y_{n+m-1} c_{n+m-1} = c,$$

where c is the column vector

$$\left(x^{m-1} f(x), \ldots, f(x), x^{n-1} g(x), \ldots, g(x) \right)^T.$$

Consider $y_0, \ldots, y_{n+m-1}$ as a system of linear equations for $y_0, \ldots, y_{n+m-1}$ and make use of Cramer's rule in order to find y_{n+m-1}. We obtain

$$y_{n+m-1} \det(c_0, \ldots, c_{n+m-1}) = \det(c_0, \ldots, c_{n+m-2}, c). \tag{1}$$

It remains to notice that $y_{n+m-1} = 1$, $\det(c_0, \ldots, c_{n+m-1}) = R(f, g)$ and the determinant on the right-hand side of (1) can be represented in the form desired, i.e., as $\varphi(x, a, b)f(x) + \psi(x, a, b)g(x)$. $\square$

1.3.2 The discriminant

Let $x_1, \ldots, x_n$ be the roots of the polynomial $f(x) = a_0 x^n + \cdots + a_n$, where $a_0 \neq 0$. The quantity

$$D(f) = a_0^{2n-2} \prod_{i<j} (a_i - a_j)^2$$

is called the *discriminant* of f.

Theorem 1.3.3. $R(f, f') = \pm a_0 D(f)$.

Proof. By Theorem 1.3.1 we have $R(f, f') = a_0^{n-1} \prod_i f'(x_i)$. It is easy to verify that $f'(x_i) = a_0 \prod_{j \neq i} (x_j - x_i)$. Therefore

$$R(f, f') = a_0^{2n-1} \prod_{j \neq i} (x_j - x_i) = \pm a_0^{2n-1} \prod_{i<j} (x_j - x_i)^2. \quad \Box$$

Remark. It is not difficult to show that

$$R(f, f') = -R(f', f) = (-1)^{n(n-1)/2} a_0 D(f).$$

Corollary. *The discriminant of f is a polynomial in $a_0, \ldots, a_n$ with integer coefficients.*

Theorem 1.3.4. *Let f, g, and h be monic polynomials. Then*

$$D(fg) = D(f)D(g)R^2(f,g)$$
$$D(fgh) = D(f)D(g)D(h)R^2(f,g)R^2(g,h)R^2(h,f).$$

Proof. Let $x_1, \ldots, x_n$ be the roots of f, and $y_1, \ldots, y_m$ the roots of g. Then

$$D(fg) = \prod (x_i - x_j)^2 \prod (y_i - y_j)^2 \prod (x_i - y_j)^2 = D(f)D(g)R^2(f,g).$$

The second formula is proved similarly. $\Box$

Theorem 1.3.5. *Let f be a real polynomial of degree n without real roots. Then $\operatorname{sgn} D(f) = (-1)^{n/2}$.*

Proof. Making use of the factorization

$$f(x) = a_0(x - x_1) \cdot \ldots \cdot (x - x_n)$$

it is easy to verify that

$$D\left((x - a)f(x)\right) = D\left(f(x)\right)\left(f(a)\right)^2.$$

Let a and $\bar{a}$ be a pair of conjugate roots of f, i.e., $f(x) = (x-a)(x-\bar{a})g(x)$. Then

$$D\left(f(x)\right) = D\left(g(x)\right)(a - \bar{a})^2 \left(f(a)f(\bar{a})\right)^2.$$

Clearly, $\operatorname{sgn}(a - \bar{a})^2 = -1$ and $\left(f(a)f(\bar{a})\right)^2 = |f(a)|^4 > 0$. Therefore $\operatorname{sgn} D(f) = -\operatorname{sgn} D(g)$. Now it is easy to obtain the statement required by induction on n. $\Box$

Theorem 1.3.6. *Let $f(x) = x^n + a_1 x^{n-1} + \cdots + a_n$ be a polynomial with integer coefficients. Then its discriminant $D(f)$ is equal to either $4k$ or $4k+1$, where k is an integer.*

Proof. Let $x_1, \ldots, x_n$ be the roots of f. Then

$$D(f) = \delta^2(f), \quad \text{where } \delta(f) = \prod_{i<j}(x_i - x_j).$$

Consider an auxiliary polynomial $\delta_1(f) = \prod_{i<j}(x_i + x_j)$. Clearly, $\delta_1(f)$ is a symmetric function of the roots of f, and hence $\delta_1(f)$ is an integer. Moreover,

$$\delta_1^2(f) - \delta^2(f) = \prod_{i<j}\left((x_i - x_j)^2 + 4x_i x_j\right) - \prod_{i<j}(x_i - x_j)^2 = 4U(x_1, \ldots, x_n),$$

where U is a symmetric polynomial in $x_1, \ldots, x_n$ with integer coefficients. Therefore $D(f) = \delta_1^2(f) + 4k_1$, where k_1 is an integer. It is also clear that $\delta_1^2(f) = 4k_2$ or $4k_2 + 1$. $\square$

1.3.3 Computing certain resultants and discriminants

In this section we give several examples on how to compute resultants and discriminants.

Example 1.3.7. $D(x^n + a) = (-1)^{n(n-1)/2} n^n a^{n-1}$.

Proof. Let us make use of the fact that

$$D(f) = (-1)^{n(n-1)/2} R(f, f') = (-1)^{n(n-1)/2} \prod_{i=1}^{n} f'(x_i),$$

where $x_1, \ldots, x_n$ are the roots of f. In our case $f'(x) = nx^{n-1}$ and $\prod x_i = (-1)^n a$, and therefore $\prod x_i^{n-1} = (-1)^{n(n-1)} a^{n-1} = a^{n-1}$. $\square$

Example 1.3.8. Let $\varphi(x) = x^{n-1} + \cdots + 1$. Then $D(\varphi) = (-1)^{(n-1)(n-2)/2} n^{n-2}$.

Proof. Since $(x-1)\varphi(x) = x^n - 1$, it follows that

$$D(\varphi)\left(\varphi(1)\right)^2 = D\left((x-1)\varphi(x)\right) = D(x^n - 1) = (-1)^{(n-1)(n-2)/2} n^n.$$

It remains to observe that $\varphi(1) = n$. $\square$

Example 1.3.9. Let $f_n(x) = 1 + x + \frac{x^2}{2!} + \cdots + \frac{x^n}{n!}$. Then

$$D(n! f_n) = (-1)^{n(n-1)/2} (n!)^n.$$

Proof. The polynomial $g_n = n! f_n$ is monic, and hence

$$D(g) = (-1)^{n(n-1)/2} R(g, g') = (-1)^{n(n-1)/2} \prod_{i=1}^{n} g'(\alpha_i),$$

where $\alpha_1, \ldots, \alpha_n$ are the roots of f_n. Clearly,

$$g'(\alpha_i) = n! f_n'(\alpha_i) = n! f_{n-1}(\alpha_i) = n! \left(f_n(\alpha_i) - \frac{\alpha_i^n}{n!} \right) = -\alpha_i^n.$$

Therefore

$$D(g) = (-1)^{n(n-1)/2} \prod_{i=1}^{n} (-\alpha_i^n).$$

It remains to observe that $\prod \alpha_i = (-1)^n g(0) = (-1)^n n!$. $\square$

Example 1.3.10. Let $d = (r, s)$, $r_1 = \dfrac{r}{d}$ and $s_1 = \dfrac{s}{d}$. Then

$$R(x^r - a, x^s - b) = (-1)^s \left(a^{s_1} - b^{r_1} \right)^d.$$

Proof. The relation $R(g, f) = (-1)^{\deg f \deg g} R(f, g)$ shows that if the desired statement holds for a pair (r, s), then it also holds for a pair (s, r). Indeed, $(-1)^{rs+d+r} = (-1)^s$. We may therefore assume that $r \geq s$.

For $s = 0$, the statement is obvious. If $s > 0$, then having divided $x^r - a$ by $x^s - b$ we get the residue $bx^{r-s} - a$. Hence

$$R(x^r - a, x^s - b) = R(bx^{r-s} - a, x^s - b) =$$
$$= R(b, x^s - b) R\left(x^{r-s} - \frac{a}{b}, x^s - b \right) =$$
$$= b^s R\left(x^{r-s} - \frac{a}{b}, x^s - b \right).$$

It is easy to see that if $R(x^{r-s} - \frac{a}{b}, x^s - b) = (-1)^s \left((\frac{a}{b})^{s_1} - b^{r_1-s_1} \right)$, then

$$R(x^r - a, x^s - b) = (-1)^s \left(a^{s_1} - b^{r_1} \right)^d.$$

It remains to use induction on $r + s$. $\square$

Example 1.3.11. Let $n > k > 0$, $d = (n, k)$, $n_1 = \dfrac{n}{d}$ and $k_1 = \dfrac{k}{d}$. Then

$$D(x^n + ax^k + b) =$$
$$(-1)^{n(n-1)/2} b^{k-1} \left(n^{n_1} b^{n_1-k_1} + (-1)^{n_1+1} (n-k)^{n_1-k_1} k^{k_1} a^{n_1} \right)^d.$$

Proof. [Sw] The formula $D(f) = (-1)^{n(n-1)/2} R(f, f')$ gives

$$D(x^n + ax^k + b) = (-1)^{n(n-1)/2} R(x^n + ax^k + b, nx^{n-1} + kax^{k-1}) =$$
$$= (-1)^{n(n-1)/2} n^n R(x^n + ax^k + b, x^{n-1} + n^{-1} kax^{k-1}).$$

Using the fact that

$$R(f, x^m g) = R(f, x^m) R(f, g) = (f(0))^m R(f, g),$$

we obtain

$$D(x^n + ax^k + b) = (-1)^{n(n-1)/2} n^n b^{k-1} R(x^n + ax^k + b, x^{n-k} + n^{-1}ka).$$

The residue after the division of $x^n + ax^k + b$ by $x^{n-k} + n^{-1}ka$ is equal to $a(1 - n^{-1}k)x^k + b$, and hence

$$R(x^n + ax^k + b, x^{n-k} + n^{-1}ka) = R\left(a(1 - n^{-1}k)x^k + b, x^{n-k} + n^{-1}ka\right).$$

The resultant of a pair of two binomials is computed in Example 1.3.10. □

1.4 Separation of roots

Here we discuss various theorems which allow us to compute, or at least estimate from above, the number of real roots of a polynomial on a given segment (a, b). Formulations of such theorems often use the notion the *number of sign changes* in the sequence $a_0, a_1, \ldots, a_n$, where $a_0 a_n \neq 0$. This number is determined as follows: all the zero terms of the sequence considered are deleted and, for the remaining non-zero terms, one counts the number of pairs of neighboring terms of different sign.

1.4.1 The Fourier–Budan theorem

Theorem 1.4.1 (Fourier–Budan). *Let $N(x)$ be the number of sign changes in the sequence $f(x), f'(x), \ldots, f^{(n)}(x)$, where f is a polynomial of degree n. Then the number of roots of f (multiplicities counted) between a and b, where $f(a) \neq 0$, $f(b) \neq 0$ and $a < b$, does not exceed $N(a) - N(b)$. Moreover, the number of roots can differ from $N(a) - N(b)$ by an even number only.*

Proof. Let x be a point which moves along the segment $[a, b]$ from a to b. The number $N(x)$ varies only if x passes through a root of the polynomial $f^{(m)}$ for some $m \leq n$.

Consider first the case when x passes through a root x_0 of multiplicity r of $f(x)$. In a neighborhood of x_0, the polynomials $f(x), f'(x), \ldots, f^{(r)}(x)$ behave approximately as

$$(x - x_0)^r g(x_0), \quad (x - x_0)^{r-1} r g(x_0), \quad \ldots, \quad r! g(x_0),$$

respectively. Therefore, for $x < x_0$, there are r sign changes in this sequence and for $x > x_0$ there are no sign changes (assuming that x is sufficiently close to x_0).

Now suppose that x passes through a root x_0 of multiplicity r of $f^{(m)}$; let x_0 be not a root of $f^{(m-1)}$. (Of course, x_0 can be a root of f as well, as it can be not a root of f.) We have to prove that under the passage through x_0 the number of sign changes in the sequence $f^{(m-1)}(x), f^{(m)}(x), \ldots, f^{(m+r)}(x)$

changes by a non-negative even integer. Indeed, in a vicinity of x_0 these polynomials behave approximately as

$$F(x_0), \ (x - x_0)^r G(x_0), \ (x - x_0)^{r-1} rG(x_0), \ \ldots, \ r!G(x_0). \tag{1}$$

Excluding $F(x_0)$, we see that the remaining system has exactly r sign changes for $x < x_0$ and no sign changes for $x > x_0$. Concerning the first two terms, $F(x_0)$ and $(x - x_0)^r G(x_0)$, of the sequence (1) we see that if r is even, then the number of sign changes is the same for $x < x_0$ and $x > x_0$ whereas if r is odd, then the number of sign changes for $x < x_0$ is by 1 greater or less than for $x > x_0$ (depending whether $F(x_0)$ and $G(x_0)$ have the same sign or the opposite sign). Thus, for r even, the difference in the number of sign changes is equal to r and, for r odd, the difference of the number of sign changes is equal to $r \pm 1$. In both these cases this difference is even and non-negative. $\square$

Corollary 1. (The Descartes Rule) *The number of positive roots of the polynomial* $f(x) = a_0 x^n + a_1 x^{n-1} + \cdots + a_n$ *does not exceed the number of sign changes in the sequence* $a_0, a_1, \ldots, a_n$.

Proof. Since $f^{(r)}(0) = r!a_{n-r}$, it follows that $N(0)$ coincides with the number of sign changes in the sequence of coefficients of f. It is also clear that $N(+\infty) = 0$. $\square$

Remark. Jacobi showed that the Descartes Rule can be used also to estimate the number of roots between α and β. To this end one should make the change of variables $y = \dfrac{x - \alpha}{\beta - x}$, i.e., set $x = \dfrac{\alpha + \beta y}{1 + y}$, and consider the polynomial

$$(1 + y)^n f\left(\frac{\alpha + \beta y}{1 + y}\right) = b_0 y^n + b_1 y^{n-1} + \cdots + b_n.$$

The Descartes Rule applied to this polynomial yields an estimate of the number of roots between α and β. Indeed, y varies from 0 to ∞, as x varies from α to β.

Corollary 2. (de Gua) *If the polynomial lacks $2m$ consecutive terms (i.e., the coefficients of these terms vanish), then this polynomial has no less than $2m$ imaginary roots. If $2m + 1$ consecutive terms are missing, then if they are between terms of different signs, the polynomial has no less than $2m$ imaginary roots, whereas if the missing terms are between terms of the same sign the polynomial has no less than $2m + 2$ imaginary roots.*

In certain cases the comparison of the sign changes in two sequences allows one to sharpen the estimate of the number of roots as compared with the estimate given by the Fourier-Budan theorem. The first to formulate this type of theorem was Newton but it was proved (by Sylvester) much later, in 1871. Let us replace the sequence $f(x), f'(x), \ldots, f^{(n)}(x)$ by the sequence $f_0(x), f_1(x), \ldots, f_n(x)$, where

$$f_i(x) = \frac{(n-i)!}{n!} f^{(i)}(x), \tag{2}$$

and consider one more sequence $F_0(x), F_1(x), \ldots, F_n(x)$, where $F_0(x) = F(x)$, $F_n(x) = f_n^2(x)$ and

$$F_i(x) = f_i^2(x) - f_{i-1}(x)f_{i+1}(x), \quad i = 1, \ldots, n-1. \tag{3}$$

Convention 1.4.1 *Let us take into account only the pairs* $f_i(x)$, $f_{i+1}(x)$ *for which* $\operatorname{sgn} F_i(x) = \operatorname{sgn} F_{i+1}(x)$.

Let $N_+(x)$ be the number of pairs for which $\operatorname{sgn} f_i(x) = \operatorname{sgn} f_{i+1}(x)$ and let $N_-(x)$ be the number of pairs for which $\operatorname{sgn} f_i(x) = -\operatorname{sgn} f_{i+1}(x)$.

Theorem 1.4.2 (Newton-Sylvester). *Let f be a polynomial of degree n without multiple roots. Then the number of roots of f between a and b, where $a < b$ and $f(a)f(b) \neq 0$, does not exceed either $N_+(b) - N_+(a)$ or $N_-(a) - N_-(b)$.*

Proof. First consider the case when f satisfies the following conditions:

    1) no two consecutive polynomials f_i have common roots;

    2) no two consecutive polynomials F_i have common roots;

    3) the roots of f_i and F_i are distinct from a and b.

In this case formula (3) implies that f_i and F_i have no common roots. It is easy to derive from (2) and (3) that

$$f_i' = (n-i)f_{i+1}, \tag{4}$$

$$f_i F_i' = (n-i-1)(F_i f_{i+1} + F_{i+1} f_i). \tag{5}$$

Let x move from a to b. The numbers $N_\pm(x)$ only vary if x passes either through a root of f_i or through a root of F_i. Consider separately the following three cases.

Case 1: the passage through a root x_0 of $f_0 = f$. If $f_0(x_0) = 0$, then

$$F_1(x_0) = f_1^2(x_0) - f_0(x_0)f_2(x_0) = f_1^2(x_0) > 0.$$

Therefore the passage through x_0 does not involve a change of sign in the sequence $F_0(x) = 1$, $F_1(x)$. Formula (4) implies that $\operatorname{sgn} f'(x) = \operatorname{sgn} f_1(x)$. Therefore, if $f_1(x_0) > 0$, then $f_0(x_0 - \varepsilon) < 0$ and $f_0(x_0 + \varepsilon) > 0$, whereas if $f_1(x_0) < 0$, then $f_0(x_0 - \varepsilon) > 0$ and $f_0(x_0 + \varepsilon) < 0$. In both cases

$$f_0(x_0 - \varepsilon)f_1(x_0 - \varepsilon) < 0 \quad \text{and} \quad f_0(x_0 + \varepsilon)f_1(x_0 + \varepsilon) > 0.$$

Thus, the passage through x_0 increases N_+ by 1 and decreases N_- by 1. (We only consider the contribution to $N_\pm$ of the pair f_0, f_1.)

Case 2: the passage through a root x_0 of the polynomial f_i, where $i \geq 1$. In this case the change of signs occurs in the sequence f_{i-1}, f_i, f_{i+1}. The possible variants of the signs of the polynomials considered at $x = x_0 \pm \varepsilon$ are considerably restricted by the following relations:

1) $\operatorname{sgn} f_{i+1} = \operatorname{sgn} f_i'$ due to (4);

2) $\operatorname{sgn} F_i(x_0) = \operatorname{sgn} \left(f_i^2(x_0) - f_{i-1}(x_0)f_{i+1}(x_0) \right) =$
$- \operatorname{sgn} \left(f_{i-1}(x_0)f_{i+1}(x_0) \right)$ due to (3);

3) $\operatorname{sgn} F_{i\pm 1} = \operatorname{sgn} f_{i\pm 1}^2 = 1.$

If $F_i(x_0) < 0$ the sign changes occur in the pairs F_{i-1}, F_i and F_i, F_{i+1} but, by Convention 1 just before the theorem, we do not consider such pairs. If $F_i(x_0) > 0$, then $f_{i-1}(x_0)f_{i+1}(x_0) < 0$. The signs of the polynomials f_{i-1}, f_i, f_{i+1} considered at $x = x_0 \pm \varepsilon$ are completely determined by the sign of $f_{i+1}(x_0)$. For both values of the signs, the pairs $f_{i-1}(x_0 - \varepsilon)$, $f_i(x_0 - \varepsilon)$ and $f_i(x_0 - \varepsilon)$, $f_{i+1}(x_0 - \varepsilon)$ contribute to N_+ and N_-, respectively, and then the pairs $f_{i-1}(x_0 + \varepsilon)$, $f_i(x_0 + \varepsilon)$ and $f_i(x_0 + \varepsilon)$, $f_{i+1}(x_0 + \varepsilon)$ contribute the other way round to N_- and N_+, respectively. Thus, their total contribution to N_+ as well as to N_- does not vary.

Case 3: passage through a root x_0 of F_i. In this case the signs of the polynomials satisfy the following relations:

1) $f_{i-1}(x_0)f_{i+1}(x_0) = f_i^2(x_0) - F_i(x_0) = f_i^2(x_0) > 0;$

2) $\operatorname{sgn} f_i' = \operatorname{sgn} f_{i+1};$

3) formula (5) implies that $\operatorname{sgn} F_i' = \operatorname{sgn} f_{i-1}f_{i+1}F_{i+1}.$

An easy perusal of the possible scenarios shows that either both N_+ and N_- do not vary, or N_+ increases by 2, or N_- decreases by 2.

It remains to explain how to get rid of conditions 1)–3) imposed on f. If some of these conditions are not satisfied, then after a small variation of the coefficients of f these conditions will be satisfied. But the roots of f are simple ones, and therefore the number of roots of f lying strictly inside the segment $[a, b]$ does not vary under a small variation of the coefficients. $\square$

Remark. For the polynomial f with multiple roots, one should make use of a slightly more subtle argument. Namely, one should consider not arbitrary small variations but only those for which the real root of multiplicity r splits into r distinct real roots. To produce such a small variation, it is more convenient to modify the roots of the polynomial rather than its coefficients.

1.4.2 Sturm's Theorem

Consider the polynomials $f(x)$ and $f_1(x) = f'(x)$. Let us seek the greatest common divisor of f and f_1 with the help of Euclid's algorithm:

$$f = q_1 f_1 - f_2,$$
$$f_1 = q_2 f_2 - f_3,$$
$$\cdots\cdots\cdots\cdots$$
$$f_{n-2} = q_{n-1} f_{n-1} - f_n,$$
$$f_{n-1} = q_n f_n.$$

The sequence $f, f_1, \ldots, f_{n-1}, f_n$ is called the *Sturm sequence* of the polynomial f.

Theorem 1.4.3 (Sturm). *Let $w(x)$ be the number of sign changes in the sequence*

$$f(x), \quad f_1(x), \quad \ldots, \quad f_n(x).$$

The number of the roots of f (without taking multiplicities into account) confined between a and b, where $f(a) \neq 0$, $f(b) \neq 0$ and $a < b$, is equal to $w(a) - w(b)$.

Proof. First, consider the case when the roots of f are simple (i.e., the polynomials f and f' have no common roots). In this case f_n is a nonzero constant.

Let us verify first of all that when we pass through one of the roots of polynomials $f_1, \ldots, f_{n-1}$ the number of sign changes does not vary. In the case considered, the neighboring polynomials have no common roots, i.e., if $f_r(\alpha) = 0$, then $f_{r\pm1}(\alpha) \neq 0$. Moreover, the equality $f_{r-1} = q_{r-1} f_r - f_{r+1}$ implies that $f_{r-1}(\alpha) = -f_{r+1}(\alpha)$. But in this case the number of sign changes in the sequence $f_{r-1}(\alpha)$, ε, $f_{r+1}(\alpha)$ is equal to 2 both for $\varepsilon > 0$ and for $\varepsilon < 0$.

Let us move from a to b. If we pass through a root x_0 of f, then first the numbers $f(x)$ and $f'(x)$ are of different signs and then they are of the same sign. Therefore the number of sign changes in the Sturm sequence diminishes by 1. All the other sign changes, as we have already shown, are preserved during the passage through x_0.

Now consider the case when x_0 is a root of multiplicity m of f. In this case f and f_1 have a common divisor $(x - x_0)^{m-1}$, and hence the polynomials are divisible by $(x - x_0)^{m-1}$. Having divided f, f_1, $\ldots$, f_r by $(x - x_0)^{m-1}$ we obtain the Sturm sequence $\varphi, \varphi_1, \ldots, \varphi_r$ for the polynomial $\varphi(x) = \dfrac{f(x)}{(x - x_0)^{m-1}}$. The root x_0 is a simple one for φ, and hence the passage through x_0 increases the number of sign changes in the sequence $\varphi, \varphi_1, \ldots, \varphi_r$ by 1. But for a fixed x the sequence $f, f_1, \ldots, f_r$ is obtained from $\varphi, \varphi_1, \ldots, \varphi_r$ by multiplication by a constant, and therefore the numbers of sign changes in these sequences coincide. $\square$

1.4.3 Sylvester's theorem

To compute the Sturm sequence is rather a laborious task. Sylvester suggested the following more elegant method for computing the number of the real roots of the polynomial. Let f be a real polynomial of degree n with simple roots $\alpha_1, \ldots, \alpha_n$. Set $s_k = \alpha_1^k + \cdots + \alpha_n^k$. (Clearly, to calculate s_k one does not have to know the roots of the polynomial because s_k, being a symmetric function, is expressed in terms of the coefficients of the polynomial.)

Theorem 1.4.4 (Sylvester). a) *The number of the real roots of f is equal to the signature of the quadratic form with the matrix*

$$\begin{pmatrix} s_0 & s_1 & \cdots & s_{n-1} \\ s_1 & s_2 & \cdots & s_n \\ \vdots & \vdots & \ddots & \vdots \\ s_{n-1} & s_n & \cdots & s_{2n} \end{pmatrix}.$$

b) *All the roots of f are positive if and only if the matrix*

$$\begin{pmatrix} s_1 & s_2 & \cdots & s_n \\ s_2 & s_3 & \cdots & s_{n+1} \\ \vdots & \vdots & \ddots & \vdots \\ s_n & s_{n+1} & \cdots & s_{2n+1} \end{pmatrix}.$$

is positive definite.

Proof. (Hermite) Let ρ be a real parameter. Consider the quadratic form

$$F(x_1, \ldots, x_n) = \frac{y_1^2}{\alpha_1 + \rho} + \cdots + \frac{y_n^2}{\alpha_n + \rho}, \tag{1.1}$$

$$\text{where } y_r = x_1 + \alpha_r x_2 + \cdots + \alpha_r^{n-1} x_n. \tag{1.2}$$

The coefficients of the polynomial F are symmetric functions in the roots of f, and hence they are real. In particular, this means that the form F can be represented as

$$h_1^2 + \cdots + h_p^2 - h_{p+1}^2 - \cdots - h_n^2,$$

where $h_1, \ldots, h_n$ are linear forms in $x_1, \ldots, x_n$ with real coefficients.

To the real root α_r there corresponds the summand

$$\frac{y_r^2}{\alpha_r + \rho} = \frac{(x_1 + \alpha_r x_2 + \cdots + \alpha_r^{n-1} x_n)^2}{\alpha_r + \rho}.$$

This summand can be represented in the form $\pm h_r^2$, where the plus sign is taken if $\alpha_r + \rho > 0$ and the minus sign otherwise.

The contribution of a pair of conjugate roots α_r and α_s is equal to

$$F_{r,s} = \frac{y_r^2}{\alpha_r + \rho} + \frac{y_s^2}{\alpha_r + \rho}.$$

Let $y_r = u + iv$ and $\dfrac{y}{\alpha_r + \rho} = \lambda + i\mu$, where u, v, λ, μ are real numbers. Then $y_s = u - iv$ and $\dfrac{y}{\alpha_s + \rho} = \lambda - i\mu$. Therefore

$$F_{r,s} = 2\lambda(u^2 - v^2) - 4\mu uv.$$

For $u = 0$ and for $v = 0$, the values of $F_{r,s}$ have opposite signs. Hence after a change of variables we may assume that $F_{r,s} = u_1^2 - v_1^2$.

As a result we see that all the roots of f are real and satisfy the inequality $\alpha_r > -\rho$ if and only if the form (1) is positive definite. The matrix elements of this form are

$$a_{ij} = \frac{\alpha_1^{i+j-2}}{\alpha_1 + \rho} + \cdots + \frac{\alpha_n^{i+j-2}}{\alpha_n + \rho}.$$

Statements a) and b) are obtained by going to the limit as $\rho \longrightarrow +\infty$ and taking $\rho = 0$, respectively. $\square$

The quadratic form that appears in Sylvester's theorem has quite an interesting interpretation. This interpretation will enable us to obtain another proof of Sylvester's theorem; moreover, even for polynomials with multiple roots.

Consider the linear space $V = \mathbb{R}[x]/(f)$ consisting of polynomials considered modulo a polynomial $f \in \mathbb{R}[x]$. We assume that f is monic and $\deg f = n$. The polynomials $1, x, \ldots, x^{n-1}$ form a basis of V. To every $a \in V$, we may assign a linear map $V \to V$ given by the formula $v \mapsto av$ (since the elements of V are polynomials we can multiply them). Let $\mathrm{tr}(a)$ be the trace of this map. Consider the symmetric bilinear form

$$\varphi(v, w) = \mathrm{tr}(vw).$$

Theorem 1.4.5. a) *Let* $f(x) = (x - \alpha_1) \cdots\cdots (x - \alpha_n) \in \mathbb{R}[x]$ *and* $s_k = \alpha_1^k + \cdots + \alpha_n^k$. *The matrix of* φ *in the basis* $1, x, \ldots, x^{n-1}$ *has the form*

$$\begin{pmatrix} s_0 & s_1 & \cdots & s_{n-1} \\ s_1 & s_2 & \cdots & s_n \\ \vdots & \vdots & \ddots & \vdots \\ s_{n-1} & s_n & \cdots & s_{2n} \end{pmatrix}.$$

b) *The signature of the form* φ *is equal to the number of distinct real roots of* f.

Proof. a) Over $\mathbb{C}$, the polynomial f can be factorized into the product of relatively prime linear factors $f = f_1^{m_1} \cdot \ldots \cdot f_r^{m_r}$. Thanks to the Chinese remainder theorem (Lemma on p. 69) the map

$$h \pmod{f} \mapsto (h \pmod{f_1^{m_1}}, \ldots, h \pmod{f_r^{m_r}})$$

determines a canonical isomorphism

$$\mathbb{C}[x]/(f) \cong \mathbb{C}[x]/(f_1^{m_1}) \times \cdots \times \mathbb{C}[x]/(f_r^{m_r}).$$

In this decomposition the factors are orthogonal with respect to φ. Indeed, let polynomials h_i and h_j correspond to factors with distinct numbers i and j, i.e., $h_i \equiv 0 \pmod{f/f_i^{m_i}}$ and $h_j \equiv 0 \pmod{f/f_j^{m_j}}$. Then $h_i h_j \equiv 0 \pmod{f}$, and therefore the map $v \mapsto h_i h_j v$ is the zero one. Hence its trace vanishes. Therefore $\varphi = \varphi_1 + \cdots + \varphi_r$, where φ_i is the restriction of φ onto the subspace $\mathbb{C}[x]/(f_i^{m_i}) = \mathbb{C}[x]/(x - \alpha_i)^{m_i}$. It remains to verify that $\varphi_i(1, x^k) = m_i \alpha_i^k$.

It is easy to calculate the matrix of the form φ_i in the basis

$$1, \; x - \alpha_i, \; \ldots, \; (x - \alpha_i)^{m_i - 1}.$$

Indeed, in this basis the map $v \mapsto (x - \alpha_i)^k v$ has a triangular matrix; and the trace of this matrix is equal to m_i if $k = 0$ and to 0 if $k > 0$. Since

$$0 = \varphi_i(1, x - \alpha_i) = \varphi_i(1, x) - \alpha_i \varphi_i(1, 1) = \varphi_i(1, x) - m_i \alpha_i,$$

it follows that $\varphi_i(1, x) = m_i \alpha_i$. Next, with the help of the equality

$$\varphi_i \left(1, (x - \alpha_i)^k\right) = 0$$

and induction on k we see that $\varphi_i(1, x^k) = m_i \alpha_i^k$.

b) Computing the signature we must remain in $\mathbb{R}$, and therefore we decompose f over $\mathbb{R}$ into the product of relatively prime linear or quadratic factors: $f = f_1^{m_1} \cdot \ldots \cdot f_r^{m_r}$. Again consider the decomposition

$$\mathbb{R}[x]/(f) \cong \mathbb{R}[x]/(f_1^{m_1}) \times \cdots \times \mathbb{R}[x]/(f_r^{m_r}).$$

It suffices to verify that the signature of the restriction of φ onto $\mathbb{R}[x]/(f_i^{m_i})$ is equal to 1 if $\deg f_i = 1$ and to 0 if f_i is an irreducible over $\mathbb{R}$ polynomial of degree 2. As we have already established, in the basis $1, x - \alpha_i, (x - \alpha_i)^{m_i - 1}$, the matrix of φ_i is equal to

$$\begin{pmatrix} m_i & 0 & \ldots & 0 \\ 0 & 0 & \ldots & 0 \\ \vdots & \vdots & \ddots & \vdots \\ 0 & 0 & \ldots & 0 \end{pmatrix}.$$

Therefore if $\deg f_i = 1$ the signature of φ_i is equal to 1.

If f_i is an irreducible over $\mathbb{R}$ polynomial of degree 2, then $\mathbb{R}[x]/(f_i^{m_i}) \cong \mathbb{R}[x]/(x^2 + 1)^{m_i}$. Here we mean an isomorphism over $\mathbb{R}$. Therefore it suffices to calculate the signature of φ on $\mathbb{R}[x]/(x^2 + 1)^m$. It is convenient to calculate the matrix of φ in the basis

$$1, \; x^2, \; x^2 + 1, \; x(x^2 + 1), \; (x^2 + 1)^2, \ldots, x(x^2 + 1)^{m-1}, \; (x^2 + 1)^{m-1}.$$

In this basis, the operators of multiplication by x and x^2 have matrices

$$\begin{pmatrix} 0 & 1 & 0 & 0 & 0 & \dots \\ -1 & 0 & 1 & 0 & 0 & \dots \\ 0 & 0 & 0 & 1 & 0 & \dots \\ 0 & 0 & -1 & 0 & 1 & \dots \\ \vdots & \vdots & \vdots & \vdots & \vdots & \ddots \end{pmatrix} \quad \text{and} \quad \begin{pmatrix} -1 & 0 & 1 & 0 & 0 & \dots \\ 0 & -1 & 0 & 1 & 0 & \dots \\ 0 & 0 & -1 & 0 & 1 & \dots \\ 0 & 0 & 0 & -1 & 0 & \dots \\ \vdots & \vdots & \vdots & \vdots & \vdots & \ddots \end{pmatrix},$$

respectively. Therefore the trace of the operator of multiplication by x is equal to 0 and the trace of the operator of multiplication by x^2 is equal to $-2m$. The operators of multiplication by $x^a(x^2+1)^k$, where $a = 0, 1, 2$ and $k \geq 1$, are represented by diagonal matrices with zero main diagonals; their traces vanish. As a result, we see that the matrix of the form φ is equal to

$$\begin{pmatrix} 2m & 0 & 0 & \dots & 0 \\ 0 & -2m & 0 & \dots & 0 \\ 0 & 0 & 0 & \dots & 0 \\ \vdots & \vdots & \vdots & \ddots & 0 \\ 0 & 0 & 0 & \dots & 0 \end{pmatrix}.$$

The signature of such a form is equal to zero. $\square$

1.4.4 Separation of complex roots

Sturm's theorem enables one to indicate algorithmically a set of segments that contain all the real roots of a real polynomial and, moreover, each such segment contains precisely one root. In a series of papers (1869–1878), Kronecker developed a theory with an algorithm to indicate a set of disks which contain all the complex roots of a complex polynomial so that each disk contains exactly one root. More exactly, Kronecker showed that the number of complex roots inside the given disk can be computed with the help of Sturm's theorem.

Let $z = x + iy$. Let us represent the polynomial $P(z)$ in the form $P(z) = \varphi(x, y) + i\psi(x, y)$. We will assume that P has no multiple roots, i.e., if $P(z) = 0$, then $P'(z) \neq 0$.

To every root of P, there corresponds the intersection point of the curves $\varphi = 0$ and $\psi = 0$. Therefore the number of roots of P lying inside a closed non-self-intersecting curve γ is equal to the number of the intersection points of the curves $\varphi = 0$ and $\psi = 0$ lying inside γ. This number can be calculated as follows. Let us circumscribe the curve γ in the positive direction, i.e., counterclockwise, and to each intersection point of the curves γ and $\varphi = 0$ we assign the number $\varepsilon_i = \pm 1$ according to the following rule: $\varepsilon_i = 1$ if we move from the domain $\varphi\psi > 0$ to the domain $\varphi\psi < 0$, or $\varepsilon_i = -1$ if, the other way round, we move from the domain $\varphi\psi < 0$ to the domain $\varphi\psi > 0$.

In the general position the number of intersection points of the curves γ and $\varphi = 0$ is even (since at every intersection point the function φ changes its sign), and hence $\sum \varepsilon_i = 2k$, where k is an integer.

Theorem 1.4.6 (Kronecker). a) *The number k is equal to the number of intersection points of the curves $\varphi = 0$ and $\psi = 0$ lying inside the curve γ.*

b) *If γ is a circle of given radius with given center, then for the given polynomial P the number k can be algorithmically computed.*

Proof. a) Clearly, $dP(z) = (\varphi_x + i\psi_x)dx + (\psi_y - i\varphi_y)i\,dy$. Hence

$$\varphi_x + i\psi_x = P'(z) = \psi_y - i\varphi_y,$$

and therefore

$$\psi_y = \varphi_x \text{ and } \psi_x = -\varphi_y$$

(the Cauchy-Riemann relations). Therefore

$$\begin{vmatrix} \phi_x & \phi_y \\ \psi_x & \psi_y \end{vmatrix} = \begin{vmatrix} \phi_x & \phi_y \\ \phi_y & -\phi_x \end{vmatrix} = \phi_x^2 + \varphi_y^2 > 0.$$

This means that the rotation from the vector $\operatorname{grad}\varphi = (\varphi_x, \varphi_y)$ to the vector $\operatorname{grad}\psi = (\psi_x, \psi_y)$ is a counterclockwise one. Geometrically this means that the domains $\varphi\psi > 0$ and $\varphi\psi < 0$ are positioned as shown in Fig. 1.1.

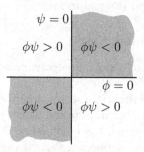

FIGURE 1.1

Let us contract the curve γ into a point. Under the passage through the intersection point of the curves $\varphi = 0$ and $\psi = 0$ the number k diminishes by 1 (Fig. 1.2) and under the reconstruction depicted on Fig. 1.3 the number k does not vary. It is also clear that when the curve becomes sufficiently small it does not intersect the curves $\varphi = 0$ and $\psi = 0$, and in this case $k = 0$.

b) The circle of radius r and center (a, b) can be parameterized with the real parameter t as follows:

$$x = a + r\frac{1 - t^2}{1 + t^2}, \quad y = b + r\frac{2t}{1 + t^2}.$$

Having substituted these expressions into $\varphi(x, y)$ we obtain a polynomial $\varPhi(t)$ with real coefficients. The real roots of this polynomial correspond to the

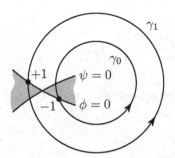

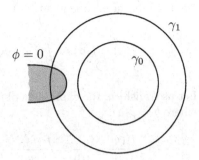

FIGURE 1.2   FIGURE 1.3

intersection points of the curves γ and $\varphi = 0$. By Sturm's theorem, for every root, we can find a segment that contains it. Having calculated the sign of the function $\varphi\psi$ at the endpoints of this segment one can find the corresponding numbers ε_i. $\square$

1.5 Lagrange's series and estimates of the roots of a given polynomial

1.5.1 The Lagrange-Bürmann series

Recall that if $f(z) = \sum\limits_{n=-\infty}^{\infty} c_n(z-a)^n$, then

$$\frac{1}{2\pi i} \int_{\gamma} f(z)\, dz = c_{-1},$$

where γ is any curve circumscribing point a. We will use this fact to obtain the expansion of the function $f(z)$ into a series in powers of $\varphi(z) - b$, where $b = \varphi(a)$. To be able to do so, the function $\varphi(z)$ should be invertible in a neighborhood of a, i.e., $\varphi'(a) \neq 0$. If $\varphi(z)$ is invertible, then

$$\frac{f'(z)\varphi'(a)}{\varphi(z) - \varphi(a)} = \frac{f'(z)\varphi'(a)}{\varphi'(a)(z-a) + \cdots} = \frac{f'(a)}{z-a} + \cdots,$$

and hence

$$\frac{1}{2\pi i} \int_{\gamma} \frac{f'(z)\varphi'(a)}{\varphi(z) - \varphi(a)}\, dz = f'(a).$$

Having integrated this identity, we obtain

$$f(z) - f(a) = \int_a^z f'(\zeta)\,d\zeta = \frac{1}{2\pi i}\int\!\!\int_{a\,\gamma}^z \frac{f'(w)\varphi'(\zeta)}{\varphi(w) - \varphi(\zeta)}\,dw\,d\zeta.$$

Let us transform the expression obtained having separated the terms $\varphi(z) - b$, where $b = \varphi(a)$:

$$\frac{f'(w)\varphi'(\zeta)}{\varphi(w) - \varphi(\zeta)} = \frac{f'(w)\varphi'(\zeta)}{\varphi(w) - b} \cdot \frac{\varphi(w) - b}{\varphi(w) - \varphi(\zeta)},$$

$$\frac{\varphi(w) - b}{\varphi(w) - \varphi(\zeta)} = \left(1 - \frac{\varphi(\zeta) - b}{\varphi(w) - b}\right)^{-1} = \sum_{m=0}^{\infty}\left(\frac{\varphi(\zeta) - b}{\varphi(w) - b}\right)^m.$$

By changing the order of integration we obtain

$$f(z) - f(a) = \frac{1}{2\pi i}\int_\gamma \left(\int_a^z \frac{f'(w)\varphi'(\zeta)}{\varphi(w) - b}\sum_{m=0}^{\infty}\left(\frac{\varphi(\zeta) - b}{\varphi(w) - b}\right)^m d\zeta\right)dw.$$

When we calculate the integral over ζ we only need the factors depending on ζ:

$$\int_a^z \varphi'(z)\,(\varphi(\zeta) - b)^m\,d\zeta = \int_{\varphi(a)}^{\varphi(z)} (\varphi(\zeta) - b)^m\,d\varphi(\zeta) = \frac{(\varphi(\zeta) - b)^{m+1}}{m+1}$$

(we have taken into account that $\varphi(a) - b = 0$).

Thus,

$$f(z) - f(a) = \sum_{m=0}^{\infty} \frac{(\varphi(\zeta) - b)^{m+1}}{m+1}\frac{1}{2\pi i}\int_\gamma \frac{f'(w)\,dw}{(\varphi(w) - b)^{m+1}}.$$

Consider a function $\psi(w)$ such that $\dfrac{1}{\varphi(w) - b} = \dfrac{\psi(w)}{w - a}$, i.e.,

$$\psi(w) = \frac{w - a}{\varphi(w) - b}. \tag{1}$$

For this function $\psi(w)$, we have

$$\frac{1}{2\pi i}\int_\gamma \frac{f'(w)\,dw}{(\varphi(w) - b)^{m+1}} = \frac{1}{2\pi i}\int_\gamma \frac{f'(w)\,(\psi(w))^{m+1}\,dw}{(w - a)^{m+1}} =$$

$$= \frac{1}{m!}\cdot\frac{d^m}{dw^m}\left(f'(w)\,(\psi(w))^{m+1}\right)_{w=a}.$$

Indeed,

$$f'(w) \, (\psi(w))^{m+1} = \sum_{k=0}^{\infty} c_k (w - a)^k,$$

where

$$c_k = \frac{1}{k!} \cdot \frac{d^k}{dw^k} \left(f'(w) \, (\psi(w))^{k+1} \right)_{w=a}.$$

The integral we are interested in is equal to c_m — the coefficient of $(w - a)^{-1}$ in the series $\sum_{k=0}^{\infty} c_k (w - a)^{k-m-1}$.

As a result, we obtain the following expansion of $f(z)$ into powers of $\varphi(z) - b$:

$$f(z) - f(a) = \sum_{n=1}^{\infty} \frac{(\varphi(z) - b)^n}{n!} \cdot \frac{d^{n-1}}{dw^{n-1}} \left(f'(w) \, (\psi(w))^n \right)_{w=a}, \qquad (2)$$

where $\psi(w)$ is given by formula (1). The series (2) is called *Bürmann's series*. Bürmann obtained it in 1799 while generalizing a series Lagrange obtained in 1770. The *Lagrange series* can be obtained from Bürmann's series for $\varphi(z) = \dfrac{z - a}{h(z)}$, where $h(z)$ is a function. In this case $b = \varphi(a) = 0$ and

$$\psi(z) = \frac{z - a}{\varphi(z) - b} = h(z).$$

Therefore

$$f(z) = f(a) + \sum_{n=0}^{\infty} \frac{s^n}{n!} \cdot \frac{d^{n-1}}{da^{n-1}} \left(f'(a) \, (h(a))^n \right),$$

where $s = \varphi(z)$. In particular,

$$z = a + \sum_{n=0}^{\infty} \frac{s^n}{n!} \cdot \frac{d^{n-1}}{da^{n-1}} \, (h(a))^n. \qquad (3)$$

Thus, if the series (3) converges, it enables one to calculate the roots of the equation

$$z = a + s \, h(z).$$

Example. Let $h(z) = \dfrac{1}{z}$. In this case the series (3) has the form

$$z = a + \sum_{n=1}^{\infty} \frac{(-1)^{n-1}(2n-2)!}{n! \, (n-1)! \, a^{2n-1}} \, s^n. \qquad (4)$$

Series (4) converges for $|s| < \dfrac{|a|^2}{4}$. The equation under consideration,

$$z = a + \frac{s}{z},$$

has two roots

$$\frac{a}{2}\left(1+\sqrt{1+\frac{4s^2}{a^2}}\right) \quad \text{and} \quad \frac{a}{2}\left(1-\sqrt{1+\frac{4s^2}{a^2}}\right).$$

The series (4) represents only the first of these roots.

1.5.2 Lagrange's series and estimation of roots

Lagrange's series enables one in certain cases to estimate the roots of polynomials. Consider, for example, the polynomial

$$f(z) = a_0 + a_1(z-c) + a_2(z-c)^2 + \cdots + a_k(z-c)^k.$$

The equation $f(z) = 0$ can be expressed in the form

$$z = c + s\, h(z),$$

where $s = -\dfrac{1}{a_1}$ and $h(z) = a_0 + a_2(z-c)^2 + a_3(z-c)^3 + \cdots + a_k(z-c)^k$.
Lagrange's series for this equation is of the form

$$z = c + \sum_{n=1}^{\infty} \frac{s^n}{n!} \cdot \frac{d^{n-1}}{dz^{n-1}} \left(h^n(z)\right)_{z=c}.$$

In our case

$$h^n(z) = \sum_{\nu_0+\nu_2+\cdots+\nu_k=n} a_0^{\nu_0} a_2^{\nu_2} \cdot \ldots \cdot a_k^{\nu_k} \frac{n!}{\nu_0!\,\nu_2!\cdot\ldots\cdot\nu_k!} (z-c)^{2\nu_2+\cdots+k\nu_k},$$

and hence

$$\frac{d^{n-1}}{dz^{n-1}}\left(h^n(z)\right)_{z=c} = \sum \frac{(n-1)!}{\nu_0!\,\nu_2!\cdot\ldots\cdot\nu_k!} a_0^{\nu_0} a_2^{\nu_2}\cdot\ldots\cdot a_k^{\nu_k}, \qquad (1)$$

where the sum runs over the collections $\{\nu_0, \nu_2, \ldots, \nu_k\}$ such that

$$\nu_0 + \nu_2 + \cdots + \nu_k = n, \quad 2\nu_2 + \cdots + k\nu_k = n-1.$$

These relations are equivalent to the relations

$$n - 1 = 2\nu_2 + \cdots + k\nu_k, \quad \nu_0 = \nu_2 + 2\nu_3 + \cdots + (k-1)\nu_k + 1.$$

Since $s = -\dfrac{1}{a_1}$, we obtain

$$z = c - \frac{a_0}{a_1} \sum \frac{(2\nu_2 + \cdots + k\nu_k)!}{\nu_0!\,\nu_2!\cdot\ldots\cdot\nu_k!} \left(\frac{a_0 a_2}{(-a_1)^2}\right)^{\nu_2} \cdot\ldots\cdot \left(\frac{a_0^{k-1} a_k}{(-a_1)^k}\right)^{\nu_k}, \qquad (2)$$

where $\nu_0 = \nu_2 + 2\nu_3 + \cdots + (k-1)\nu_k + 1$.

   If the series (2) converges, the number z so determined is one of the roots of the equation $f(z) = 0$.

Theorem 1.5.1 ([Be3]). *Let* $|a_0| + |a_2| + \cdots + |a_k| < |a_1|$. *Then the series* (2) *converges and the root* z *determined by the series satisfies*

$$|z - c| \leq -\ln\left(1 - \frac{1}{|a_1|}\left(|a_0| + |a_2| + \cdots + |a_k|\right)\right).$$

Proof. Formula (1) implies that

$$\left|\frac{1}{n!} \cdot \frac{d^{n-1}}{dz^{n-1}}\left(h^n(z)\right)_{z=c}\right| \leq \frac{1}{n}\left(|a_0| + |a_2| + \cdots + |a_k|\right)^n.$$

Hence

$$|z - c| \leq \sum_{n=1}^{\infty} \frac{|a_1|^{-n}}{n}\left(|a_0| + |a_2| + \cdots + |a_k|\right)^n =$$

$$- \ln\left(1 - \frac{1}{|a_1|}\left(|a_0| + |a_2| + \cdots + |a_k|\right)\right). \quad \square$$

1.6 Problems to Chapter 1

1.1 Prove that a polynomial $f(x)$ is divisible by $f'(x)$ if and only if $f(x) = a_0(x - x_0)^n$.

1.2 Prove that the polynomial

$$a_0 + a_1 x^{m_1} + a_2 x^{m_2} + \cdots + a_n x^{m_n}$$

has at most n positive roots.

1.3 [Newton] Prove that if all the roots of the polynomial

$$P(x) = a_0 x^n + a_1 x^{n-1} + \cdots + a_n$$

with real coefficients are real and distinct, then

$$a_i^2 > \frac{n - i + 1}{n - 1} \cdot \frac{i + 1}{i} a_{i-1} a_{i+1} \quad \text{for } i = 1, 2, \ldots, n - 1.$$

1.4 Prove that the polynomial

$$a_1 x^{m_1} + a_2 x^{m_2} + \cdots + a_n x^{m_n}$$

has no nonzero roots of multiplicity greater than $n - 1$.

1.5 Find the number of real roots of the following polynomials:

$$\text{a) } 1 + x + \frac{x^2}{2} + \cdots + \frac{x^n}{n};$$

$$\text{b) } nx^n - x^{n-1} - \cdots - 1.$$

1.6 Let $0 = m_0 < m_1 < \cdots < m_n$ and $m_i \equiv i \pmod 2$. Prove that the polynomial

$$a_0 + a_1 x^{m_1} + a_2 x^{m_2} + \cdots + a_n x^{m_n}$$

has at most n real roots.

1.7 Let x_0 be a root of the polynomial $x^n + a_1 x^{n-1} + \cdots + a_n$. Prove that for any $\varepsilon > 0$ there exists a $\delta > 0$ such that if $|a_i - a_i'| < \delta$ for $i = 1, \ldots, n$, then the polynomial $x^n + a_1' x^{n-1} + \cdots + a_n'$ has a root x_0' such that $|x_0 - x_0'| < \varepsilon$.

1.8 Let the numbers $a_1, \ldots, a_n$ be distinct and let the numbers $b_1, \ldots, b_n$ be positive. Prove that all the roots of the equation

$$\sum \frac{b_k}{x - a_k} = x - c, \quad \text{where } c \in \mathbb{R},$$

are real.

1.9 Find all the roots of the equation

$$\frac{(x^2 - x + 1)^3}{x^2 (x - 1)^2} = \frac{(a^2 - a + 1)^3}{a^2 (a - 1)^2}.$$

1.10 Find the number of roots of the polynomial $x^n + x^m - 1$ whose absolute values are less than 1.

1.11 Let $f(z) = z^n + a_1 z^{n-1} + \cdots + a_n$, where $a_1, \ldots, a_n \in \mathbb{C}$. Prove that any root z of f satisfies $-\beta \leq \operatorname{Re} z \leq \alpha$, where α is the only positive root of the polynomial

$$x^n + (\operatorname{Re} a_1) x^{n-1} - |a_2| x^{n-2} - \cdots - |a_n|$$

and β is the only positive root of the polynomial

$$x^n - (\operatorname{Re} a_1) x^{n-1} - |a_2| x^{n-2} - \cdots - |a_n|.$$

1.12 [Su1] Let $f(z)$ be a polynomial of degree n with complex coefficients. Prove that the polynomial $F = f \cdot f' \cdot f'' \cdot \ldots \cdot f^{(n-1)}$ has at least $n+1$ distinct roots.

1.7 Solutions of selected problems

1.3. Set $Q(y) = y^n P(y^{-1})$. The roots of $Q(y)$ are also real and distinct. Hence the roots of the quadratic polynomial

$$Q^{(n-2)}(y) = (n-2) \cdot (n-3) \cdots \cdots 4 \cdot 3 \left(n(n-1) a_n y^2 + 2(n-1) a_{n-1} y + 2 a_{n-2} \right)$$

are real and distinct. Therefore

$$(n-1)^2 a_{n-1}^2 > 2n(n-1)a_n a_{n-2}.$$

If $i = n - 1$ the desired inequality is proved.

Now consider the polynomial

$$P^{(n-i-1)}(x) = b_0 x^{i+1} + b_1 x^i + \cdots + b_{i+1} x^2 + b_i x + b_{i-1}.$$

Applying to it the inequality already proved we obtain

$$b_i^2 > \frac{2(i+1)}{i} b_{i-1} b_{i+1}.$$

Since

$$b_{i+1} = (n-i+1) \cdot \ldots \cdot 4 \cdot 3\, a_{i+1},$$
$$b_i = (n-i) \cdot \ldots \cdot 3 \cdot 2\, a_i,$$
$$b_{i-1} = (n-i-1) \cdot \ldots \cdot 2 \cdot 1\, a_{i-1},$$

it follows that

$$\left(2(n-i)a_i\right)^2 > \frac{2(i+1)}{i} 2(n-i+1)(n-i)a_{i-1}a_{i+1}.$$

After simplification we obtain the desired inequality.

1.11. As x grows from 0 to $+\infty$, the function $x^n \pm \operatorname{Re} a_1$ monotonically increases, whereas the function

$$\frac{|a_2|}{x} + \frac{|a_3|}{x^2} + \cdots + \frac{|a_n|}{x^{n-1}}$$

monotonically decreases. Therefore each of the polynomials considered has only one positive root.

Let $f(z) = 0$ and $\operatorname{Re} z > \alpha$. Then

$$\alpha + \operatorname{Re} a_1 < \operatorname{Re}(z + a_1) \le |z + a_1| = \left| \frac{a_2}{z} + \frac{a_3}{z^2} + \cdots + \frac{a_n}{z^{n-1}} \right| \le$$
$$\le \frac{|a_2|}{|z|} + \cdots + \frac{|a_n|}{|z|^{n-1}} < \frac{|a_2|}{\alpha} + \cdots + \frac{|a_n|}{\alpha^{n-1}}$$

(the last inequality follows since $|z| \ge \operatorname{Re} z > \alpha$). On the other hand, by the hypothesis

$$\alpha + \operatorname{Re} a_1 = \frac{|a_2|}{\alpha} + \cdots + \frac{|a_n|}{\alpha^{n-1}}:$$

a contradiction.

The estimate of $\operatorname{Re} z$ from below is obtained as the estimate from above of the real part of the root z of $(-1)^n f(-z)$.

1.12. Let $z_1, \ldots, z_m$ be the distinct roots of F, and let $\mu_j(r)$ be the multiplicity of z_j as a root of $f^{(r)}$, where $r = 0, 1, \ldots, n-1$. Consider the symmetric functions

$$s_k(r) = \sum_{j=1}^{k} \mu_j(r) z_j^k, \tag{1}$$

i.e., $s_k(r)$ is the sum of the kth powers of the roots of $f^{(r)}$. The elementary symmetric functions in the roots of $f^{(r)}$ will be denoted by $\sigma_k(r)$ (for $k > n-r$ we set $\sigma_k(r) = 0$).

It is easy to verify that, if $f(z) = \sum_{k=0}^{n} (-1)^k a_k z^{n-k}$, then

$$f^{(r)}(z) = \sum_{k=0}^{n-r} (-1)^k a_k \frac{(n-k)!}{(n-k-r)!} z^{n-k-r}.$$

Hence

$$\sigma_k(r) = \frac{a_k}{a_0} \cdot \frac{(n-k)!(n-r)!}{n!(n-k-r)!} = \frac{a_k}{a_0} \cdot \frac{(n-k)!}{n!} \cdot (n-r) \cdots \cdots (n-k-r+1).$$

Therefore $\sigma_k(r)$ is a polynomial of degree k in r and $\sigma_k(n) = 0$.

On p. 79, for $k \geq 1$, the identity

$$s_k = \begin{vmatrix} \sigma_1 & 1 & 0 & \ldots & 0 \\ 2\sigma_2 & \sigma_1 & 1 & \ldots & 0 \\ \vdots & \vdots & \vdots & \ddots & \vdots \\ k\sigma_k & \sigma_{k-1} & \sigma_{k-2} & \ldots & \sigma_1 \end{vmatrix}$$

is proved. This identity implies, in particular, that $s_k(r)$, where $k > 0$, can be represented as a linear combination of expressions $\sigma_{k_1}(r) \cdots \sigma_{k_p}(r)$, where $k_1 + \cdots + k_p = k$, and the coefficients of this linear combination do not depend on r. Therefore, if $k \geq 1$, then $s_k(r)$ is a polynomial in r of degree not greater than k. It is also clear that $s_0(r) = \sum_{j=1}^{m} \mu_j(r) = n - r$ and $s_k(n) = 0$ for all $k \geq 0$.

Consider the relation (1) for $k = 0, 1, \ldots, m-1$ as a system of linear equations for unknowns $\mu_j(r)$, where $j = 1, \ldots, m$. By the hypothesis, the numbers $z_1, \ldots, z_m$ are distinct, and therefore the determinant of the system considered does not vanish (this determinant is a it Vandermond determinant, see [Pr1]). Having solved this system of linear equations via Cramer's algorithm we obtain a representation of $\mu_j(r)$ in the form of a linear combination of the $s_k(r)$, where $k = 0, \ldots, m-1$, with coefficients independent of r. Hence $\mu_j(r)$ is a polynomial in r of degree $d_j \leq m - 1$. Since $s_k(n) = 0$ for all k, we have $\mu_j(n) = 0$.

Let the number of distinct roots of F be strictly less than $n + 1$, i.e., $m < n + 1$. Then $d_j \leq m - 1 < n$, i.e., $\mu_j(r)$ is a polynomial in r of degree $\leq n - 1$. In this case

$$\deg \Delta^1 \mu_j(r) = \mu_j(r+1) - \mu_j(r) \leq n - 2,$$
$$\deg \Delta^2 \mu_j(r) = \Delta^1 \mu_j(r+1) - \Delta^1 \mu_j(r) \leq n - 3, \ldots,$$

$\Delta^{n-1} \mu_j(r)$ is a constant, and $\Delta^n \mu_j(r)$ is identically zero. In particular,

$$\Delta^n \mu_j(0) = \sum_{r=0}^{n} (-1)^r \binom{n}{r} \mu_j(r) = 0.$$

To arrive at a contradiction, it suffices to show that $\Delta^n \mu_1(0) \neq 0$.

Consider the convex hull of the roots of f. By the Gauss-Lucas theorem (Theorem 1.2.1 on p. 13), this convex hull coincides with the convex hull of the points $z_1, \ldots, z_m$. We may assume that z_1 is a vertex of the convex hull of the roots of f. Then z_1 lies outside the convex hull of the points $z_2, \ldots, z_m$. Let $\mu = \mu_1(0)$ be the multiplicity of z_1 as of a root of f. Then for $0 \leq r \leq \mu - 1$ the number z_1 is a root of multiplicity $\mu - r$ of $f^{(r)}$ and $f^{(\mu)}(z_1) \neq 0$. The convex hull of the roots of $f^{(\mu)}$ does not contain z_1, and hence $f^{(r)}(z_1) \neq 0$ for $r \geq \mu$. Therefore

$$\mu_1(r) = \begin{cases} \mu - r & \text{for } 0 \leq r \leq \mu - 1; \\ 0 & \text{for } r \geq \mu. \end{cases}$$

It is also clear that $\mu \leq n - 1$, since f has at least one root distinct from z_1. Hence

$$\Delta^2 \mu_1(r) = \begin{cases} 0 & \text{for } 0 \leq r \leq n - 1, \, r \neq \mu - 1; \\ 1 & \text{for } r = \mu - 1. \end{cases}$$

Therefore, for $n > 2$, we obtain

$$\Delta^n \mu_1(0) = \Delta^{n-2} \left(\Delta^2 \mu_1 \right)(0) = \sum_{r=0}^{n-2} (-1)^r \binom{n-2}{r} \Delta^2 \mu_1(r) = (-1)^{\mu-1} \binom{n-2}{\mu-1},$$

and, for $n = 2$, we obtain $\mu = 1$ and $\Delta^2 \mu_1(0) = 1$. In both cases $\Delta^n \mu_1(0) \neq 0$, as was required.

2

Irreducible Polynomials

2.1 Main properties of irreducible polynomials

2.1.1 Factorization of polynomials into irreducible factors

Let f and g be polynomials in one variable with coefficients from a field k. We say that f is *divisible* by g if $f = gh$, where h is a polynomial (with coefficients in k).

The polynomial d is called a *common divisor* of f and g if both f and g are divisible by d. The common divisor d of f and g is called the *greatest common divisor* if it is divisible by any common divisor of f and g. Clearly, the greatest common divisor is defined uniquely up to multiplication by a nonzero element of k.

One can find the greatest common divisor $d = (f, g)$ of f and g with the help of the following *Euclid's algorithm*. Suppose, for the sake of definiteness, that $\deg f \geq \deg g$. Let r_1 be the remainder after division of f by g, let r_2 be the remainder after division of g by r_1, and generally let r_{k+1} be the remainder after division of r_{k-1} by r_k. Since the degrees of the polynomials r_i strictly decrease, it follows that, for some n, we have $r_{n+1} = 0$, i.e., r_{n-1} is divisible by r_n. We see that both f and g are also divisible by r_n because r_n divides all the polynomials $r_{n-1}, r_{n-2}, \ldots$. Moreover, if f and g are divisible by a polynomial h, then r_n is also divisible by h since h divides $r_1, r_2, \ldots$. Therefore $r_n = (f, g)$.

Euclid's algorithm directly implies important corollaries which we formulate as a separate theorem.

Theorem 2.1.1. a) *If d is the greatest common divisor of polynomials f and g, then there exist polynomials a and b such that $d = af + bg$.*

b) *Let f and g be polynomials over the field $k \subset K$. If f and g have a non-trivial common divisor over K, they have a nontrivial common divisor over k also.*

A polynomial f with coefficients from a ring k is called *reducible* over k if $f = gh$, where g and h are polynomials of positive degree with coefficients from k. Otherwise f is called *irreducible* over k.

Let $f = f_1 \cdot \ldots \cdot f_s$ be a factorization of a polynomial f over a field k into factors $f_1, \ldots, f_s$ which are polynomials over k. From the factorization into the product of factors with arbitrary coefficients we can pass to factorization into monic polynomials. Indeed, if $f_i(x) = a_i x^i + \cdots$ is a polynomial over the field k, then $g_i = \dfrac{f_i}{a_i}$ is a monic polynomial over k. Hence, we can replace the factorization $f = f_1 \cdot \ldots \cdot f_s$ by the factorization $f = a g_1 \cdot \ldots \cdot g_s$, where $a = a_1 \cdot \ldots \cdot a_s$. We will not distinguish two factorizations of such a form that differ only by the order of factors.

Theorem 2.1.2. *Let k be a field. Then the polynomial $f \in k[x]$ can be factorized into irreducible factors and this factorization is unique.*

Proof. The existence of the factorization is easy to prove by induction on $n = \deg f$. First of all, observe that, for irreducible f, the desired factorization consists of f itself.

For $n = 1$, the polynomial f is irreducible. Let the factorization exist for any polynomial of degree $< n$ and let $\deg f = n$. We may assume that f is reducible, i.e., $f = gh$, where $\deg g < n$ and $\deg h < n$. But the factorizations for g and h exist by the induction hypothesis.

Let us prove now the uniqueness of the factorization. Let $a g_1 \cdot \ldots \cdot g_s = b h_1 \cdot \ldots \cdot h_t$, where $a, b \in k$ and $g_1, \ldots, g_s, h_1, \ldots, h_t$ are irreducible monic polynomials over k. Clearly, in this case $a = b$. The polynomial $g_1 \cdots g_s$ is divisible by the irreducible polynomial h_1. This means that one of the polynomials $g_1, \ldots, g_s$ is divisible by h_1. To see this, it suffices to prove the following auxiliary statement.

Lemma. *If the polynomial qr is divisible by an irreducible polynomial p, then either q or r is divisible by p.*

Proof. Suppose that q is not divisible by p. Then $(p, q) = 1$, i.e., there exist polynomials a and b such that $ap + bq = 1$. Having multiplied both sides of this identity by r we get $apr + bqr = r$. But pr and qr are divisible by p, and so r is also divisible by p. $\square$

For definiteness, let g_1 be divisible by h_1. Taking into account that g_1 and h_1 are monic irreducible polynomials, we deduce that $g_1 = h_1$. Let us simplify the equality $g_1 \cdot \ldots \cdot g_s = h_1 \cdot \ldots \cdot h_t$ having divided by $g_1 = h_1$. After several such operations we deduce that $s = t$ and $g_1 = h_{i_1}, \ldots, g_s = h_{i_s}$, where $\{i_1, \ldots, i_s\}$ is a permutation of the set $\{1, \ldots, s\}$. $\square$

Irreducibility of polynomials over the ring of integers $\mathbb{Z}$ is defined exactly as for polynomials over fields, i.e., $f \in \mathbb{Z}[x]$ is irreducible over $\mathbb{Z}$ if it cannot be represented as a product of polynomials of positive degree with integer coefficients. But when the coefficients of the polynomial belong to a ring one cannot always divide the coefficients by the highest coefficient; one can only divide the coefficients by the greatest common divisor of all the coefficients. This complication leads to the following definition. Let $f(x) = \sum a_i x^i$, where $a_i \in \mathbb{Z}$. The greatest common divisor of the coefficients $a_0, \ldots, a_n$ is called the *content* of f and denoted by $\mathrm{cont}(f)$. Clearly, $f(x) = \mathrm{cont}(f)g(x)$, where g is a polynomial over $\mathbb{Z}$ with content 1.

Lemma. $\mathrm{cont}(fg) = \mathrm{cont}(f)\,\mathrm{cont}(g)$.

Proof. It suffices to consider the case where $\mathrm{cont}(f) = \mathrm{cont}(g) = 1$. Indeed, the coefficients of the polynomials f and g can be divided by $\mathrm{cont}(f)$ and $\mathrm{cont}(g)$ respectively.

Let $f(x) = \sum a_i x^i$, $g(x) = \sum b_i x^i$, $fg(x) = \sum c_i x^i$. Suppose that $\mathrm{cont}(fg) = d > 1$ and p is a prime divisor of d. Then all the coefficients of fg are divisible by p whereas f and g have coefficients not divisible by p. Let a_r be the first coefficient of f not divisible by p, and b_s the first coefficient of g not divisible by p. Then

$$c_{r+s} = a_r b_s + a_{r+1} b_{s-1} + a_{r+2} b_{s-2} + \cdots + a_{r-1} b_{s+1} + a_{r-2} b_{s+2} + \cdots$$
$$\equiv a_r b_s \not\equiv 0 \pmod{p},$$

since

$$b_{s-1} \equiv b_{s-2} \equiv \cdots \equiv b_0 \equiv 0 \pmod{p},$$
$$a_{r-1} \equiv a_{r-2} \equiv \cdots \equiv a_0 \equiv 0 \pmod{p}.$$

Thus we have a contradiction. $\square$

Corollary. *A polynomial with integer coefficients is irreducible over $\mathbb{Z}$ if and only if it is irreducible over $\mathbb{Q}$.*

Proof. Let $f \in \mathbb{Z}[x]$ and $f = gh$, where $g, h \in \mathbb{Q}[x]$. We may assume that $\mathrm{cont}(f) = 1$. For g, select a positive integer m such that $mg \in \mathbb{Z}[x]$. Let $n = \mathrm{cont}(mg)$. Then the rational $r = \dfrac{m}{n}$ is such that $rg \in \mathbb{Z}[x]$ and $\mathrm{cont}(rg) = 1$. Similarly, select a positive rational number s for h. Let us show that in this case $rs = 1$, i.e., the factorization $f = (rg)(sh)$ is a factorization over $\mathbb{Z}$. Indeed, thanks to Gauss's lemma, $\mathrm{cont}(rg)\,\mathrm{cont}(sh) = \mathrm{cont}(rsgh)$, i.e., $1 = \mathrm{cont}(rsf)$. Since $\mathrm{cont}(f) = 1$ we deduce that $rs = 1$. $\square$

Kronecker suggested the following algorithm for factorization of any polynomial $f \in \mathbb{Z}[x]$ into irreducible factors (*Kronecker's algorithm*). Let $\deg f = n$ and $r = \left[\dfrac{n}{2}\right]$. If $f(x)$ is reducible, it has a divisor $g(x)$ of degree not higher than r.

To find this divisor $g(x)$, consider the numbers $c_j = f(j)$, where $j = 0, 1, \ldots, r$. If $c_j = 0$, then $x - j$ divides $f(x)$. If on the contrary $c_j \neq 0$, then $g(j)$ divides c_j. To every set $d_0, \ldots, d_r$ of divisors of the numbers $c_0, \ldots, c_r$, respectively, there corresponds precisely one polynomial $g(x)$, of degree not higher than r, such that $g(j) = d_j$ for $j = 0, 1, \ldots, r$. Namely,

$$g(x) = \sum_{j=0}^{r} d_j g_j(x), \quad \text{where} \quad g_j(x) = \prod_{0 \le k \le r, k \ne j} \left(\frac{x - k}{j - k} \right).$$

For each such polynomial, one has to verify if its coefficients are integers and if it actually divides $f(x)$.

Other, more effective, algorithms for factorization of polynomials into irreducible factors are given below (see p. 71–73 and 279–288).

2.1.2 Eisenstein's criterion

One of the best known irreducibility criteria of polynomials is the following *Eisenstein's criterion*.

Theorem 2.1.3 (Eisenstein's criterion). *Let $f(x) = a_0 + a_1 x + \cdots + a_n x^n$ be a polynomial with integer coefficients such that the coefficient a_n is not divisible by a prime p, while the coefficients $a_0, \ldots, a_{n-1}$ are divisible by p but a_0 is not divisible by p^2. Then f is irreducible over $\mathbb{Z}$.*

Proof. Suppose that

$$f = gh = \left(\sum b_k x^k \right) \left(\sum c_l x^l \right),$$

where g and h are polynomials of positive degree with integer coefficients. The number $b_0 c_0 = a_0$ is divisible by p, and hence one of the numbers b_0 and c_0 is divisible by p. Let, for the sake of definiteness, b_0 be divisible by p. Then c_0 is not divisible by p since $a_0 = b_0 c_0$ is not divisible by p^2. If all the numbers b_i are divisible by p, then so is a_n. Therefore b_i is not divisible by p for some i, where $0 < i \le \deg g < n$; we may assume that i is the least index of the numbers b_i not divisible by p.

On the one hand, by assumption the number a_i is divisible by p. On the other hand, $a_i = b_i c_0 + b_{i-1} c_1 + \cdots + b_0 c_i$, where all the summands $b_{i-1} c_1, \ldots, b_0 c_i$ are divisible by p while $b_i c_0$ is not divisible by p: a contradiction. $\square$

Example 2.1.4. Let p be a prime and let q be not divisible by p. Then $x^m - pq$ is irreducible over $\mathbb{Z}$.

Example 2.1.5. If p is a prime, then $f(x) = x^{p-1} + x^{p-2} + \cdots + x + 1$ is irreducible.

Indeed, one can apply Eisenstein's criterion to the polynomial

$$f(x+1) = \frac{(x+1)^p - 1}{(x+1) - 1} = x^{p-1} + \binom{p}{1} x^{p-2} + \cdots + \binom{p}{p-1}.$$

Example 2.1.6. For any positive integer n, the polynomial

$$f(x) = 1 + x + \frac{x^2}{2!} + \cdots + \frac{x^n}{n!}$$

is irreducible.

Proof. We have to prove that the polynomial

$$n!f(x) = x^n + nx^{n-1} + n(n-1)x^{n-2} + \cdots + n!$$

is irreducible over $\mathbb{Z}$. To this end, it suffices to find the prime p such that $n!$ is divisible by p but is not divisible by p^2, i.e., $p \leq n < 2p$.

Let $n = 2m$ or $n = 2m + 1$. *Bertrand's postulate* (for its proof, see, e.g., [Ch1]) states that

there exists a prime p such that $m < p \leq 2m$.

For $n = 2m$, the inequalities $p \leq n < 2p$ are obvious. For $n = 2m + 1$, we obtain the inequalities $p \leq n - 1$ and $n - 1 < 2p$. But in this case the number $n - 1$ is even, and hence the inequality $n - 1 < 2p$ implies $n < 2p$. It is also clear that $p \leq n - 1 < n$. $\square$

2.1.3 Irreducibility modulo p

Let $\mathbb{F}_p$ be the residue field modulo p. Every polynomial with integer coefficients can be also considered as a polynomial with coefficients from $\mathbb{F}_p$. A polynomial irreducible over $\mathbb{Z}$ can become reducible over $\mathbb{F}_p$ for all p, and the construction of an example to show this is based on the following theorem.

Theorem 2.1.7. *The polynomial $P(x) = x^4 + ax^2 + b^2$, where $a, b \in \mathbb{Z}$, is reducible over $\mathbb{F}_p$ for all primes p.*

Proof. For $p = 2$, there are only 4 polynomials of the form indicated, namely,

$$x^4, \quad x^4 + x^2 = x^2(x^2 + 1), \quad x^4 + 1 = (x+1)^4, \quad x^4 + x^2 + 1 = (x^2 + x + 1)^2.$$

All these polynomials are reducible.

Let p be an odd prime. Then we can select an integer s such that $a \equiv 2s$ (mod p). We have

$$P(x) = x^4 + ax^2 + b^2 \equiv (x^2 + s)^2 - (s^2 - b^2) \equiv$$
$$\equiv (x^2 + b)^2 - (2b - 2s)x^2 \equiv$$
$$\equiv (x^2 - b)^2 - (-2b - 2s)x^2 \quad (\text{mod } p).$$

Thus it suffices to prove that one of the numbers $s^2 - b^2$, $2b - 2s$, $-2b - 2s$ is a quadratic residue modulo p.

Let us recall the basic notions of the theory of quadratic residues. Under the map $x \mapsto x^2$ the elements x and $-x$ turn into the same element. Therefore the image of the set of nonzero elements of $\mathbb{F}_p$ under this map consists of $\dfrac{p-1}{2}$ elements. On the other hand, if $x = y^2$, then $x^{(p-1)/2} = y^{p-1} = 1$, i.e., all the $\dfrac{p-1}{2}$ elements of the image satisfy the equation $x^{(p-1)/2} = 1$, which cannot have more than $\dfrac{p-1}{2}$ solutions. The elements that do not lie in the image of the map $x \mapsto x^2$ satisfy the equation $x^{(p-1)/2} = -1$. Therefore, if two integers are not perfect squares modulo p, then their product is a perfect square modulo p.

Suppose that $2b - 2s$ and $-2b - 2s$ are not squares modulo p. Then their product $4(s^2 - b^2)$ is a square modulo p, and hence so is $s^2 - b^2$. $\square$

Example 1. The polynomial $x^4 + 1$ is irreducible over $\mathbb{Z}$ but reducible modulo p for all primes p.

Proof. It suffices to prove that $x^4 + 1$ is irreducible over $\mathbb{Z}$. The roots of this polynomial are $\dfrac{\pm 1 \pm i}{2}$. Every polynomial with real coefficients can have non-real roots only if they occur in complex conjugate pairs. Therefore the only nontrivial real divisors of $x^4 + 1$ are the polynomials $x^2 \pm \sqrt{2}x + 1$ whose roots are $\dfrac{1 \pm i}{2}$ and $\dfrac{-1 \pm i}{2}$. Both these polynomials do not lie in $\mathbb{Z}[x]$. $\square$

Example 2. Let $c \in \mathbb{N}$ and $\sqrt{c} \notin \mathbb{Q}$. Then the polynomial $P(x) = x^4 + 2(1 - c)x^2 + (1 + c)^2$ is irreducible over $\mathbb{Z}$ but reducible modulo any prime p.

Proof. It suffices to prove that $P(x)$ is irreducible over $\mathbb{Z}$. It is easy to verify that the roots of P are equal to $\pm\sqrt{-1 + c \pm 2i\sqrt{c}} = \pm i \pm \sqrt{c}$. Combining the complex conjugate roots into pairs we obtain the polynomials $x^2 \pm 2\sqrt{c}x + 1 + c$. These polynomials do not lie in $\mathbb{Z}[x]$. $\square$

2.2 Irreducibility criteria

2.2.1 Dumas's criterion

Let p be a fixed prime, and let $f(x) = \sum\limits_{i=0}^{n} A_i x^i$ be a polynomial with integer coefficients such that $A_0 A_n \neq 0$. Let us represent the nonzero coefficients of f in the form $A_i = a_i p^{\alpha_i}$, where a_i is an integer not divisible by p. To every nonzero coefficient $a_i p^{\alpha_i}$ we assign a point in the plane with coordinates (i, α_i). These points give rise to the *Newton diagram of the polynomial f* (corresponding to p). The construction of the diagram is as follows.

Let $P_0 = (0, \alpha_0)$ and $P_1 = (i_1, \alpha_{i_1})$, where i_1 is the largest integer for which there are no points (i, α_i) below the line $P_0 P_1$. Further, let $P_2 = (i_2, \alpha_{i_2})$, where i_2 is the largest integer for which there are no points (i, α_i) below the line P_1P_2, etc. (fig. 2.1). The very last segment is of the form $P_{r-1}P_r$, where $P_r = (n, \alpha_n)$. If some segments of the broken line $P_0 \ldots P_r$ pass through points with integer coordinates, then such points will be also considered as vertices of the broken line. In this way, to the vertices $P_0, \ldots, P_r$, we add $s \geq 0$ more vertices. The resulting broken line $Q_0 \ldots Q_{r+s}$ is called *the Newton diagram* (here $Q_0 = P_0$ and $Q_{r+s} = P_r$). The segments P_lP_{l+1} and Q_iQ_{i+1} will be called *sides* and *segments* of the Newton diagram respectively, and the vectors $\overrightarrow{Q_iQ_{i+1}}$ will be called the *vectors of the segments* of the Newton diagram.

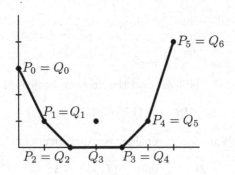

FIGURE 2.1

Consider the system of vectors of the segments for the Newton diagram, taking each vector with its multiplicity, i.e., as many times as it enters the set of vectors of segments.

Theorem 2.2.1 (Dumas, [Du2]). *Let $f = gh$, where f, g, and h are polynomials with integer coefficients. Then the system of vectors of the segments for f is the union of the systems of vectors of the segments for g and h (provided p is the same for all the polynomials).*

Proof. [W] Let

$$f(x) = \sum_{i=0}^{n} a_i p^{\alpha_i} x^i, \quad g(x) = \sum_{j=0}^{m} b_j p^{\beta_j} x^j, \quad h(x) = \sum_{k=0}^{n-m} c_k p^{\gamma_k} x^k,$$

where the numbers a_i, b_j, c_k are not divisible by p. Take a side of the Newton diagram for f (recall that a side P_lP_{l+1} may consist of several segments of the Newton diagram). Let the coordinates of P_l and P_{l+1} be (i_-, α_{i_-}) and (i_+, α_{i_+}) respectively. The slope of P_lP_{l+1} is

$$M = \frac{\alpha_{i_+} - \alpha_{i_-}}{i_+ - i_-}.$$

Let $\alpha_{i_+} - \alpha_{i_-} = At$ and $i_+ - i_- = It$, where $t > 0$ is the greatest common divisor of $\alpha_{i_+} - \alpha_{i_-}$ and $i_+ - i_-$. Then $M = A/I$, where $(A, I) = 1$.

The side $P_l P_{l+1}$ of the Newton diagram belongs to the straight line

$$I\alpha - Ai = F, \quad \text{where } F = I\alpha_{i_+} - Ai_+ = I\alpha_{i_-} - Ai_-.$$

By assumption all the points (i, α_i), where $i = 0, 1, \ldots, n$, lie on or above this line, i.e., $I\alpha_i - Ai \geq F$, where this inequality is strict for $i < i_-$ and $i > i_+$. The number $I\alpha_i - Ai$ will be called the *weight* of the monomial $ap^\alpha x^i$, where $(a, p) = 1$. The numbers i_- and i_+ are uniquely determined as the least and the greatest exponents of the power of x for the monomials entering f with the minimal weight.

For the polynomial g, consider the quantity

$$G = \min_{j=0,\ldots,m} \{I\beta_j - Aj\},$$

and define j_- and j_+ as the least and the greatest indices for which

$$G = I\beta_{j_-} - Aj_- = I\beta_{j_+} - Aj_+.$$

Similarly, for the polynomial h, consider the quantity

$$H = \min_{k=0,\ldots,n-m} \{I\gamma_k - Ak\},$$

and define k_- and k_+ as the least and the greatest indices for which

$$H = I\gamma_{k_-} - Ak_- = I\gamma_{k_+} - Ak_+.$$

Clearly,

$$a_{j_-+k_-} p^{\alpha_{j_-+k_-}} = \sum_{j+k=j_-+k_-} (b_j p^{\beta_j} x^j)(c_k p^{\gamma_k} x^k).$$

The weight of the product of two terms is equal to the sum of their weights, and therefore the weight of the summand with $j = j_-$ and $k = k_-$ is equal to $G + H$. The weights of all the other summands are strictly greater than $G + H$ since, for them, either $j < j_-$ or $k < k_-$.

Indeed, let, for example, $j < j_-$. Then the weight of $b_j p^{\beta_j} x^j$ is strictly greater than G and the weight of $c_k p^{\gamma_k} x^k$ is not less than H.

The weight of $(b_j p^{\beta_j} x^j)(c_k p^{\gamma_k} x^k)$ for $j+k = \text{const}$ increases monotonically as $\beta_j + \gamma_k$ grows since $I > 0$. In the case considered, $j + k = j_- + k_-$, and therefore the sum $\beta_j + \gamma_k$ is strictly minimal at $j = j_-$ and $k = k_-$. Therefore the weight of $a_{j_-+k_-} p^{\alpha_{j_-+k_-}}$ is equal to $G + H$.

It is also clear that for $i < j_- + k_-$ the weight of $a_i p^{\alpha_i} x^i$ is strictly greater than $G + H$, whereas for $i \geq j_- + k_-$ the weight of $a_i p^{\alpha_i} x^i$ is not less than $G + H$. Therefore $G + H = F$ and $j_- + k_- = i_-$. We similarly prove that $j_+ + k_+ = i_+$. Thus,

$$i_+ - i_- = (j_+ - j_-) + (k_+ - k_-). \tag{1}$$

In particular, one of the numbers $j_+ - j_-$ and $k_+ - k_-$ is nonzero.

If both the numbers $j_+ - j_-$ and $k_+ - k_-$ are nonzero, then the segment with the end points (j_-, β_{j_-}) and (j_+, β_{j_+}) is a side of the Newton diagram for g and the segment with the end points (k_-, γ_{k_-}) and (k_+, γ_{k_+}) is a side of the Newton diagram for h. The slope of both segments is equal to $M = A/I$ since

$$\frac{\beta_{j_+} - \beta_{j_-}}{j_+ - j_-} = \frac{A}{I} = \frac{\gamma_{k_+} - \gamma_{k_-}}{k_+ - k_-}.$$

Relation (1) shows that the sum of the lengths of the sides with slope M of the Newton diagrams for g and h is equal to the length of the side with the same slope M of the Newton diagram for f.

If one of the numbers $j_+ - j_-$ and $k_+ - k_-$ vanishes, then the Newton diagram of one of the polynomials g or h has a side with slope M and its length is equal to the length of the side of the Newton diagram for f, whereas the Newton diagram of the other polynomial has no side with slope M.

Thus, the vector of the side with slope M of the Newton diagram for f is equal to the sum of the vectors of the sides with the same slope M of the Newton diagrams for g and h. Relation (1) shows that if one of the Newton diagrams for g and h possesses a side with a certain slope M, then the Newton diagram for f should also possess a side with the same slope. $\square$

Corollary. *If, for a prime p, the Newton diagram for f consists of precisely one segment, i.e., consists of a segment containing no points with integer coordinates, then f is irreducible.*

Let us give now three examples of the application of Dumas's criterion to the proof of irreducibility of polynomials.

Example 2.2.2 (Eisenstein's criterion). Let $f = a_0 + a_1 x + \cdots + a_n x^n$ be a polynomial with integer coefficients such that, for a prime p, the coefficient a_n is not divisible by p, the coefficients $a_0, \ldots, a_{n-1}$ are divisible by p and a_0 is not divisible by p^2. Then f is irreducible.

Proof. The Newton diagram for f consists of one segment with the end points $(0, 1)$ and $(n, 0)$; inside this segment there are no points with integer coordinates. $\square$

Example 2.2.3. Let p be a prime, $(c, p) = 1$ and $(m, n) = 1$. Then the polynomial $x^n + cp^m$ is irreducible.

Proof. The Newton diagram for the polynomial considered is a segment with the end points $(0, m)$ and $(n, 0)$. Since $(m, n) = 1$, there are no points with integer coordinates inside this segment. $\square$

Example 2.2.4. Let p be a prime. If the polynomial $f(x) = x^n + px + bp^2$, where $(b, p) = 1$, has no integer roots, then this polynomial is irreducible.

Proof. The Newton diagram for f is the union of the segment with the end points $(0,2)$ and $(1,1)$ and the segment with the end points $(1,1)$ and $(n,0)$. Inside these segments, there are no points with integer coordinates. Therefore the nontrivial factorization of f over $\mathbb{Z}$ can consist only of a linear factor and a factor of degree $n - 1$. □

2.2.2 Polynomials with a dominant coefficient

In certain situations one can ensure that a polynomial with a sufficiently large coefficient will necessarily be irreducible. Among criteria of this type, the best known one is the following *Perron's criterion*.

Theorem 2.2.5 ([Pe]). *Let* $f(x) = x^n + a_1 x^{n-1} + \cdots + a_n$ *be a polynomial with integer coefficients such that* $a_n \neq 0$.
a) *If* $|a_1| > 1 + |a_2| + \cdots + |a_n|$, *then* f *is irreducible.*
b) *If* $|a_1| \geq 1 + |a_2| + \cdots + |a_n|$ *and* $f(\pm 1) \neq 0$, *then* f *is irreducible.*

Proof. a) Let us prove first that all the roots of f, except precisely one root, lie inside the unit disk $|z| \leq 1$. Clearly, the polynomial

$$g(x) = x^n + a_1 x^{n-1}$$

satisfies this property, i.e., all the roots of g, except precisely one root, lie inside the unit disk $|z| \leq 1$. Hence by Rouché's theorem (see p. 1) it suffices to prove that for $|z| = 1$ we have

$$\left| f(z) - g(z) \right| < \left| f(z) \right| + \left| g(z) \right|.$$

But for $|z| = 1$ we have, on the one hand,

$$\left| f(z) - g(z) \right| = |a_2 z^{n-2} + \cdots + a_n| \leq |a_2| + \cdots + |a_n| < |a_1| - 1, \quad (1)$$

and, on the other hand,

$$\left| f(z) \right| + \left| g(z) \right| \geq \left| g(z) \right| = |z^n + a_1 z^{n-1}| = |z + a_1| \geq |a_1| - 1. \quad (2)$$

Suppose now that, on the contrary, that f can be represented as the product of polynomials f_1 and f_2 of positive degree with integer coefficients. The product of the roots of each of the polynomials f_1 and f_2 is a non-zero integer, and therefore each of these polynomials has a root whose absolute value is not less than 1. But f has only one such root, and we have a contradiction.

b) If $|a_1| = 1 + |a_2| + \cdots + |a_n|$, then inequality (1) becomes non-strict. But if $f(\pm 1) \neq 0$, inequality (2) becomes strict. Indeed, for $|z| = 1$ the equality

$$\left| f(z) \right| + \left| g(z) \right| = |a_1| - 1$$

is only possible when simultaneously $\left| f(z) \right| = 0$ and $|z + a_1| = |a_1| - 1$. The latter equality can only hold if $z \in \mathbb{R}$. Since $|z| = 1$, it follows that $z = \pm 1$. □

Theorem 2.2.6 ([Br]). *Let $a_1 \geq a_2 \geq \cdots \geq a_n$ be positive integers and $n \geq 2$. Then the polynomial $p(x) = x^n - a_1 x^{n-1} - a_2 x^{n-2} - \cdots - a_n$ is irreducible over $\mathbb{Z}$.*

Proof. Consider the polynomial $f(x) = (x - 1)p(x)$. Clearly,

$$f(x) = x^{n+1} - b_1 x^n + b_2 x^{n-1} + \cdots + b_{n+1},$$

where $b_1 = a_1 + 1, b_2 = a_1 - a_2, \ldots, b_n = a_{n-1} - a_n, b_{n+1} = a_n$. The numbers $b_1, \ldots, b_{n+1}$ are positive integers and $b_1 = 1 + b_2 + \cdots + b_{n+1}$. Therefore $f(x)$ satisfies one of the conditions of Theorem 2.2.5 (b). But it does not satisfy the second condition since $f(1) = 0$. We must therefore apply a subtler argument.

Let

$$h(z) = b_1 z^n - b_2 z^{n-1} - \cdots - b_{n+1}.$$

First we show that, for all sufficiently small $\varepsilon > 0$, we have

$$|h(z)| > |z^{n+1}| = |f(z) + h(z)|$$

everywhere on the circle $|z| = 1 + \varepsilon$. Indeed, if $|z| = 1 + \varepsilon$, then

$$\left|h(z)\right| - |z^{n+1}| \geq b_1(1 + \varepsilon)^n - b_2(1 + \varepsilon)^{n-1} - \cdots - b_{n+1} - (1 + \varepsilon)^{n+1} =$$
$$= \varepsilon(nb_1 - (n-1)b_2 - \cdots - 2b_{n-1} - b_n - (n+1)) + \cdots =$$
$$= \varepsilon(b_2 + 2b_3 + \cdots + (n-1)b_n + nb_{n+1} - 1) + \cdots$$

The coefficient of ε is positive, and therefore, for sufficiently small $\varepsilon > 0$, we have $\left|h(z)\right| - |z^{n+1}| > 0$. In this case

$$\left|f(z) + h(z)\right| = |z^{n+1}| < \left|h(z)\right| \leq \left|f(z)\right| + \left|h(z)\right|.$$

Therefore, by Rouché's theorem, the polynomial $f(z)$ has as many roots inside the disk $|z| \leq 1 + \varepsilon$ as $h(z)$ does. But all the roots of $h(z)$ lie strictly inside the unit disk $|z| \leq 1$. Indeed, if $|z| \geq 1$, then

$$\left|h(z)\right| \geq b_1|z|^n - b_2|z|^{n-1} - \cdots - b_{n+1} \geq$$
$$\geq |z|^n(b_1 - b_2 - \cdots - b_{n+1}) = |z|^n > 0.$$

Letting $\varepsilon \longrightarrow 0$, we see that inside and on the boundary of the unit disk there are exactly n roots of the polynomial $f(x) = (x - 1)p(x)$. Hence exactly $n - 1$ roots of $p(x)$ lie inside the unit disk and at least one of its roots lies outside it. Hence p is irreducible. $\square$

A criterion similar to Perron's criterion but with a condition on the constant term instead of a_1 also holds. It holds, however, only if the constant term is a prime.

Theorem 2.2.7 ([Os1]). *Let $f(x) = x^n + a_1 x^{n-1} + \cdots + a_{n-1}x \pm p$ be a polynomial with integer coefficients, where p is a prime.*

a) If $p > 1 + |a_1| + \cdots + |a_{n-1}|$, then f is irreducible.

b) If $p \geq 1 + |a_1| + \cdots + |a_{n-1}|$ and among the roots of f there are no roots of unity, then f is irreducible.

Proof. Suppose that $f(x) = g(x)h(x)$, where g and h are polynomials of positive degree with integer coefficients. The product of the constant terms of g and h is equal to $\pm p$. Since p is prime, one of these constant terms is equal to ± 1. Therefore the product of the absolute values of the roots of one of the polynomials g and h is equal to 1. This polynomial must therefore possess a root α such that $|\alpha| \leq 1$. Since α is also a root of f, it follows from $f(\alpha) = 0$ that

$$p = |\alpha^n + a_1\alpha^{n-1} + \cdots + a_{n-1}\alpha| \leq 1 + |a_1| + \cdots + |a_{n-1}|.$$

In case a) we arrive at a contradiction.

In case b), α is not a root of unity. Hence $|\alpha| < 1$, and therefore

$$p < 1 + |a_1| + \cdots + |a_{n-1}|.$$

A contradiction again. $\square$

2.2.3 Irreducibility of polynomials attaining small values

Theorem 2.2.8 (Pólya). *Let f be a polynomial of degree n with integer coefficients and define $m = \left[\frac{n+1}{2}\right]$. Suppose that, for n different integers $a_1, \ldots, a_n$, we have $|f(a_i)| < 2^{-m}m!$ and the numbers $a_1, \ldots, a_n$ are not roots of f. Then f is irreducible.*

Proof. We will need the following auxiliary statement.

Lemma. *Let g be a polynomial of degree k with integer coefficients, and let $d_0 < d_1 < \cdots < d_k$ be integers. Then $|g(d_i)| \geq k!2^{-k}$ for some i.*

Proof. Consider the polynomial

$$G(x) = (x - d_0) \cdot \ldots \cdot (x - d_k) \sum_{i=0}^{k} \frac{g(d_i)}{x - d_i} \prod_{j \neq i} \frac{1}{d_i - d_j}.$$

It is easy to see that $G(d_i) = g(d_i)$ for $i = 0, \ldots, k$, and $\deg G \leq k$. Hence $G(x) = g(x)$.

The highest coefficient of G is equal to

$$\sum_{i=0}^{k} g(d_i) \prod_{j \neq i} \frac{1}{d_i - d_j}.$$

By assumption this coefficient is a nonzero integer, and hence its absolute value is ≥ 1. Therefore one of the numbers $\left|g(d_i)\right|$ is not less than

$$\cfrac{1}{\cfrac{1}{\displaystyle\sum_{0\leq i\leq k}\prod_{j\neq i}\frac{1}{|d_i-d_j|}}} \geq \cfrac{1}{\cfrac{1}{\displaystyle\sum_{0\leq i\leq k}\prod_{j\neq i}\frac{1}{|i-j|}}} =$$

$$= \cfrac{1}{\displaystyle\sum_{0\leq i\leq k}\frac{1}{i!(k-i)!}} = \cfrac{k!}{\displaystyle\sum_{0\leq i\leq k}\binom{k}{i}} = \frac{k!}{2^k}. \quad \square$$

Returning to the proof of the theorem, we suppose on the contrary that $f = gh$, where g and h are polynomials with integer coefficients. We may assume that $\deg h \leq \deg g = k$. Then $m \leq k < n$. Clearly, $g(a_i) \neq 0$, and $g(a_i)$ divides $f(a_i)$. Hence

$$\left|g(a_i)\right| \leq \left|f(a_i)\right| < \frac{m!}{2^m}.$$

On the other hand, by Pólya's lemma, we have $\left|g(a_i)\right| \geq 2^{-k}k!$ for one of the a_i (we apply Pólya's lemma to $d_i = a_{i-1}$). It remains to notice that, since $k \geq m$, it follows that $2^{-k}k! \geq 2^{-m}m!$. Indeed, if $m = k + r$,

$$\frac{m!}{k!} = (k+1) \cdot (k+2) \cdot \ldots \cdot (k+r) \leq 2^r = \frac{2^m}{2^k}. \quad \square$$

Example. The polynomial $(x-1) \cdot (x-2) \cdot \ldots \cdot (x-n) + 1$ is irreducible.

For other irreducibility criteria for polynomials attaining small values, see [Tv].

2.3 Irreducibility of trinomials and fournomials

2.3.1 Irreducibility of polynomials of the form $x^n \pm x^m \pm x^p \pm 1$

Let $f(x) = x^n + \varepsilon_1 x^m + \varepsilon_2 x^p + \varepsilon_3$, where $n > m > p \geq 1$ and $\varepsilon_i = \pm 1$. Let us find out following [Lj] when f is irreducible. We first show that it suffices to consider the case where $m + p \geq n$. Clearly, f is irreducible if and only if the polynomial

$$x^n f\left(\frac{1}{x}\right) = 1 + \varepsilon_1 x^{n-m} + \varepsilon_2 x^{n-p} + \varepsilon_3 x^n$$

is irreducible and then, if $m + p < n$, we have $(n - m) + (n - p) > n$.

We may also exclude from further considerations the trivial case $f(x) = (x^m + \varepsilon_2)(x^p + \varepsilon_1)$, i.e., when $n = m + p$ and $\varepsilon_3 = \varepsilon_1\varepsilon_2$.

The polynomial $\varphi(x)$ of degree s is said to be *recursive* if

$$\varphi(x) = \pm x^s \varphi\left(\frac{1}{x}\right).$$

Lemma 2.3.1. *Let $f(x) = \varphi(x)\psi(x)$, where $\varphi(x)$ and $\psi(x)$ are monic polynomials of positive degree with integer coefficients. Then at least one of the polynomials $\varphi(x)$ and $\psi(x)$ is recursive.*

Proof. Let $r = \deg \varphi$ and $s = n - r = \deg \psi$. Consider the polynomials

$$f_1(x) = x^r \varphi\left(\frac{1}{x}\right)\psi(x) = \sum_{i=0}^{n} c_i x^{n-i},$$

$$f_2(x) = x^s \psi\left(\frac{1}{x}\right)\varphi(x) = x^n f_1\left(\frac{1}{x}\right) = \sum_{i=0}^{n} c_{n-i} x^{n-i}.$$

Clearly,

$$f_1(x)f_2(x) = x^n f\left(\frac{1}{x}\right) =$$

$$(x^n + \varepsilon_1 x^m + \varepsilon_2 x^p + \varepsilon_3)(\varepsilon_3 x^n + \varepsilon_2 x^{n-p} + \varepsilon_1 x^{n-m} + 1).$$

Comparison of the coefficients of x^{2n} shows that $c_0 c_n = \varepsilon_3$, and hence $c_0 = \pm 1$ and $c_n = \pm 1$. Comparison of the coefficients of x^n shows that

$$c_0^2 + c_1^2 + \cdots + c_{n-1}^2 + c_n^2 = 1 + \varepsilon_1^2 + \varepsilon_2^2 + \varepsilon_3^2 = 4,$$

i.e., $c_1^2 + \cdots + c_{n-1}^2 = 2$. Thus, $c_0 = \pm 1$, $c_n = \pm 1$, $c_\alpha = \pm 1$ and $c_\beta = \pm 1$ for some $1 \leq \alpha < \beta \leq n - 1$, all the other coefficients c_i being zero. Hence $f_1(x)f_2(x)$ can be expressed in the following two forms:

$$c_0 c_n x^{2n} + c_\alpha c_n x^{2n-\alpha} + c_\beta c_n x^{2n-\beta} + c_0 c_\alpha x^{n+\alpha} + \tag{1}$$
$$c_0 c_\beta x^{n+\beta} + c_\alpha c_\beta x^{n+\beta-\alpha} + 4x^n + \cdots$$

and

$$\varepsilon_3 x^{2n} + \varepsilon_2 x^{2n-p} + \varepsilon_1 x^{2n-m} + \varepsilon_1 \varepsilon_3 x^{n+m} + \varepsilon_2 \varepsilon_3 x^{n+p} + \varepsilon_1 \varepsilon_2 x^{n+m-p} + \tag{2}$$
$$4x^n + \cdots$$

In order to compare (1) and (2), let us order the monomials with respect to the size of the degrees taking into account only the three highest monomials. For (1), we obtain the four possibilities:

$$\beta \leq \frac{n}{2} \quad : \quad 2n > 2n - \alpha > 2n - \beta,$$

$$\beta > \frac{n}{2}, \quad \alpha \leq n - \beta \quad : \quad 2n > 2n - \alpha \geq n + \beta,$$

$$\beta > \frac{n}{2}, \quad \frac{n}{2} \geq \alpha > n - \beta \quad : \quad 2n > n + \beta > 2n - \alpha,$$

$$\beta > \frac{n}{2}, \quad \alpha > \frac{n}{2} \quad : \quad 2n > n + \beta > n + \alpha.$$

For (2), we obtain the two possibilities:

$$n \geq 2m \ : \ 2n > 2n - p > 2n - m,$$
$$2m > n \geq n + p \ : \ 2n > 2n - p > n + m.$$

Comparing the three highest monomials in (1) and (2) we obtain for the pair (α, β) the following four possibilities:

$$(\alpha, \beta) = (p, m), \ (p, n - m), \ (m, n - p) \text{ or } (n - m, n - p).$$

If $(\alpha, \beta) = (p, m)$, comparison of (1) with (2) shows that

$$c_0 c_n = \varepsilon_3, \quad c_p c_n = \varepsilon_2, \quad c_m c_n = \varepsilon_1.$$

Hence

$$f_1(x) = c_n(\varepsilon_3 x^n + \varepsilon_2 x^{n-p} + \varepsilon_1 x^{n-m} + 1) = c_n x^n f\left(\frac{1}{x}\right).$$

Therefore $\psi(x) = c_n x^s \psi\left(\frac{1}{x}\right).$

If $(\alpha, \beta) = (n - m, n - p)$, we similarly see that

$$c_0 c_n = \varepsilon_3, \quad c_0 c_{n-m} = \varepsilon_1, \quad c_0 c_{n-p} = \varepsilon_2.$$

Hence

$$f_1(x) = c_0(x^n + \varepsilon_1 x^m + \varepsilon_2 x^p + \varepsilon_3) = c_0 f(x).$$

Therefore $\varphi(x) = c_0 x^r \varphi\left(\frac{1}{x}\right).$

If $(\alpha, \beta) = (p, n - m)$, then we encounter in (1) monomials of degrees

$$2n, \ 2n - p, \ n + m, \ n + p, \ 2n - m, \ 2n - m - p, \ n,$$

and in (2) monomials of degrees

$$2n, \ 2n - p, \ 2n - m, \ n + m, \ n + p, \ n + m - p, \ n.$$

Therefore the number $2n - m - p$ is equal to one of the three numbers: $n + m$, $n + p$, $n + m - p$. The equalities $2n - m - p = n + m$ and $2n - m - p = n + p$ contradict the assumption that $n \leq m + p$. and hence $2n - m - p = n + m - p$, i.e., $n = 2m$. Therefore $(\alpha, \beta) = (p, m)$.

If $(\alpha, \beta) = (m, n - p)$, we similarly see that $n = 2m$, i.e., $(\alpha, \beta) = (n - m, n - p)$. $\square$

Lemma 2.3.2. *Let λ and λ^{-1} be the roots of $f(x)$. Then one of the following three pairs of conditions hold:*

$$\text{(I)} \quad \lambda^n = -\varepsilon_3 \quad \text{and} \quad \lambda^{n-p} = -\varepsilon_1\varepsilon_2,$$

$$\text{(II)} \quad \lambda^m = -\varepsilon_1\varepsilon_3 \ \text{and} \ \lambda^{n-p} = -\varepsilon_2,$$

$$\text{(III)} \ \lambda^p = -\varepsilon_2\varepsilon_3 \ \text{and} \ \lambda^{n-m} = -\varepsilon_1.$$

Proof. The conditions $f(\lambda) = 0$ and $f(\lambda^{-1}) = 0$ can be expressed as

$$\lambda^n + \varepsilon_1\lambda^m + \varepsilon_2\lambda^p + \varepsilon_3 = 0, \quad \lambda^n + \varepsilon_2\varepsilon_3\lambda^{n-p} + \varepsilon_1\varepsilon_3\lambda^{n-m} + \varepsilon_3 = 0.$$

By subtracting one equation from the other one we get

$$\varepsilon_2\varepsilon_3\lambda^{n-p} + \varepsilon_1\varepsilon_3\lambda^{n-m} - \varepsilon_1\lambda^m - \varepsilon_2\lambda^p = 0,$$

i.e.,

$$(\varepsilon_2\lambda^{m-p} + \varepsilon_1)(\varepsilon_3\lambda^{n-m} - \varepsilon_1\varepsilon_2\lambda^p) = 0.$$

and hence either $\lambda^p = -\varepsilon_1\varepsilon_2\lambda^m$ or $\lambda^p = \varepsilon_1\varepsilon_2\varepsilon_3\lambda^{n-m}$. Substituting these values of λ^p into the relation $f(\lambda) = 0$ we obtain accordingly either $\lambda^n = -\varepsilon_3$ or

$$(\lambda^m + \varepsilon_1\varepsilon_2)(\lambda^{n-m} - \varepsilon_1) = 0. \ \square$$

With the help of Lemmas 2.3.1 and 2.3.2 it is easy to prove the following two theorems which in turn lead to a complete description of irreducible polynomials of the form $x^n + \varepsilon_1 x^m + \varepsilon_2 x^p + \varepsilon_3$. In both theorems (as well as in Lemmas 2.3.1 and 2.3.2) we assume that $n \le m+p$ and $f(x) \ne (x^m + \varepsilon_2)(x^p + \varepsilon_1)$.

Theorem 2.3.3. a) *If the polynomial $f(x)$ has no roots which are roots of unity, then $f(x)$ is irreducible.*

b) *If the polynomial $f(x)$ has exactly q roots which are roots of unity, then $f(x)$ can be represented as the product of two polynomials with integer coefficients one of which is of degree q and its roots are the given roots of unity, while the other polynomial is irreducible.*

Proof. Let $f(x) = \varphi(x)\psi(x)$, where $\varphi, \psi \in \mathbb{Z}[x]$. By Lemma 2.3.1 we may assume that if λ is a root of φ, then λ^{-1} is also a root of φ. It then follows from Lemma 2.3.2 that λ is a root of unity. If not all the roots of f are the roots of unity, then either ψ is irreducible over $\mathbb{Z}$ or $\psi = \psi_1\psi_2$, where $\psi_1, \psi_2 \in \mathbb{Z}[x]$ and all the roots of ψ_1 are the roots of unity whereas ψ_2 has a root which is not a root of unity. In this case all the roots of $\varphi\psi_1$ are the roots of unity. By continuing the same arguments now applied to ψ_2 we obtain the factorization desired of f. $\square$

It remains to determine, when f has roots which are roots of unity. The answer is given by the following theorem.

Theorem 2.3.4. *Let d be the greatest common divisor of n, m, p. Set*

$$n_1 = \frac{n}{d}, \quad m_1 = \frac{m}{d}, \quad p_1 = \frac{p}{d},$$
$$d_1 = (n_1, m_1 - p_1), \quad d_2 = (m_1, n_1 - p_1), \quad d_3 = (p_1, n_1 - m_1).$$

Then any root of unity which is a root of f satisfies one of the equations

$$x^{dd_1} = \pm 1, \quad x^{dd_2} = \pm 1, \quad x^{dd_3} = \pm 1,$$

and it is a simple root of f.

Proof. Let λ be a root of unity which is a root of f. Then λ^{-1} is also a root of f. Lemma 2.3.2 provides three possibilities for conditions on λ. Consider, e.g., case (I): $\lambda^n = -\varepsilon_3$ and $\lambda^{m-p} = -\varepsilon_1\varepsilon_2$. Clearly, $(n, m - p) = dd_1$, and hence there exist integers u and v such that $dd_1 = nu + (m - p)v$. Therefore $\lambda^{dd_1} = (\lambda^n)^u(\lambda^{m-p})^v = \pm 1$ since $\lambda^n = -\varepsilon_3 = \pm 1$ and $\lambda^{m-p} = -\varepsilon_1\varepsilon_3 = \pm 1$. Cases (II) and (III) are similarly considered.

It remains to prove that λ is a simple root of f, i.e.,

$$\lambda f'(\lambda) = n\lambda^n + \varepsilon_1 m\lambda^m + \varepsilon_2 p\lambda^p \neq 0.$$

Substituting (I), (II) and (III) into $n\lambda^n + \varepsilon_1 m\lambda^m + \varepsilon_2 p\lambda^p = 0$ we respectively obtain

$$\varepsilon_2\lambda^p(p - m) = n\varepsilon_3, \quad \varepsilon_2\lambda^p(p - n) = m\varepsilon_3, \quad \varepsilon_1\lambda^m(m - n) = p\varepsilon_2\varepsilon_3.$$

The equality $|\lambda| = 1$ cannot occur in the first case whereas in the second and third ones it means that $n = m + p$. If $n = m + p$, the relations (II) take the form $\lambda^m = -\varepsilon_1\varepsilon_3$ and $\lambda^n = -\varepsilon_2$ whereas relations (III) take the form $\lambda^p = -\varepsilon_2\varepsilon_3$ and $\lambda^p = -\varepsilon_1$. In both cases $\varepsilon_3 = \varepsilon_1\varepsilon_2$ which corresponds to the excluded polynomial $f(x) = (x^m + \varepsilon_2)(x^p + \varepsilon_1)$. $\square$

2.3.2 Irreducibility of certain trinomials

Making use of the results obtained in the preceding section it is not difficult to determine which of the trinomials $x^n \pm x^m \pm 1$ are irreducible.

Theorem 2.3.5 ([Lj]). *Let $n \geq 2m$, $d = (n, m)$, $n_1 = \dfrac{n}{d}$ and $m_1 = \dfrac{m}{d}$. Then the trinomial*

$$g(x) = x^n + \varepsilon x^m + \varepsilon', \quad where \ \varepsilon = \pm 1 \ and \ \varepsilon' = \pm 1,$$

is irreducible except for the three cases in which $n_1 + m_1 \equiv 0 \pmod{3}$:
  a) *n_1 and m_1 are odd and $\varepsilon = 1$.*
  b) *n_1 is even and $\varepsilon' = 1$.*
  c) *m_1 is even and $\varepsilon' = \varepsilon$.*
    In all these cases $g(x)$ is a product of $x^{2d} + \varepsilon^m\varepsilon'^n x^d + 1$ by an irreducible polynomial.

Proof. The case when $n = 2m$ and $\varepsilon' = 1$ is obvious. We will therefore assume that either $n = 2m$ and $\varepsilon' = -1$ or $n > 2m$. For such n and ε', we can apply Theorems 2.3.3 and 2.3.4 to the product

$$(x^n + \varepsilon x^m + \varepsilon')(x^n - \varepsilon') = x^{2n} + \varepsilon x^{n+m} - \varepsilon\varepsilon' x^m - 1$$

since $2n > n+m > m$, and if $2n = (n+m)+m$, i.e., $n = 2m$, then $\varepsilon_3 \neq \varepsilon_1\varepsilon_2$. In the notation of Theorem 2.3.4 we have

$$(2n, n + m, m) = (n, m) = d,$$
$$d_1 = (2n_1, n_1) = n_1,$$
$$d_3 = (m_1, n_1 - m_1) = 1$$

and $d_2 = (n_1 + m_1, 2n_1 - m_1) = (n_1 + m_1, 3n_1)$. Therefore

$$d_2 = \begin{cases} 1 & \text{if } n_1 + m_1 \not\equiv 0 \pmod 3 \\ 3 & \text{if } n_1 + m_1 \equiv 0 \pmod 3. \end{cases}$$

By Theorem 2.3.4 the roots of unity which are roots of g satisfy one of the equations $x^{dd_1} = \pm1$, $x^{dd_2} = \pm1$, $x^{dd_3} = \pm1$. The first of these equations is of the form $x^n = \pm1$, the third one is of the form $x^d = \pm1$ because $d_3 = 1$.

If $x^n = \pm1$, then $g(x) = \pm1 + \varepsilon x^m \pm 1 \neq 0$ and, if $x^d = \pm1$, then $g(x) = \pm1 \pm 1 \pm 1 \neq 0$.

It remains to consider the case when $d_2 = 3$. In Lemma 2.3.2, case (I) leads to the relations

$$\lambda^{2n} = \pm1 \text{ and } \lambda^n = \pm1$$

whereas case (III) leads to the relations

$$\lambda^m = \pm1 \text{ and } \lambda^{n-m} = \pm1.$$

In both cases we get $\lambda^n = \pm1$, and hence $g(\lambda) \neq 0$. Case (II) leads to the relations $\lambda^{n+m} = \varepsilon$, $\lambda^{2n-m} = \varepsilon\varepsilon'$, i.e., $\lambda^{3n} = \varepsilon'$, $\lambda^{3m} = \varepsilon\varepsilon'$. Therefore $(\lambda^{3d})^{n_1} = \varepsilon'$ and $(\lambda^{3d})^{m_1} = \varepsilon\varepsilon'$, where $(n_1, m_1) = 1$ and $n_1 + m_1 \equiv 0 \pmod 3$. From the condition $(n_1, m_1) = 1$ it follows that $n_1 u + m_1 v = 1$ for some integers u and v. Hence

$$\lambda^{3d} = \lambda^{3dn_1 u + 3dm_1 v} = (\varepsilon')^u (\varepsilon\varepsilon')^v = \pm1.$$

If n_1 and m_1 are odd, then $\lambda^{3d} = \varepsilon' = \varepsilon\varepsilon'$, and hence $\varepsilon = 1$ and $\lambda^{3d} = \varepsilon'$.
If n_1 is even, then $\varepsilon' = 1$.
If m_1 is even, then $\varepsilon\varepsilon' = 1$. $\square$

By Perron's criterion (Theorem 2.2.5 on page 56) the trinomial $x^n \pm ax^{n-1} \pm 1$, where $a \geq 3$ is an integer, is irreducible. For $a = 2$, this trinomial is irreducible if it has no roots equal to ±1. All these statements also hold for the trinomial $x^n \pm ax \pm 1$.

Irreducibility of trinomials $x^n \pm 2x^m \pm 1$ is studied in [Sc2].

In conclusion we formulate two further theorems on the irreducibility of trinomials.

Theorem 2.3.6 ([Mi2]). *Let the trinomial* $x^n \pm px^m \pm 1$, *where* $n > m$ *and* p *is a prime, be reducible. Then* $\dfrac{n}{(n,m)} \leq 4^{p^2}$.

Theorem 2.3.7 ([Ra2]). a) *The trinomial* $x^5 + x + n$ *factorizes into the product of irreducible quadratic and cubic polynomials if and only if* $n = \pm 1$ *or* $n = \pm 6$.

b) *The trinomial* $x^5 - x + n$ *factorizes into the product of irreducible quadratic and cubic polynomials if and only if* $n = \pm 15$, $n = \pm 22\,440$ *or* $n = \pm 2\,759\,640$.

2.4 Hilbert's irreducibility theorem

Let $f(t, x) \in \mathbb{Q}[t, x]$ be a polynomial in two variables. The polynomial f is called *reducible* if $f = gh$, where $g, h \in \mathbb{Q}[t, x]$ are polynomials of positive degree.

Theorem 2.4.1 (Hilbert, [Hi3]). *If* $f(t, x)$ *is an irreducible polynomial over* $\mathbb{Q}$, *then there exist infinitely many rationals* t_0 *for which the polynomial* $f(t_0, x)$ *in one indeterminate is irreducible over* $\mathbb{Q}$.

We give the proof of Hilbert's theorem due to Dörge [Do2] using the exposition of this proof given in [La3] and [Se2]. A proof using more modern language can be found in [Fr].

Let us start by establishing a relation between reducibility of the polynomial $f(t_0, x)$ and existence of point (t_0, y_0) with rational coordinates on a certain algebraic curve. Let $f(t, x) \in \mathbb{Q}[t, x]$ be an irreducible polynomial. Let us represent it in the form

$$f(t, x) = a_n(t)x^n + \cdots + a_0(t), \text{ where } a_i(t) \in \mathbb{Q}[t],$$

and we define the polynomial $F(x) = f(t, x)$ with coefficients from the field $k = \mathbb{Q}(t)$. Let $\overline{k}$ be the algebraic closure of k. Then

$$F(x) = a_n(t) \cdot (x - \alpha_1) \cdot \ldots \cdot (x - \alpha_n),$$

where $\alpha_1, \ldots, \alpha_n \in \overline{k}$ are the roots of $F(x)$ which is irreducible over $k = \mathbb{Q}(t)$. If $a_n(t_0) \neq 0$ there correspond to them the roots $\alpha_1', \ldots, \alpha_n' \in \overline{\mathbb{Q}}$ of $f(t_0, x)$.

Suppose that $f(t_0, x)$ is reducible over $\mathbb{Q}$. After a renumeration of the roots we may assume that $f(t_0, x) = a_n(t_0)g_0(x)h_0(x)$, where

$$g_0(x) = (x - \alpha_1') \cdot \ldots \cdot (x - \alpha_s') \in \mathbb{Q}[x],$$
$$h_0(x) = (x - \alpha_{s+1}') \cdot \ldots \cdot (x - \alpha_n') \in \mathbb{Q}[x].$$

Set

$$g(x) = (x - \alpha_1) \cdot \ldots \cdot (x - \alpha_s) \text{ and } h(x) = (x - \alpha_{s+1}) \cdot \ldots \cdot (x - \alpha_n).$$

Then $F(x) = a_n(t)g(x)h(x)$, where $g(x)$, $h(x) \in \overline{k}[x]$. By assumption $F(x)$ is irreducible over $k[x]$, and hence $g(x)$ has a coefficient y which belongs to $\overline{k} \setminus k$. Recall that $k = \mathbb{Q}(t)$, so that y is algebraic over $\mathbb{Q}(t)$, i.e.,

$$b_m(t)y^m + b_{m-1}(t)y^{m-1} + \cdots + b_0(t) = 0, \quad \text{where} \quad b_i(t) \in \mathbb{Q}(t).$$

As a result we obtain an algebraic curve C (with rational coefficients) in the plane (t, y). To the coefficient y of g there corresponds a coefficient $y_0 \in \mathbb{Q}$ of the polynomial g_0 and this coefficient satisfies the relation

$$b_m(t_0)y_0^m + b_{m-1}(t_0)y_0^{m-1} + \cdots + b_0(t_0) = 0,$$

i.e., C has a rational point (t_0, y_0).

Thus, consider all polynomials of the form $(x - \alpha_{i_1}) \cdot \ldots \cdot (x - \alpha_{i_k})$, where $1 \le k \le n - 1$, and, for each of these polynomials, select the coefficient that does not belong to $\mathbb{Q}[t]$. To these coefficients we assign plane algebraic curves $C_1, \ldots, C_M$ with rational coefficients. If $t_0 \in \mathbb{Q}$ is such that none of the curves $C_1, \ldots, C_M$ has rational points (t_0, y_0), the polynomial $f(t_0, x)$ is irreducible.

Let us now study rational points of the plane algebraic curve C given by the equation

$$b_m(t)y^m + b_{m-1}(t)y^{m-1} + \cdots + b_0(t) = 0, \quad \text{where} \quad b_i(t) \in \mathbb{Z}(t).$$

First we make the change of variable $\widetilde{y} = b_m(t)y$. As a result we obtain the curve

$$\widetilde{y}^m + b_{m-1}(t)\widetilde{y}^{m-1} + b_{m-2}(t)b_m(t)\widetilde{y}^{m-2} + \cdots + b_0(t)\big(b_m(t)\big)^{m-1} = 0.$$

If $(t_0, \widetilde{y}_0)$ is a rational point on this curve and $t_0 \in \mathbb{Z}$, then $\widetilde{y}_0 \in \mathbb{Z}$. In what follows we confine ourselves to the study of integer points on the curve.

Thus, we may assume that $b_m(t) = 1$, i.e., the curve is given by the equation

$$y^m + b_{m-1}(t)y^{m-1} + \cdots + b_0(t) = 0, \quad \text{where} \quad b_i(t) \in \mathbb{Z}(t).$$

Let us show that in a vicinity of the point $t = \infty$ the algebraic function $y(t)$ can be expanded as

$$y(t) = a(\sqrt[k]{t})^n + \cdots + b + c(\sqrt[k]{t})^{-1} + \cdots,$$

where $\sqrt[k]{t}$ is one of the branches of the k-th root of t (to be definite, we select the branch for which $\sqrt[k]{t} > 0$ for $t > 0$). The map $(y, t) \mapsto t$ determines a ramified covering $M^2 \to \mathbb{C}P^1$ where M^2 is the Riemann surface of the algebraic function $y(t)$. We are interested in the branches of this covering over ∞. Take one of the branches and consider its intersection with the preimage of a neighborhood of ∞. The restriction of the ramified covering onto this set is of the form $z \mapsto z^k$. This means that $y(z)$ is single-valued and $z^k = t$. It is also clear that $z = \infty$ is not an essentially singular point of $y(z)$.

We are interested in the case where there exists an infinite increasing sequence of positive integers t_i for which $y(t_i)$ is a real (and moreover integer) number. Let us show that in this case all the coefficients of the expansion $y(t)$ are real. Suppose that these coefficients are not all real. Let $\xi t^{s/k}$ be the term of highest degree s/k with non-real ξ. Then, for real values of t, the terms of higher degree do not affect the imaginary part of the sum of the series, and as $t \to +\infty$ the terms of lesser degree are small as compared with $\xi t^{s/k}$ and cannot cancel its imaginary part.

It remains to perform the last step — proving that the numbers $t_i \in \mathbb{N}$ for which $y(t_i) \in \mathbb{Z}$ constitute a set of zero density. This easily follows from the next statement.

Theorem 2.4.2. *Let*

$$\varphi(t) = a(\sqrt[k]{t})^n + \cdots + b + c(\sqrt[k]{t})^{-1} + \cdots,$$

where t is real and the series is real and converges for $t \geq R$. Suppose that $\varphi(t)$ is not a polynomial. Then there exist constants $C > 0$ and $\varepsilon \in (0,1)$ such that the number of positive integers $t \leq N$ for which $\varphi(t) \in \mathbb{Z}$ does not exceed CN^ε.

Proof. Observe first of all that in the expansion of the m-th derivative of $\varphi(t)$ there are no terms of the form t^ν, where $\nu > \dfrac{n}{k} - m$. Therefore we can select an integer $m \geq 1$ such that $\varphi^{(m)}(t) \sim ct^{-\mu}$ as $t \to \infty$, where $\mu > 0$ and $c \neq 0$ (the latter property is ensured by the fact that φ is not a polynomial).

Lemma. *There exist positive constants c_1 and α such that, if T is sufficiently large, then the interval $[T, T+c_1T^\alpha]$ contains no more than m positive integers t for which $\varphi(t) \in \mathbb{Z}$.*

Proof. Let $t_1 < \cdots < t_{m+1}$. Consider Lagrange's interpolation polynomial

$$f(t) = \sum_{i=1}^{m+1} \varphi(t_i) \frac{(t - t_1) \cdot \ldots \cdot (t - t_{i-1})(t - t_{i+1}) \cdot \ldots \cdot (t - t_{m+1})}{(t_i - t_1) \cdot \ldots \cdot (t_i - t_{i-1})(t_i - t_{i+1}) \cdot \ldots \cdot (t_i - t_{m+1})}.$$

The function $\varphi - f$ vanishes at $t_1, \ldots, t_{m+1}$, and hence by Rolle's theorem there exists a point $\xi \in [t_1, t_{m+1}]$ such that

$$\varphi^{(m)}(\xi) = f^{(m)}(\xi) =$$
$$m! \sum_{1 \leq i \leq m+1} \frac{\varphi(t_i)}{(t_i - t_1) \cdot \ldots \cdot (t_i - t_{i-1})(t_i - t_{i+1}) \cdot \ldots \cdot (t_i - t_{m+1})}.$$

Hence, $\varphi^{(m)}(\xi)$ is a rational number whose denominator does not exceed

$$\prod_{1 \leq i < j \leq m+1} (t_j - t_i) < (t_{m+1} - t_1)^{m(m+1)/2}.$$

On the other hand, if t_1 is sufficiently large, then $0 < \left|\varphi^{(m)}(\xi)\right| \leq c_2 t_1^{-\mu}$.

Set $\Delta T = t_{m+1} - t_1$. Then $\left|f^{(m)}(\xi)\right| \geq \Delta T^{-m(m+1)/2}$, and so $c_2 t_1^{-\mu} \geq \Delta T^{-m(m+1)/2}$, i.e., $\Delta T > c_1 T^\alpha$, where $\alpha = \dfrac{2\mu}{m(m+1)}$ and $c_1 = c_2^{-2/m(m+1)}$. $\square$

Returning to the proof of the theorem, we select ε so that $1 - \alpha\varepsilon = \varepsilon$, i.e., $\varepsilon = \dfrac{1}{1+\alpha}$. Then $0 < \varepsilon < 1$. Let us split the interval $[1, N]$ into subintervals $[1, N^\varepsilon]$ and $[N^\varepsilon, N]$. By the Lemma any segment of length $c_1(N^\varepsilon)^\alpha$ that lies in $[N^\varepsilon, N]$ contains no more than m positive integers t for which $\varphi(t) \in \mathbb{Z}$. and hence the total number of such positive integers in $[1, N]$ does not exceed

$$N^\varepsilon + n\frac{N - N^\varepsilon}{c_1 N^{\alpha\varepsilon}} < N^\varepsilon + \frac{m}{c_1}N^{1-\alpha\varepsilon} = N^\varepsilon + \frac{m}{c_1}N^\varepsilon.$$

Therefore we can set $C = 1 + \dfrac{m}{c_1}$. $\square$

Let $B(N)$ be the total number of positive integers $t \leq N$ for which $\varphi(t) \in \mathbb{Z}$. Then $\lim\limits_{N\to\infty} \dfrac{CN^\varepsilon}{N} = \lim\limits_{N\to\infty} \dfrac{C}{N^{1-\varepsilon}} = 0$. This means in particular that there exist infinitely many positive integers t for which $\varphi(t) \notin \mathbb{Z}$.

2.5 Algorithms for factorization into irreducible factors

2.5.1 Berlekamp's algorithm

The most effective algorithms for factorizing a polynomial with integer coefficients into irreducible factors uses the factorization of this polynomial over fields $\mathbb{F}_p$ for a prime p. Therefore we first discuss one of the algorithms for factorizing polynomials modulo p suggested by Berlekamp [Be4].

Let f be a polynomial with coefficients from $\mathbb{F}_p$. We may assume that the polynomial is monic. Before we apply Berlekamp's algorithm we have to get rid of multiple irreducible factors of f. This is done as follows. Let $f = f_1^{n_1} \cdot \ldots \cdot f_k^{n_k}$, where $f_1, \ldots, f_k$ are distinct irreducible monic polynomials. It is easy to verify that

$$d = (f, f') = \prod_{p\nmid n_i} f_i^{n_i-1} \prod_{p\mid n_i} f_i^{n_i}, \quad \text{i.e.,} \quad \frac{f}{d} = \prod_{p\nmid n_i} f_i.$$

The polynomial f factorizes into the product of d and $\dfrac{f}{d}$, and in the factorization of $\dfrac{f}{d}$ there are no multiple irreducible factors. If $\deg d < \deg f$, then we can apply the same procedure to d. If $0 < \deg d = \deg f$, then $d = \prod\limits_{p\mid n_i} f_i^{n_i} = g^p$. Clearly, $\deg g < \deg f$ and from factorization of g one can recover the factorization of d.

Berlekamp's algorithm is based on the following theorem.

Theorem 2.5.1. *Let $f \in \mathbb{F}_p[x]$ be a monic polynomial of positive degree n.*

a) *If $h \in \mathbb{F}_p[x]$ satisfies the relation $h^p \equiv h \pmod{f}$, i.e., $h^p - h$ is divisible by f, then*

$$f(x) = \prod_{a \in \mathbb{F}_p} \left(f(x), h(x) - a \right).$$

b) *Let $f = f_1 \cdot \ldots \cdot f_k$, where $f_1, \ldots, f_k$ are different irreducible monic polynomials. In this case h satisfies the relation $h^p \equiv h \pmod{f}$ if and only if $h(x) \equiv a_i \pmod{f_i}$, where $a_i \in \mathbb{F}_p$. To each collection $(a_1, \ldots, a_k)$ there corresponds exactly one polynomial h whose degree is less than that of f.*

Proof. a) Set $F(x) = \prod_{a \in \mathbb{F}_p} \left(f(x), h(x) - a \right)$. Polynomials $h(x) - a$ with distinct a are relatively prime, and hence polynomials $\left(f(x), h(x) - a \right)$ are relatively prime divisors of $f(x)$. Hence $f(x)$ is divisible by their product $F(x)$. On the other hand, in $\mathbb{F}_p$ the polynomial identity $\prod_{a \in \mathbb{F}_p} (y - a) = y^p - y$

holds, and hence the polynomial

$$\prod_{a \in \mathbb{F}_p} \left(h(x) - a \right) = \left(h(x) \right)^p - h(x)$$

is divisible by $f(x)$, and therefore $F(x)$ is divisible by $f(x)$. Thus, the polynomials f and F are divisible by each other and are monic. Hence $F = f$.

b) If $h(x) \equiv a_i \pmod{f_i}$, then $\left(h(x) \right)^p \equiv a_i^p \equiv a_i \equiv h(x) \pmod{f_i}$, and therefore $\left(h(x) \right)^p \equiv h(x) \pmod{f_1 \cdot \ldots \cdot f_k}$. Conversely, if the polynomial

$$\left(h(x) \right)^p - h(x) = \prod_{a \in \mathbb{F}_p} \left(h(x) - a \right)$$

is divisible by f, then it is divisible by all the polynomials $f_1, \ldots, f_k$. It is also clear that if the irreducible polynomial f_i divides the product of pairwise relatively prime factors $h(x) - a$, then it divides one of these factors, i.e., $h(x) \equiv a_i \pmod{f_i}$.

The existence and uniqueness of a polynomial h corresponding to a given collection $(a_1, \ldots, a_k)$ obviously follows from the next statement called the Chinese remainder theorem for polynomials.

Lemma. *Let $f_1, \ldots, f_k$ be relatively prime irreducible polynomials over a field F, and let $g_1, \ldots, g_k$ be arbitrary polynomials over the same field. Then there exists a polynomial h such that $h(x) \equiv g_i \pmod{f_i}$ and this polynomial is uniquely determined modulo $f = f_1 \cdot \ldots \cdot f_k$.*

Proof. The polynomials f_i and $F_i = \dfrac{f}{f_i}$ are relatively prime, and hence there exist polynomials a_i and b_i such that $a_i f_i + b_i F_i = 1$. Further, $b_i F_i \equiv 1 \pmod{f_i}$ and $b_i F_i \equiv 0 \pmod{f_j}$ for $j \neq i$.

Set $h = \sum g_i b_i F_i$. Then $h \equiv g_i b_i F_i \pmod{f_i} \equiv g_i \pmod{f_i}$. The existence of the required polynomial h is thereby proved.

The uniqueness of h follows from the fact that, if $h_1 - g_i$ and $h_2 - g_i$ are divisible by f_i, then $h_1 - h_2$ is divisible by $f_1 \cdot \ldots \cdot f_k = f$. $\square$ $\square$

The relation
$$\big(h(x)\big)^p \equiv h(x) \pmod{f}$$
is equivalent to a system of linear equations over $\mathbb{F}_p$. Indeed, recall that $\deg h < \deg f = n$ and let
$$h(x) = t_0 + t_1 x + \cdots + t_{n-1} x^{n-1}.$$

Then
$$h(x)^p = h(x^p) = t_0 + t_1 x^p + \cdots + t_{n-1} x^{p(n-1)}.$$

We find the residue of each monomial x^{pj}, where $j = 0, 1, \ldots, n-1$, after division by f:
$$x^{pj} \equiv \sum_{i=0}^{n-1} q_{ij} x^i \pmod{f}.$$

As a result we obtain a system of linear equations:
$$\sum_{i=0}^{n-1} t_j q_{ij} = t_i, \quad i = 1, \ \ldots, \ n-1.$$

The dimension of the space of solutions of this system is equal to k, the number of irreducible factors of f. Clearly, $q_{00} = 1$ and $q_{i0} = 0$ for $i > 0$. Therefore the system has a trivial solution $t_0 = c$, $t_1 = \cdots = t_{n-1} = 0$. This solution corresponds to the polynomial h of degree zero.

Let $h_1 = 1, h_2, \ldots, h_k$ be a basis in the space of solutions. If $k = 1$, then f is irreducible. If $k > 1$, let us find the greatest common divisors of polynomials $f(x)$ and $h_2(x) - a$ for all $a \in \mathbb{F}_p$. As a result we obtain a collection of divisors $g_1, \ldots, g_s$ of f. If $s < k$, then, for each g_i, we compute $\big(g_i, h_3(x) - a\big)$, and so on, until we obtain all k divisors.

It is easy to verify that at the end we necessarily obtain all the k divisors. Indeed, let f_1 and f_2 be distinct irreducible divisors of f. Consider a collection $(a_1, a_2, \ldots, a_k)$, where $a_1 \neq a_2$. Then there is a corresponding polynomial h for which
$$h(x) \equiv a_1 \pmod{f_1} \text{ and } h(x) \equiv a_2 \pmod{f_2}.$$

Therefore, for a certain basic polynomial h_i, we should have $h_i(x) \equiv a_{1i} \pmod{f_1}$ and $h_i(x) \equiv a_{2i} \pmod{f_2}$, where $a_{1i} \neq a_{2i}$. Such a polynomial distinguishes factors f_1 and f_2.

Remark. The efficiency of Berlekamp's algorithm can be essentially enhanced by using the factorization algorithm due to *Cantor and Zassenhaus* [Ca2]. Namely, instead of the polynomials $h_i(x) - a = h_i(x) - ah_1(x)$, we take the polynomials

$$H(x) = a_1 h_1(x) + \cdots + a_k h_k(x),$$

where $a_1, \ \ldots, \ a_k$ is a random collection of elements from $\mathbb{F}_p$, and then compute the greatest common divisors of f and $H^{(p-1)/2} - 1$. If f is reducible and $p \geq 3$, then with probability $\geq \dfrac{4}{9}$ we immediately obtain a nontrivial factorization.

2.5.2 Factorization with the help of Hensel's lemma

On page 49 we gave Kronecker's algorithm for factorizing a polynomial with integer coefficients into irreducible factors, but this algorithm requires too much computation. Much more effective algorithms are known. The *triple-L algorithm* due to Lenstra–Lenstra–Lovász which we discuss in the Appendix (see p. 279) is of the greatest theoretical interest. In practice, however, it often turns out to be slower than the factorization algorithm we describe now.

Let f be a polynomial with integer coefficients. If $\text{cont}(f) \neq 1$, then having divided f by $\text{cont}(f)$ we get a polynomial with content 1. Let $f = f_1^{n_1} \cdot \ldots \cdot f_k^{n_k}$ be a factorization of f over $\mathbb{Z}$ into irreducible factors. Then

$$f' = f_1^{n_1-1} \cdot \ldots \cdot f_k^{n_k-1} g, \quad \text{where } g \in \mathbb{Z}[x].$$

Therefore, over $\mathbb{Z}$, the greatest common divisor of f and f' is equal to $f_1^{n_1-1} \cdot \ldots \cdot f_k^{n_k-1}$, the same as over $\mathbb{Q}$. Therefore, over $\mathbb{Z}$, we can also divide f by (f, f') and obtain a polynomial without multiple roots. So in what follows we will assume that $\text{cont}(f) = 1$ and polynomials f and f' are relatively prime.

Since $(f, f') = 1$, there exist polynomials $u, \ v \in \mathbb{Q}[x]$ for which

$$uf + vf' = 1.$$

Hence there exist polynomials $\overline{u}, \overline{v} \in \mathbb{Z}[x]$ for which $\overline{u}f + \overline{v}f' = n$, where $n \in \mathbb{N}$. If a prime p is relatively prime to n, then, over $\mathbb{F}_p$, the greatest common divisor of f and f' is equal to 1. Let us consecutively calculate (f, f') over $\mathbb{F}_p$ for $p = 2, 3, 5, \ldots$ until we get $(f, f') = 1$ and simultaneously the highest coefficient of f will be relatively prime to p. Fix this p in what follows.

Modulo p, the polynomial f has no multiple irreducible factors. Hence we can apply Berlekamp's algorithm and obtain a factorization $f \equiv a f_1 \cdot \ldots \cdot f_k$ modulo p, where $f_1, \ldots, f_k \in \mathbb{Z}[x]$ are monic polynomials, a is the highest coefficient of f and $\deg f = \deg f_1 + \cdots + \deg f_k$. Hensel's lemma given below enables us to construct a factorization of f modulo p^m starting from the above factorization.

It suffices to consider the following situation:

$f \equiv f_1 f_2 \pmod{p^m}$, where $f, f_1, f_2 \in \mathbb{Z}[x]$, $\deg f = \deg f_1 + \deg f_2$, f_1 is monic, the leading coefficient of f is relatively prime to p, and $\quad$ (∗) the polynomials f_1 and f_2 are relatively prime modulo p.

The last line implies that there exist polynomials $u, v \in \mathbb{Z}[x]$ for which $uf_1 + vf_2 \equiv 1 \pmod{p}$. If u', v' are some other such polynomials, then $u' \equiv u + wf_2 \pmod{p}$ and $v' \equiv v - wf_1 \pmod{p}$, where $w \in \mathbb{Z}[x]$. Therefore the conditions $\deg u < \deg f_2$, $\deg v < \deg f_1$ uniquely determine u and v modulo p.

Hensel's continuation of the factorization $f \equiv f_1 f_2 \pmod{p^m}$ is a factorization $f \equiv \overline{f}_1 \overline{f}_2 \pmod{p^{m+1}}$, where the polynomials $\overline{f}_1$ and $\overline{f}_2$ satisfy the same conditions as f_1 and f_2, and where $\overline{f}_i \equiv f_i \pmod{p^m}$ and $\deg \overline{f}_i = \deg f_i$.

Lemma. *If $m \geq 1$, then, for any factorization $f \equiv f_1 f_2 \pmod{p^m}$ satisfying the above condition (*), there exists Hensel's extension $f \equiv \overline{f}_1 \overline{f}_2 \pmod{p^{m+1}}$ for which the polynomials $\overline{f}_1$ and $\overline{f}_2$ are uniquely determined modulo p^{m+1}.*

Proof. We are looking for polynomials $\overline{f}_1$, $\overline{f}_2 \in \mathbb{Z}[x]$ such that

$$\overline{f}_i = f_i + p^m g_i, \quad \deg \overline{f}_i = \deg f_i \quad \text{for } i = 1, 2,$$

$\overline{f}_1$ is monic and the congruence $f \equiv \overline{f}_1 \overline{f}_2 \pmod{p^{m+1}}$ holds, i.e.,

$$f_1 f_2 + p^m (g_2 f_1 + g_1 f_2) + p^{2m} g_1 g_2 \equiv f \pmod{p^{m+1}}.$$

Clearly, $p^{2m} g_1 g_2 \equiv 0 \pmod{p^{m+1}}$, and so we come to the congruence

$$g_2 f_1 + g_1 f_2 \equiv d \pmod{p}, \tag{1}$$

where $d = p^{-m}(f - f_1 f_2) \in \mathbb{Z}[x]$. The solutions of (1) can be obtained in terms of polynomials u and v for which $uf_1 + vf_2 \equiv 1 \pmod{p}$. Namely,

$$g_1 \equiv dv + wf_1 \pmod{p} \quad \text{and} \quad g_2 \equiv du - wf_2 \pmod{p},$$

where $w \in \mathbb{Z}[x]$ is an arbitrary polynomial. Since $\overline{f}_1 = f_1 + p^m g_1$, where f_1 and $\overline{f}_1$ are monic polynomials, it follows that $\deg g_1 < \deg f_1$. Therefore, for a given v, the polynomial g_1 is uniquely determined modulo p. In this case the polynomial g_2 is also uniquely determined modulo p. Hence $\overline{f}_1$ and $\overline{f}_2$ are uniquely determined modulo p^{m+1}. $\square$

Remark. The process of deriving the polynomial factorization modulo p^m, where m is large, can be essentially speeded up if one raises factorizations modulo q to factorizations modulo qr, where $r = (p, q)$. For details, see [Co3].

To obtain a factorization of $f \in \mathbb{Z}[x]$ into irreducible factors, one can use the following procedure (we assume that $\text{cont}(f) = 1$ and f has no multiple roots). By means of Mignotte's inequality (Theorem 4.2.6 on page 152) we get an estimate M for the coefficients of the divisors of f whose degree does not exceed $\frac{1}{2} \deg f$. Next, we select m such that $p^m > 2aM$, where $a > 0$ is the leading coefficient of f. Then, using Berlekamp's algorithm and Hensel's lemma, we factorize: $f \equiv a \cdot f_1 \cdot \ldots \cdot f_k \pmod{p^m}$, where $f_1, \ldots, f_k \in \mathbb{Z}[x]$ are monic polynomials.

Let $g(x) = a_1 x^l + \cdots \in \mathbb{Z}[x]$ be a divisor of f. Then $a_2 = a/a_1 \in \mathbb{N}$, and, modulo p, the polynomial $a_2 g$ is of the form $a \cdot f_{i_1} \cdot \ldots \cdot f_{i_d}$. The condition $p^m > 2aM$ shows that the polynomial $a_2 g$ is uniquely recovered from the polynomial $a \cdot f_1 \cdot \ldots \cdot f_k \pmod{p^m}$. Indeed, the coefficients of $a_2 g$ lie strictly between $-\dfrac{mp^m}{2}$ and $\dfrac{p^m}{2}$, and hence they are uniquely recovered from their residues after division by p^m.

2.6 Problems to Chapter 2

2.1 Let $f \in \mathbb{Z}[x]$ be a polynomial with roots $\alpha_1, \ldots, \alpha_n$ and let $M = \max_i |\alpha_i|$. Prove that, if $f(x_0)$ is a prime for an integer x_0 such that $|x_0| > M + 1$, then f is irreducible.

2.2 Let p be a prime, and a a positive integer not divisible by p. Prove that $x^p - x - a$ is irreducible.

2.3 For a polynomial $f \in \mathbb{Z}[x]$, let there be an integer n such that:

1) All the roots of f lie in the half-plane $\operatorname{Re} z < n - \dfrac{1}{2}$.

2) $f(n - 1) \neq 0$.

3) $f(n)$ is a prime.

Prove that f is irreducible.

2.4 [Kl] a) Let $f(x) = f_n x^n + \cdots + f_0 \in \mathbb{Z}[x]$, where $|f_0| > 1$. Further, let $\{c_1, \ldots, c_r\}$ be the set of all divisors of $|f_0|$. Let f assume prime values $p_1, \ldots, p_n$ at n distinct integer points $a_1, \ldots, a_n$ such that $|a_i| > 2$ and a_i does not divide $c_j \pm 1$, where $i = 1, \ldots, n$ and $j = 1, \ldots, r$. Prove that f is irreducible.

b) Let $a_1, \ldots, a_n, r$ and s be integers such that $|a_k| > 2$. Let the numbers $q = (-1)^n a_1 \cdot \ldots \cdot a_n + s$ and $p_k = r a_k + s$, where $k = 1, \ldots, n$, be prime with $q \pm 1$ not divisible by a_k. Prove that the polynomial

$$f(x) = (x - a_1) \cdot \ldots \cdot (x - a_n) + rx + s$$

is irreducible.

2.5 Let $f(x) = x^n + a_1 x^{n-1} + \cdots + a_{n-1} x + p a_n$ be a polynomial with integer coefficients and p a prime. Prove that if $p > \sum\limits_{i=0}^{n-1} |a_n|^{n-1-i} |a_i|$, then f is irreducible.

2.6 Let $a_1, \ldots, a_n$ be distinct integers.

a) Prove that the polynomial $(x-a_1)(x-a_2) \cdot \ldots \cdot (x-a_n) - 1$ is irreducible.

b) Prove that the polynomial $(x-a_1)(x-a_2) \cdot \ldots \cdot (x-a_n) + 1$ is irreducible except for the following cases:

$$(x-a)(x-a-2) + 1 = (x-a-1)^2;$$
$$(x-a)(x-a-1)(x-a-2)(x-a-3) + 1 = \big((x-a-1)(x-a-2) - 1\big)^2.$$

c) Prove that the polynomial $(x-a_1)^2(x-a_2)^2 \cdot \ldots \cdot (x-a_n)^2 + 1$ is irreducible.

2.7 Prove that any polynomial with integer coefficients can be represented as the sum of two irreducible polynomials.

2.8 a) Let $f(x)$ be a polynomial with integer coefficients assuming the value $+1$ at more than three integer points. Prove that $f(n) \neq -1$ for any $n \in \mathbb{Z}$.

b) Let $a, b \in \mathbb{Z}$ and let the polynomial $ax^2 + bx + 1$ be irreducible. Let $n \geq 7$ and let $a_1, \ldots, a_n$ be distinct integers; set $\varphi(x) = (x-a_1)(x-a_2) \cdot \ldots \cdot (x-a_n)$. Prove that the polynomial $a\big(\varphi(x)\big)^2 + b\varphi(x) + 1$ is irreducible.

2.9 Let $F(x_1, \ldots, x_n) \in \mathbb{Z}[x_1, \ldots, x_n]$; set $f(x) = F(x, \ldots, x)$. Prove that if f is irreducible, then F is also irreducible.

2.10 Let $p > 3$ be a prime and let $n < 2p$. Prove that the polynomial $x^{2p} + px^n - 1$ is irreducible.

2.11 Let $p > 3$ be a prime, $a_1 + \cdots + a_p = 2p$ and $n < 2p$. Prove that the polynomial

$$x_1^{a_1} \cdot \ldots \cdot x_p^{a_p} + x_1^n + \ldots + x_p^n - 1$$

is irreducible.

2.12 Let f be an irreducible polynomial with integer coefficients, let D be its discriminant and p a prime. Suppose that modulo p the polynomial f factors into k irreducible factors. Prove that $D^{(p-1)/2} \equiv (-1)^{n-k} \pmod{p}$.

2.7 Solutions of selected problems

2.1. Let $f(x) = g(x)h(x)$, where $g, h \in \mathbb{Z}[x]$ and $\deg g \geq 1$, $\deg h \geq 1$. Since $f(x_0) = p$ is a prime, we may assume that $g(x_0) = \pm 1$ and $h(x_0) = \pm p$. On the other hand, the roots $\beta_1, \ldots, \beta_k$ of g are also roots of f, so that $|\beta_i| \leq M$, and therefore

$$\big|g(x_0)\big| = |a_0| \prod |x_0 - \beta_i|,$$

where $|a_0| \geq 1$ and $|x_0 - \beta_i| \geq |x_0| - |\beta_i| > (M+1) - M = 1$. Hence $\big|g(x_0)\big| > 1$: a contradiction.

2.2. Let $x^p - x - a$ be reducible over $\mathbb{Z}$. Then it is reducible as a polynomial over $\mathbb{Z}_p$. Thus, over $\mathbb{Z}_p$, we have

$$x^p - x - a = g(x)h(x), \quad \text{where } 1 \le \deg g \le p - 1,$$

and the polynomial g is irreducible. If $b \in \mathbb{Z}_p$, then

$$g(x - b)h(x - b) = (x - b)^p - (x - b) - a = x^p - x - a.$$

Thus, the polynomial $x^p - x - a$ is divisible by p polynomials $g_i(x) = g(x - i)$, where $i = 0, 1, \ldots, p - 1$. As $\deg g \le p - 1$, these polynomials are distinct because

$$(x - i)^k - (x - j)^k = (j - i)k x^{k-1} + \ldots .$$

Therefore $p = \deg(x^p - x - a) \ge p \deg g$. It follows that $\deg g = 1$. But if a is not divisible by p, the polynomial $x^p - x - a$ has no roots in $\mathbb{Z}_p$ because $b^p - b = 0$ for any $b \in \mathbb{Z}_p$.

2.3. Let $f(x) = g(x)h(x)$, where $g, h \in \mathbb{Z}[x]$ and $\deg g \ge 1$, $\deg h \ge 1$. Since $f(n) = p$ is prime, we may assume that $g(n) = \pm 1$ and $h(n) = \pm p$. On the other hand, if $g(\beta_i) = 0$, then $f(\beta_i) = 0$, and so $\operatorname{Re} \beta_i < n - \frac{1}{2}$, i.e., $\operatorname{Re}(n - \frac{1}{2} - \beta_i) > 0$. This means that

$$\left| n - \frac{1}{2} - \beta_i - t \right| < \left| n - \frac{1}{2} - \beta_i + t \right| \quad \text{for } t > 0,$$

and so $\left| g(n - \frac{1}{2} - t) \right| < \left| g(n - \frac{1}{2} + t) \right|$. The condition $f(n - 1) \ne 0$ implies that $\left| g(n - 1) \right| \ge 1$. Therefore

$$|g(n)| = \left| g\left(n - \frac{1}{2} + \frac{1}{2}\right) \right| > \left| g\left(n - \frac{1}{2} - \frac{1}{2}\right) \right| = |g(n - 1)| \ge 1.$$

This is a contradiction.

2.4. a) Let $f = f_1 f_2$, where $f_1, f_2 \in \mathbb{Z}[x]$. Let $f_1(0) = b_1$ and $f_2(0) = c_1$. For definiteness sake, let us assume that $|b_1| \le |c_1|$.

Case 1: $|b_1| = 1$. In this case $|c_1| = |f_0| > 1$. The conditions $f(a_k) = p_k$, where p_k is a prime, implies that either

$$f_1(a_k) = \pm p_k \text{ and } f_2(a_k) = \pm 1 \tag{1}$$

or

$$f_1(a_k) = \pm 1 \text{ and } f_2(a_k) = \pm p_k. \tag{2}$$

First, suppose that $f_2(a_k) = \pm 1$. Then $f_2(a_k) - f_2(0) = -(c_1 \pm 1)$.

On the other hand, $f_2(a_k) - f_2(0)$ is divisible by a_k. This is a contradiction.

It remains to assume that $f_1(a_k) = \pm 1 = \pm b_1 = \pm f_1(0)$. If $f_1(a_k) = -f_1(0)$, then $f_1(a_k) - f_1(0) = 2 f_1(a_k) = \pm 2$. On the other hand, $f_1(a_k) - f_1(0)$ is divisible by a_k, and so $|a_k| \le 2$ which contradicts our assumption.

Thus $f_1(a_k) = f_1(a_0) = b_1$ for $k = 1, \ldots, n$. Since $\deg f_1 < n$, we deduce that $f_1(x) = b_1$ for all x.

Case 2: $|b_1| > 1$. In this case b_1 and c_1 are on equal footing since $|c_1| \geq |b_1| > 1$. Let, for definiteness sake, $f_1(a_k) = \pm p_k$ and $f_2(a_k) = \pm 1$. Then $f_2(a_k) - f_2(0) = -(c_1 \pm 1)$ is divisible by a_k, which contradicts to the assumption.

b) Obviously follows from a).

2.5. As in the proof of Theorem 2.2.7, we deduce that the product of absolute values of the roots of one of the polynomials g and h does not exceed $|a_n|$. To obtain a contradiction, it suffices to show that for any root α of f we have $|\alpha| > |a_n|$. Suppose that $f(\alpha) = 0$ and $|\alpha| \leq |a_n|$. Then

$$|pa_n| = |\alpha^n + a_1\alpha^{n-1} + \cdots + a_{n-1}\alpha| \leq |a_n| \sum_{i=0}^{n-1} |a_n|^{n-1-i}|a_i| < p|a_n|,$$

which is impossible.

3

Polynomials of a Particular Form

3.1 Symmetric polynomials

3.1.1 Examples of symmetric polynomials

A polynomial $f(x_1, \ldots, x_n)$ is called *symmetric* if, for any permutation $\sigma \in S_n$, we have
$$f(x_{\sigma(1)}, \ldots, x_{\sigma(n)}) = f(x_1, \ldots, x_n).$$

The main examples of symmetric polynomials are the *elementary* symmetric polynomials
$$\sigma_k(x_1, \ldots, x_n) = \sum_{i_1 < \cdots < i_k} x_{i_1} \cdot \ldots \cdot x_{i_k},$$

where $1 \leq k \leq n$. It is convenient to set $\sigma_0 = 1$ and $\sigma_k(x_1, \ldots, x_n) = 0$ for $k > n$.

One can determine elementary symmetric polynomials with the help of the *generating function*

$$\sigma(t) = \sum_{k=0}^{\infty} \sigma_k t^k = \prod_{i=1}^{n} (1 + tx_i).$$

If $x_1, \ldots, x_n$ are the roots of the polynomial $x^n + a_1 x^{n-1} + \cdots + a_n$, then
$$\sigma_k(x_1, \ldots, x_n) = (-1)^k a_k.$$

Another example of symmetric polynomials is given by the *complete homogeneous* symmetric polynomials

$$p_k(x_1, \ldots, x_n) = \sum_{i_1 + \cdots + i_n = k} x_1^{i_1} \cdot \ldots \cdot x_n^{i_n}.$$

Their generating function is of the form

V.V. Prasolov, *Polynomials*, Algorithms and Computation in Mathematics 11,
DOI 10.1007/978-3-642-03980-5_3, © Springer-Verlag Berlin Heidelberg 2004

$$p(t) = \sum_{k=0}^{\infty} p_k t^k = \prod_{i=1}^{n} \frac{1}{1 - tx_i}.$$

An important example of symmetric polynomials is given by the *sums of powers*

$$s_k(x_1, \ldots, x_n) = x_1^k + \cdots + x_n^k.$$

Their generating function is of the form

$$s(t) = \sum_{k=0}^{\infty} s_k t^{k-1} = \sum_{i=1}^{n} \frac{x_i}{1 - tx_i}.$$

Sometimes one uses *monomial* symmetric polynomials:

$$m_{i_1 \ldots i_n}(x_1, \ldots, x_n) = \sum_{\sigma \in S_n} x_{\sigma(1)}^{i_1} \cdot \cdots \cdot x_{\sigma(n)}^{i_n}.$$

The generating functions $\sigma(t)$ and $p(t)$ are related by the equation

$$\sigma(t)p(-t) = 1.$$

Equating the coefficients of t^n, where $n \geq 1$, on the two sides here, we obtain

$$\sum_{r=0}^{n} (-1)^r \sigma_r p_{n-r} = 0. \tag{3.1}$$

The generating function $s(t)$ is expressed in terms of $p(t)$ and $\sigma(t)$ as follows:

$$s(t) = \frac{d}{dt} \ln p(t) = \frac{p'(t)}{p(t)}, \text{ i.e., } s(t)p(t) = p'(t);$$

$$s(-t) = -\frac{d}{dt} \ln \sigma(t) = -\frac{\sigma'(t)}{\sigma(t)}, \text{ i.e., } s(-t)\sigma(t) = -\sigma'(t).$$

Equating the coefficients of t^{n+1} on the two sides we obtain

$$np_n = \sum_{r=1}^{n} s_r p_{n-r}, \tag{3.2}$$

$$n\sigma_n = \sum_{r=1}^{n} (-1)^{r-1} s_r \sigma_{n-r}. \tag{3.3}$$

Relations (3.3) are called *Newton's formulas*.

Let us make more explicit the relations (3.1) for $n = 1, \ldots, k$. With σ_1, $\ldots$, σ_k regarded as fixed, these relations can be considered as a system of linear equations for $p_1, \ldots, p_k$ and similarly, with $p_1, \ldots, p_k$ fixed, as a system of linear equations for $\sigma_1, \ldots, \sigma_k$. Solving these systems we find that

$$\sigma_k = \begin{vmatrix} p_1 & 1 & 0 & \ldots & 0 \\ p_2 & p_1 & 1 & \ldots & 0 \\ \vdots & \vdots & \vdots & \ddots & \vdots \\ p_{k-1} & p_{k-2} & p_{k-3} & \cdots & 1 \\ p_k & p_{k-1} & p_{k-2} & \cdots & p_1 \end{vmatrix}, \quad p_k = \begin{vmatrix} \sigma_1 & 1 & 0 & \ldots & 0 \\ \sigma_2 & \sigma_1 & 1 & \ldots & 0 \\ \vdots & \vdots & \vdots & \ddots & \vdots \\ \sigma_{k-1} & \sigma_{k-2} & \sigma_{k-3} & \cdots & 1 \\ \sigma_k & \sigma_{k-1} & \sigma_{k-2} & \cdots & \sigma_1 \end{vmatrix}.$$

Similarly, from (3.2) we obtain

$$s_k = (-1)^{k-1} \begin{vmatrix} p_1 & 1 & 0 & \ldots & 0 \\ 2p_2 & p_1 & 1 & \ldots & 0 \\ \vdots & \vdots & \vdots & \ddots & \vdots \\ kp_k & p_{k-1} & p_{k-2} & \cdots & p_1 \end{vmatrix}, \quad p_k = \frac{1}{k!} \begin{vmatrix} s_1 & -1 & 0 & \ldots & 0 \\ s_2 & s_1 & -2 & \ldots & 0 \\ \vdots & \vdots & \vdots & \ddots & \vdots \\ s_{k-1} & s_{k-2} & s_{k-3} & \cdots & -k+1 \\ s_k & s_{k-1} & s_{k-2} & \cdots & s_1 \end{vmatrix}.$$

From (3.3) we deduce that

$$s_k = \begin{vmatrix} \sigma_1 & 1 & 0 & \ldots & 0 \\ 2\sigma_2 & \sigma_1 & 1 & \ldots & 0 \\ \vdots & \vdots & \vdots & \ddots & \vdots \\ k\sigma_k & \sigma_{k-1} & \sigma_{k-2} & \cdots & \sigma_1 \end{vmatrix}, \quad \sigma_k = \frac{1}{k!} \begin{vmatrix} s_1 & 1 & 0 & \ldots & 0 \\ s_2 & s_1 & 2 & \ldots & 0 \\ \vdots & \vdots & \vdots & \ddots & \vdots \\ s_{k-1} & s_{k-2} & s_{k-3} & \cdots & k-1 \\ s_k & s_{k-1} & s_{k-2} & \cdots & s_1 \end{vmatrix}.$$

3.1.2 Main theorem on symmetric polynomials

The elementary symmetric polynomials are algebraically independent and form a basis of the ring of symmetric polynomials. A more precise formulation of this statement is as follows.

Theorem 3.1.1. *Let $f(x_1, \ldots, x_n)$ be a symmetric polynomial. Then there exists a polynomial $g(y_1, \ldots, y_n)$ such that $f(x_1, \ldots, x_n) = g(\sigma_1, \ldots, \sigma_n)$. This polynomial g is unique.*

Proof. It suffices to consider the case where f is a homogeneous polynomial (a form). We will say that the *order* of a monomial $x_1^{\lambda_1} \cdots \cdot x_n^{\lambda_n}$ is greater than that of $x_1^{\mu_1} \cdot \ldots \cdot x_n^{\mu_n}$ if

$$\lambda_1 = \mu_1, \quad \ldots, \quad \lambda_k = \mu_k \text{ and } \lambda_{k+1} > \mu_{k+1}.$$

(Here $k = 0$ is possible.) Let $ax_1^{\lambda_1} \cdot \ldots \cdot x_n^{\lambda_n}$ be the highest order monomial of f. Then $\lambda_1 \geq \cdots \geq \lambda_n$. Consider the symmetric polynomial

$$f_1 = f - a\sigma_1^{\lambda_1 - \lambda_2} \cdot \sigma_2^{\lambda_2 - \lambda_3} \cdot \ldots \cdot \sigma_n^{\lambda_n}. \tag{1}$$

The highest order term of the monomial $\sigma_1^{\lambda_1 - \lambda_2} \cdot \ldots \cdot \sigma_n^{\lambda_n}$ is equal to

$$x_1^{\lambda_1 - \lambda_2} (x_1 x_2)^{\lambda_2 - \lambda_3} \cdot \ldots \cdot (x_1 \cdot \ldots \cdot x_n)^{\lambda_n} = x_1^{\lambda_1} \cdot x_2^{\lambda_2} \cdot \ldots \cdot x_n^{\lambda_n}.$$

Hence the order of the highest order monomial of f_1 is strictly lower than that of the highest order monomial of f.

Let us apply the operation (1) to f_1, and so on. Clearly, after finitely many such operations we obtain the zero polynomial.

Let us prove now the uniqueness of the representation $f(x_1, \ldots, x_n) = g(\sigma_1, \ldots, \sigma_n)$. It suffices to verify that if

$$g(y_1, \ldots, y_n) = \sum a_{i_1 \ldots i_n} y_1^{i_1} \cdot \ldots \cdot y_n^{i_n}$$

is a nonzero polynomial, then after the substitution

$$y_1 = \sigma_1 = x_1 + \cdots + x_n, \quad \ldots, \quad y_n = \sigma_n = x_1 \cdot \ldots \cdot x_n$$

this polynomial remains nonzero. Let us confine ourselves to the highest order monomials of the form

$$a_{i_1 \ldots i_n} x_1^{i_1 + \cdots + i_n} x_2^{i_2 + \cdots + i_n} \cdot \ldots \cdot x_n^{i_n},$$

obtained after the substitution. It is clear that the highest among these monomials cannot cancel with any other monomial. $\square$

It is obvious from the proof of Theorem 3.1.1 that, if $f(x_1, \ldots, x_n)$ is a symmetric polynomial with integer coefficients, then $f(x_1, \ldots, x_n) = g(\sigma_1, \ldots, \sigma_n)$, where the coefficients of g are also integers. The determinant formula for σ_k in terms of p_1, $\ldots$, p_k indicates that for complete homogeneous polynomials an analogous statement also holds.

As to sums of powers, an expression of the form $f(x_1, \ldots, x_n) = g(s_1, \ldots, s_n)$ also exists but the coefficients of g are not integers now. For example,

$$x_1 x_2 = \frac{(x_1 + x_2)^2 - (x_1^2 + x_2^2)}{2} = \frac{s_1^2 - s_2}{2}.$$

The main theorem on symmetric polynomials implies that, if x_1, $\ldots$, x_n are the roots of the polynomial

$$f(x) = x^n + a_1 x^{n-1} + \cdots + a_n,$$

then the quantity

$$D = \prod_{i<j} (x_i - x_j)^2,$$

which represents a symmetric polynomial in $x_1, \ldots, x_n$, can be polynomially expressed in terms of $a_1, \ldots, a_n$. This quantity is called the *discriminant* of f.

A polynomial $f(x_1, \ldots, x_n)$ is called *skew-symmetric* if

$$f(\ldots, x_i, \ldots, x_j, \ldots) = -f(\ldots, x_j, \ldots, x_i, \ldots),$$

i.e., under transposition of any two of its indeterminates x_i and x_j it changes its sign. The polynomial $\Delta = \prod_{i<j}(x_i - x_j)$ is an example of a skew-symmetric polynomial. Clearly, $\Delta^2 = D$.

Theorem 3.1.2. *Any skew-symmetric polynomial $f(x_1, \ldots, x_n)$ can be represented in the form*

$$\Delta(x_1, \ldots, x_n)g(x_1, \ldots, x_n),$$

where g is a symmetric polynomial.

Proof. It suffices to verify that f is divisible by Δ. Indeed, if $\dfrac{f}{\Delta}$ is a polynomial, then, for obvious reasons, this polynomial is symmetric. Let us show, for example, that f is divisible by $x_1 - x_2$. We make the change of variables $x_1 = u + v$, $x_2 = u - v$. As a result we obtain

$$f(x_1, x_2, x_3, \ldots, x_n) = f_1(u, v, x_3, \ldots, x_n).$$

If $x_1 = x_2$, then $u = 0$ and so $f_1(0, v, x_3, \ldots, x_n) = 0$. This means that f_1 is divisible by u, i.e., f is divisible by $x_1 - x_2$. We similarly prove that f is divisible by $x_i - x_j$ for all $i < j$. $\square$

The equality $\Delta^2 = D$ shows that the representation $f = \Delta g$ is not unique.

3.1.3 Muirhead's inequalities

Let $\lambda = (\lambda_1, \ldots, \lambda_n)$ be a *partition*, i.e., an ordered set of non-negative integers $\lambda_1 \geq \lambda_2 \geq \cdots \geq \lambda_n \geq 0$. Set $|\lambda| = \lambda_1 + \cdots + \lambda_n$. We say that $\lambda \geq \mu$ if $\lambda_1 + \cdots + \lambda_k \geq \mu_1 + \cdots + \mu_k$ for $k = 1, 2, \ldots, n$.

To every partition λ one can assign a homogeneous symmetric polynomial

$$M_\lambda(x_1, \ldots, x_n) = \frac{1}{n!} \sum_{\sigma \in S_n} x_1^{\lambda_{\sigma(1)}} \cdot \ldots \cdot x_n^{\lambda_{\sigma(n)}}. \tag{1}$$

Clearly, $\deg M_\lambda = |\lambda|$.

Example 1. If $\lambda = (1, \ldots, 1)$, then $M_\lambda(x_1, \ldots, x_n) = x_1 \cdot \ldots \cdot x_n$.

Indeed, the sum (1) in this case consists of $n!$ summands $x_1 \cdot \ldots \cdot x_n$.

Example 2. If $\lambda = (n, 0, \ldots, 0)$, then $M_\lambda(x_1, \ldots, x_n) = \frac{1}{n}(x_1^n + \ldots + x_n^n)$.

Indeed, the sum (1) consists in this case of $(n-1)!$ summands x_1^n, $(n-1)!$ summands x_2^n, and so on.

For positive $x_1, \ldots, x_n$, the inequality

$$\frac{x_1^n + \cdots + x_n^n}{n} \geq x_1 \cdot \ldots \cdot x_n$$

holds (this is the well-known inequality between the arithmetic and geometric means). The following statement is a generalization of this inequality.

Theorem 3.1.3 (Muirhead, [Mu1]). *The inequality*

$$M_\lambda(x) \geq M_\mu(x) \tag{2}$$

holds for all vectors $x = (x_1, \ldots, x_n)$ *with positive coordinates* $x_1, \ldots, x_n$ *if and only if* $|\lambda| = |\mu|$ *and* $\lambda \geq \mu$. *The equality is only attained if* $\lambda = \mu$ *and* $x_1 = \cdots = x_n$.

Proof. Suppose first that (2) holds for all $x > 0$. Let $x_1 = \cdots = x_k = a$ and $x_{k+1} = \cdots = x_n = 1$. Then

$$1 \leq \lim_{a \to \infty} \frac{M_\lambda(x)}{M_\mu(x)} = \lim_{a \to \infty} \frac{a^{\lambda_1 + \cdots + \lambda_k}}{a^{\mu_1 + \cdots + \mu_k}}.$$

Therefore $\lambda_1 + \cdots + \lambda_k \geq \mu_1 + \cdots + \mu_k$.

Now take $k = n$ and $x_1 = \cdots = x_n = a$. Then

$$\frac{M_\lambda(x)}{M_\mu(x)} = \frac{a^{\lambda_1 + \cdots + \lambda_k}}{a^{\mu_1 + \cdots + \mu_k}}.$$

For $a > 1$, we deduce, as earlier, that $|\lambda| \geq |\mu|$ whereas, for $0 < a < 1$, we obtain $|\lambda| \leq |\mu|$.

The proof of the statement in the opposite direction is more complicated. It makes use of the following transformation R_{ij}. Let $\mu_i \geq \mu_j > 0$, where $i < j$. Set $R_{ij}\mu = \mu'$, where $\mu_i' = \mu_i + 1$, $\mu_j' = \mu_j - 1$ and $\mu_k' = \mu_k$ for $k \neq i, j$. It is easy to verify that $\mu' > \mu$ and $|\mu'| = |\mu|$.

Lemma 1. *If* $\lambda = R_{ij}\mu$, *then* $M_\lambda(x) \geq M_\mu(x)$ *and the equality is only obtained if* $x_1 = \cdots = x_n$ *(we assume that the numbers* $x_1, \ldots, x_n$ *are positive.)*

Proof. For every pair of indices p and q such that $1 \leq p < q \leq n$, the difference $M_\lambda(x) - M_\mu(x)$ contains a summand of the form

$$A \cdot \left(x_p^{\lambda_i} x_q^{\lambda_j} + x_q^{\lambda_i} x_p^{\lambda_j} - x_p^{\mu_i} x_q^{\mu_j} - x_q^{\mu_i} x_p^{\mu_j} \right), \tag{3}$$

where A is a positive number.

To make the presentation more readable, we write $x_p = a$, $x_q = b$, $\mu_i = \alpha$, $\mu_j = \beta$. Recall that $\lambda_i = \alpha + 1$, $\lambda_j = \beta - 1$ and $\alpha \geq \beta$. Expression (3) divided by A is equal to

$$a^{\alpha+1}b^{\beta-1} + a^{\beta-1}b^{\alpha+1} - a^\alpha b^\beta - a^\beta b^\alpha =$$
$$(ab)^{\beta-1}(a-b)(a^{\alpha+1-\beta} - b^{\alpha+1-\beta}) \geq 0,$$

where the equality is only possible when $a = b$. Thus $M_\lambda(x) - M_\mu(x) \geq 0$, and, if among the numbers $x_1, \ldots, x_n$ there are at least two distinct ones, then the inequality is strict. $\square$

Lemma 2. *If $\lambda \geq \mu$ and $|\lambda| = |\mu|$ but $\lambda \neq \mu$, then λ can be obtained from μ after a finite number of transformations R_{ij}.*

Proof. Let i be the least index for which $\lambda_i \neq \mu_i$. Then the condition $\lambda \geq \mu$ implies that $\lambda_i > \mu_i$. The equality $|\lambda| = |\mu|$ means that $\sum(\lambda_k - \mu_k) = 0$, and so $\lambda_j < \mu_j$ for some j. Clearly, $i < j$ and $\mu_j > 0$. Hence we can apply R_{ij} to μ. As a result, we obtain a sequence ν in which $\nu_i = \mu_i + 1$, $\nu_j = \mu_j - 1$, and $\nu_k = \mu_k$ for $k \neq i, j$. Taking into account that $\lambda_i > \mu_i$ and $\lambda_j < \mu_j$ we obtain

$$|\lambda_i - \mu_i| = |\lambda_i - \nu_i| + 1, \quad |\lambda_j - \mu_j| = |\lambda_j - \nu_j| + 1.$$

Therefore

$$\sum |\lambda_k - \nu_k| = \sum |\lambda_k - \mu_k| - 2,$$

i.e., using R_{ij} we have diminished $\sum |\lambda_k - \mu_k|$ by 2. Therefore, using a certain number of transformations R_{ij}, we can reduce $\sum |\lambda_k - \mu_k|$ to zero. $\square$

Lemmas 1 and 2 obviously imply Muirhead's inequality. $\square$

3.1.4 The Schur functions

Consider the infinite matrix

$$P = \begin{pmatrix} p_0 & p_1 & p_2 & \cdots \\ 0 & p_0 & p_1 & \cdots \\ 0 & 0 & p_0 & \cdots \\ \vdots & \vdots & \vdots & \ddots \end{pmatrix},$$

where $p_i = p_i(x_1, \ldots, x_n)$, is the complete homogeneous polynomial of degree n. The (i, j)th element of the matrix P is equal to p_{j-i}. The *Schur function* or *S-function* corresponding to a partition λ is the minor of P formed by the first n rows $0, 1, \ldots, n-1$ and columns $\lambda_n, \lambda_{n-1} + 1, \ldots, \lambda_1 + n - 1$. This symmetric function in $x_1, \ldots, x_n$ is denoted by s_λ. The function s_λ can be expressed as a determinant as follows:

$$s_\lambda = |p_{\lambda_{n+1-j}+j-i}|_1^n = |p_{\lambda_j+i-j}|_1^n.$$

Considering transposed matrix we get $s_\lambda = |p_{\lambda_i+j-i}|_1^n$.

The skew Schur function $s_{\lambda,\mu}$ corresponding to a pair of partitions λ and μ is the minor of the matrix P formed by the rows with numbers $\mu_n, \mu_{n-1} + 1$,

..., $\mu_1 + n - 1$ and the columns with numbers λ_n, $\lambda_{n-1} + 1$, ..., $\lambda_1 + n - 1$. Clearly, $s_\lambda = s_{\lambda,0}$. A partition μ is called a *subpartition* of λ if $\lambda_i \geq \mu_i$ for $i = 1, \ldots, n$. One can prove that, if μ is not a subpartition of λ, then $s_{\lambda,\mu} = 0$.

The Schur functions were introduced by Jacobi long before Schur as quotients of skew-symmetric functions of a certain type. Let $\alpha = (\alpha_1, \ldots, \alpha_n)$ be a partition and a_α the anti-symmetrization of the monomial $x_1^{\alpha_1} \cdot \ldots \cdot x_n^{\alpha_n}$, i.e.,

$$a_\alpha = \sum_{w \in S_n} (-1)^w w(x^\alpha),$$

where $(-1)^w$ is the sign of the permutation w and $w(x^\alpha) = x_{w(1)}^{\alpha_1} \cdot \ldots \cdot x_{w(n)}^{\alpha_n}$. It is easy to verify that the polynomial $a_\alpha(x_1, \ldots, x_n)$ is equal to the determinant $|x_i^{\alpha_j}|_1^n$; in particular, this polynomial is skew-symmetric. Hence, if $\alpha_i = \alpha_{i+1}$ for some i, then $a_\alpha = 0$. Thus, we may assume that $\alpha = \lambda + \delta$, where

$$\delta = (n - 1, n - 2, \ldots, 1, 0).$$

Theorem 3.1.4 (Jacobi-Trudi identity). *Let* $\delta = (n - 1, \ldots, 1, 0)$. *Then*

$$s_\lambda = \frac{a_{\lambda+\delta}}{a_\delta}, \quad i.e., \quad |p_{\lambda_i+j-i}|_1^n = \frac{|x_i^{\lambda_j+n-j}|_1^n}{|x_i^{n-j}|_1^n}.$$

Proof. Let $\alpha = (\alpha_1, \ldots, \alpha_n)$ be a partition. Consider the matrices

$$A_\alpha = \|x_j^{\alpha_i}\|_1^n, \quad H_\alpha = \|p_{\alpha_i-n+j}\|_1^n \quad \text{and} \quad M = \|(-1)^{n-i}\sigma_{n-i}(\widehat{x}_j)\|_1^n,$$

where $\widehat{x}_j = (x_1, \ldots, x_{j-1}, x_{j+1}, \ldots, x_n)$. Let us show that these three matrices are related by the formula

$$H_\alpha M = A_\alpha. \tag{1}$$

Let

$$\sigma^{(j)}(t) = \sum_{k=0}^{n-1} \sigma_k(\widehat{x}_j) t^k = \prod_{l \neq j} (1 + x_l t)$$

and

$$p(t) = \sum_{k=0}^{\infty} p_k t^k = \prod_{l=1}^{n} (1 - x_l t)^{-1}.$$

Then

$$p(t)\sigma^{(j)}(-t) = (1 - x_j t)^{-1}.$$

By comparing the coefficients of t^{α_i} on both sides of this equality we deduce that

$$\sum_{l=1}^{n} p_{\alpha_i-n+l}(-1)^{n-l}\sigma_{n-l}(\widehat{x}_j) = x_j^{\alpha_i}.$$

This is precisely the relation (1) required.

Relation (1) implies, in particular, that

$$\det H_\alpha \det M = \det A_\alpha. \tag{2}$$

To compute $\det M$, we set $\alpha = \delta = (n-1, \ldots, 1, 0)$. In this case the matrix H_α is of the form $\|p_{j-i}\|_i^n$. This matrix is triangular with elements $p_0 = 1$ on the main diagonal. Hence $\det H_\delta = 1$, and therefore $\det M = \det A_\delta = a_\delta$. But since $\det H_\alpha = s_{\alpha-\delta}$, it follows that for $\alpha = \lambda + \delta$ equation (2) takes the form $s_\lambda a_\delta = a_{\lambda+\delta}$, i.e., $s_\lambda = \dfrac{a_{\lambda+\delta}}{a_\delta}$. $\square$

3.2 Integer-valued polynomials

3.2.1 A basis in the space of integer-valued polynomials

The polynomial $p(x)$ is called *integer-valued* if it assumes only integer values for all integers x.

By induction on k one can prove that the polynomial

$$\binom{x}{k} = \frac{x \cdot (x-1) \cdot \ldots \cdot (x-k+1)}{k!}$$

is integer-valued. Indeed, for $k = 1$, this is obvious. Suppose that $\binom{x}{k}$ is an integer-valued polynomial. It is easy to verify that

$$\binom{x+1}{k+1} - \binom{x}{k+1} = \binom{x}{k}.$$

Hence, for all integers m and n, the difference $\binom{m}{k+1} - \binom{n}{k+1}$ is an integer. It remains to observe that $\binom{0}{k+1} = 0$ for any k.

In a sense, the integer-valued polynomials are exhausted by the polynomials $\binom{x}{k}$. Moreover, the requirement that $p(n) \in \mathbb{Z}$ for all $n \in \mathbb{Z}$ for $p(x)$ to be integer-valued can be considerably weakened, as we now prove.

Theorem 3.2.1. *Let p_k be a polynomial of degree k assuming integer values at $x = n, n+1, \ldots, n+k$ for an integer n. Then*

$$p_k(x) = c_0 \binom{x}{k} + c_1 \binom{x}{k-1} + c_2 \binom{x}{k-2} + \cdots + c_k,$$

where $c_0, c_1, \ldots, c_k$ are integers.

Proof. The polynomials

$$\binom{x}{0} = 1, \quad \binom{x}{1} = x, \quad \binom{x}{2} = \frac{x^2}{2} - \frac{x}{2}, \quad \ldots, \quad \binom{x}{k} = \frac{x^k}{k!} + \cdots$$

form a basis in the space of polynomials of degree not greater than k, and hence

$$p_k(x) = c_0 \binom{x}{k} + c_1 \binom{x}{k-1} + \cdots + c_k,$$

where $c_0, c_1, \ldots, c_k$ are some numbers. It only remains to prove that these numbers are integers.

We use induction on k. For $k = 0$, the polynomial $p_0(x) = c_0$ assumes an integer value at $x = n$ for any n, and so c_0 is an integer. Suppose now the required statement is true for all polynomials of degree not greater than k. Let the polynomial

$$p_{k+1}(x) = c_0 \binom{x}{k+1}' + \cdots + c_{k+1}$$

take integer values at $x = n, n+1, \ldots, n+k+1$. Then the polynomial

$$\Delta p_{k+1}(x) = p_{k+1}(x+1) - p_{k+1}(x) = c_0 \binom{x}{k} + c_1 \binom{x}{k-1} + \cdots + c_k$$

takes integer values at $x = n, n+1, \ldots, n+k$. Therefore $c_0, c_1, \ldots, c_k$ are integers, and hence so is

$$c_{k+1} = p_{k+1}(n) - c_0 \binom{n}{k+1} - c_1 \binom{n}{k} - \cdots - c_k \binom{n}{1}. \quad \square$$

Theorem 3.2.2. *Let $R(x)$ be a rational function which takes integer values at all integer x. Then $R(x)$ is an integer-valued polynomial.*

Proof. We write $R(x) = \dfrac{f(x)}{g(x)}$, where f and g are polynomials. Having divided f by g with a residue, we see that

$$R(x) = p_k(x) + r(x),$$

where p_k is a polynomial of degree k and $r(x) \to 0$ as $x \to \infty$. Therefore, for large values of n, the values of $p_k(n)$ differ but slightly from integers. Let us show that $p_k(x)$ is an integer-valued polynomial. This is done almost by the same arguments as used for the proof of Theorem 3.2.1.

Let us express $p_k(x)$ in the form

$$p_k(x) = c_0 \binom{x}{k} + \cdots + c_k.$$

For $k = 0$, the number c_0 is arbitrarily close to an integer, and hence $c_0 \in \mathbb{Z}$. The polynomial

$$\Delta p_k(x) = p_k(x+1) - p_k(x) = c_0 \binom{x}{k-1} + \cdots + c_{k-1}$$

also assumes almost integer values at large integer x and its degree is equal to $k - 1$. Applying the induction hypothesis to it, we see that $c_0, c_1, \ldots, c_{k-1}$ are integers. It is also clear that the number

$$c_k = p_k(n) - c_0 \binom{n}{k} - \cdots - c_{k-1} \binom{n}{1}$$

is also an integer. It remains to prove that $r(x) = 0$. As we already know $r(n) \in \mathbb{Z}$ for $n \in \mathbb{Z}$ and $r(n) \to 0$ as $n \to \infty$. Therefore $r(n) = 0$ for all sufficiently large integer n. But any rational function with infinitely many zeros is identically zero. $\square$

Corollary. *Let $f(x)$ and $g(x)$ be polynomials with integer coefficients and let $f(n)$ be divisible by $g(n)$ at all integer n. Then*

$$f(x) = \left(\sum_{k=0}^{m} c_k \binom{x}{k} \right) g(x),$$

where $c_0, \ldots, c_m$ are integers.

Pólya [Po1] has shown that if an entire analytic function $f(z)$ assumes integer values at all integer or positive integer values of z and grows not too quickly, then $f(z)$ is an integer-valued polynomial. More precisely, the following statements hold:

1) If $f(\mathbb{N}) \subset \mathbb{N}$ and $|f(z)| < Ce^{k|z|}$, where $k < \ln 2$, then f is an integer-valued polynomial (for a proof of this statement, see [Ge2]).

2) If $f(\mathbb{Z}) \subset \mathbb{Z}$ and $|f(z)| < Ce^{k|z|}$, where $k < \ln \left(\frac{3+\sqrt{5}}{2} \right)$, then f is an integer-valued polynomial.

The examples of functions 2^z and $\frac{1}{\sqrt{5}} \left(\left(\frac{3+\sqrt{5}}{2} \right)^z - \left(\frac{3-\sqrt{5}}{2} \right)^z \right)$ show that both estimates are the best possible.

3.2.2 Integer-valued polynomials in several variables

The structure of the basis for the space of integer-valued polynomials in n variables is similar to the one-variable case.

Theorem 3.2.3 ([Os2]). *The polynomial $p_{d_1 \ldots d_n}(x_1, \ldots, x_n)$ of degree d_i with respect to x_i assumes integer values at $x_1 = a_1, a_1 + 1, \ldots, a_1 + d_1,$ $\ldots, x_n = a_n, a_n + 1, \ldots, a_n + d_n$ if and only if*

$$p_{d_1 \ldots d_n}(x_1, \ldots, x_n) = \sum c_{k_1 \ldots k_n} \binom{x_1}{k_1} \cdots \binom{x_n}{k_n},$$

where $c_{k_1 \ldots k_n}$ are integers. In particular, such a polynomial assumes integer values at all integer points $(x_1, \ldots, x_n)$.

Proof. Let us consider the case $n = 2$ since the general case is similar. For a fixed $x_1 \in \{a_1, \ldots, a_1 + d_1\}$, the polynomial $p_{d_1 d_2}(x_1, x_2)$ takes integer values at $x_2 = a_2, \ldots, a_2 + d_2$. Therefore, by Theorem 3.2.1 for $x_1 = a_1, \ldots, a_1 + d_1$, we have the identity

$$p_{d_1 d_2}(x_1, x_2) = \sum_{k_2=0}^{d_2} c_{k_2}(x_1) \binom{x_2}{k_2}, \qquad (1)$$

where $c_{k_2}(a_1), \ldots, c_{k_2}(a_1 + d_1)$ are integers. If we now consider (1) as a relation for polynomials in variables x_1 and x_2, then it is clear that $c_{k_2}(x_1)$ is a uniquely determined polynomial. As we have already shown, this polynomial (of degree not greater than d_1) assumes integer values at $x_1 = a_1, \ldots, a_1 + d_1$, as was required. $\square$

3.2.3 The q-analogue of integer-valued polynomials

Gauss's binomial coefficient, or the *q-binomial coefficient*, is

$$\begin{bmatrix} n \\ k \end{bmatrix}_q = \frac{(q^n - 1)(q^{n-1} - 1) \cdot \ldots \cdot (q^{n-k+1} - 1)}{(q^k - 1)(q^{k-1} - 1) \cdot \ldots \cdot (q - 1)}.$$

In the limit as $q \to 1$ Gauss's binomial coefficient becomes the usual binomial coefficient $\binom{n}{k}$. The Gauss binomial coefficient is one of the numerous q-analogues of elementary and special functions, see, e.g.; [Ki] and [Ga3].

A q-analogue of the identity $\binom{n+1}{k} = \binom{n}{k} + \binom{n}{k-1}$ is

$$\begin{bmatrix} n+1 \\ k \end{bmatrix}_q = \begin{bmatrix} n \\ k \end{bmatrix}_q + \begin{bmatrix} n \\ k-1 \end{bmatrix}_q q^{n-k+1}. \qquad (1)$$

To prove (1), it suffices to observe that after simplification this identity becomes

$$\frac{q^{n+1} - 1}{(q^k - 1)(q^{n-k+1} - 1)} = \frac{1}{q^k - 1} + \frac{q^{n-k+1}}{q^{n-k+1} - 1}.$$

In what follows we will assume that q, n and k are integers such that $q \geq 2$ and $1 \leq k \leq n$. In this case induction on n based on formula (1) shows that $\begin{bmatrix} n \\ k \end{bmatrix}_q$ is an integer.

Now consider polynomials $f_0, f_1, f_2, \ldots$, where

$$f_0 = 1 \text{ and } f_k(x) = q^{-k(k-1)2} \frac{(x-1)(x-q) \cdot \ldots \cdot (x - q^{k-1})}{(q-1)(q^2-1) \cdot \ldots \cdot (q^k - 1)} \text{ for } k \geq 1. \quad (2)$$

It is easy to verify that

$$f_k(q^n) = 0 \quad \text{for} \quad n = 0, 1, \ldots, k-1 \quad \text{and} \quad f_k(q^k) = 1. \qquad (3)$$

Moreover, $f_k(q^n) = \begin{bmatrix} n \\ k \end{bmatrix}_q$ for $n \geq k$. In particular, for any positive integer n, the number $f_k(q^n)$ is an integer.

Theorem 3.2.4. *The polynomial $p_k(x)$ of degree k assumes integer values at $x = 1, q, q^2, \ldots, q^k$ if and only if*

$$p_k(x) = c_k f_k(x) + c_{k-1} f_{k-1}(x) + \cdots + c_1 f_1(x) + c_0, \qquad (4)$$

where $c_0, c_1, \ldots, c_k$ are integers and the f_i are defined by formula (2). In particular, such a polynomial assumes integer values at all $x = q^n$ $(n \in \mathbb{N})$.

Proof. The polynomials $f_0, f_1, \ldots, f_k$ form a basis of the linear space of the polynomials of degree not greater than k, and so (4) holds for some $c_0, c_1, \ldots, c_k \in \mathbb{C}$. It remains to verify that $c_0, \ldots, c_k \in \mathbb{Z}$. Formulas (3) show that

$$
\begin{aligned}
p_k(1) &= c_0, \\
p_k(q) &= c_1 + c_0, \\
p_k(q^2) &= c_2 + c_1 f_1(q^2) + c_0, \\
&\cdots\cdots\cdots\cdots\cdots\cdots\cdots\cdots\cdots\cdots\cdots \\
p_k(q^k) &= c_k + c_{k-1} f_{k-1}(q^k) + \cdots + c_1 f_1(q^k) + c_0.
\end{aligned}
$$

Therefore we obtain recursively

$$c_0 \in \mathbb{Z} \Longrightarrow c_1 \in \mathbb{Z} \Longrightarrow \cdots \Longrightarrow c_k \in \mathbb{Z}. \quad \square$$

Lastly, observe that if an entire analytic function assumes integer values at points $1, q, q^2, \ldots$ and grows not too quickly, then this function is a polynomial. For a precise formulation and proof of this statement, see [Ge2].

3.3 The cyclotomic polynomials

3.3.1 Main properties of the cyclotomic polynomials

The polynomial

$$\Phi_n(x) = \prod (x - \varepsilon_k),$$

where $\varepsilon_1, \ldots, \varepsilon_{\varphi(n)}$ are the primitive n-th roots of unity, is called the *cyclotomic polynomial* of degree n. For example,

$$\Phi_1(x) = x - 1, \quad \Phi_2(x) = x + 1, \quad \Phi_3(x) = x^2 + x + 1, \quad \Phi_4(x) = x^2 + 1.$$

If $n > 2$, then ± 1 are not primitive roots of unity of degree n. In this case the primitive roots split into pairs of complex conjugate numbers. Therefore, for $n > 2$, the degree of Φ_n is even.

We deduce directly from the definition of the cyclotomic polynomial that

$$\prod_{d \mid n} \Phi_d(x) = x^n - 1.$$

Theorem 3.3.1. *Let $n > 1$ be odd. Then*

$$\Phi_{2n}(x) = \Phi_n(-x).$$

Proof. If $\varepsilon_1, \dots, \varepsilon_{\varphi(n)}$ are all the primitive roots of degree n, then

$$-\varepsilon_1, \dots, -\varepsilon_{\varphi(n)}$$

are all the primitive roots of degree $2n$. Indeed, for n odd, we have $\varphi(2n) = \varphi(n)$. Therefore the number of primitive n-th roots of unity is equal to the number of primitive $2n$-th roots of unity. It remains to prove that if ε is a primitive n-th root of unity, then $-\varepsilon$ is a primitive $2n$-th root of unity. If $0 < k < n$, then $\varepsilon^k \neq -1$. Indeed, if $\varepsilon^k = -1$, then $\varepsilon^{2k} = 1$ and $\varepsilon^{2n-2k} = 1$, but either $2k < n$ or $2n - 2k < n$. Thus, if $0 < k < n$, then $(-\varepsilon)^k \neq 1$ and $(-\varepsilon)^{n+k} = -(-\varepsilon)^k \neq 1$.

Therefore

$$\Phi_n(-x) = (-x - \varepsilon_1) \cdot \ldots \cdot (-x - \varepsilon_{\varphi(n)}),$$
$$\Phi_{2n}(x) = (x + \varepsilon_1) \cdot \ldots \cdot (x + \varepsilon_{\varphi(n)}).$$

It remains to recall that the degree of Φ_n is even. $\square$

3.3.2 The Möbius inversion formula

The relation

$$\prod_{d|n} \Phi_d(x) = x^n - 1$$

enables us to express $\Phi_n(x)$ in terms of $x^d - 1$, where d runs over the set of divisors of n. To this end, we have a rather general construction based on the *Möbius function*

$$\mu(n) = \begin{cases} 1 & \text{if } n = 1; \\ (-1)^k & \text{if } n = p_1 \cdots p_k; \\ 0 & \text{if } n = p^2 m, \end{cases}$$

where $p, p_1, \dots, p_k$ are primes.

Theorem 3.3.2 (Möbius). *If $F(n) = \sum\limits_{d|n} f(d)$, then*

$$f(n) = \sum_{d|n} \mu(d) F\left(\frac{n}{d}\right) = \sum_{d|n} \mu\left(\frac{n}{d}\right) F(d).$$

Proof. Let us verify first that $\sum\limits_{d|n} \mu(d) = 0$ for all $n > 1$. Let $n = p_1^{\alpha_1} \cdot \ldots \cdot p_k^{\alpha_k}$. Then

$$\sum_{d|n} \mu(d) = \sum_{d|p_1 \cdots p_k} \mu(d) = 1 - \binom{k}{1} + \binom{k}{2} + \cdots + (-1)^k = (1 - 1)^k = 0.$$

Clearly,

$$\sum_{ab=n} F(a)\mu(b) = \sum_{d_1 d_2 b=n} f(d_1)\mu(b) = \sum_{d_1|n}\left(f(d_1)\sum_{d_2 b=n/d_1}\mu(b)\right).$$

Let $m = \dfrac{n}{d_1}$. Then

$$\sum_{d_2 b=m}\mu(b) = \sum_{b|m}\mu(b) = \begin{cases} 1 & \text{if } n=d_1; \\ 0 & \text{if } n \neq d_1. \end{cases}$$

Hence

$$\sum_{d_1|n}\left(f(d_1)\sum_{d_2 b=n/d_1}\mu(b)\right) = f(n). \quad \square$$

Corollary. *If* $F(n) = \prod_{d|n} f(d)$, *then*

$$f(n) = \prod_{d|n} F\left(\frac{n}{d}\right)^{\mu(d)} = \prod_{d|n} F(d)^{\mu(n/d)}.$$

For cyclotomic polynomials, the Möbius inversion formula yields

$$\Phi_n(x) = \prod_{d|n}(x^d - 1)^{\mu(n/d)}.$$

Theorem 3.3.3. *The coefficients of* $\Phi_n(x)$ *are integers.*

Proof. In the product $\prod_{d|n}(x^d - 1)^{\mu(n/d)}$, let us group the factors with $\mu = 1$ and, separately, the factors with $\mu = -1$. As a result, we see that $\Phi_n(x) = \dfrac{P(x)}{Q(x)}$, where P and Q are monic polynomials with integer coefficients. The algorithm of polynomial division shows that Φ_n is a polynomial with rational coefficients. Therefore there exists an integer m such that $m\Phi_n$ has integer coefficients and the greatest common divisor of these coefficients is equal to 1. By Gauss's lemma the greatest common divisor of the coefficients of $mP = (m\Phi_n)Q$ is equal to the product of the greatest common divisors of the coefficients of $m\Phi_n$ and Q, i.e., is equal to 1.

On the other hand, the greatest common divisor of the coefficients of mP is equal to m. Therefore $m = \pm 1$, i.e., the coefficients of Φ_n are integers. $\square$

3.3.3 Irreducibility of cyclotomic polynomials

In the preceding subsection we have shown that the coefficients of $\Phi_n(x)$ are integers.

Theorem 3.3.4. *The polynomial Φ_n is irreducible over $\mathbb{Z}$.*

Proof. Suppose that $\Phi_n = fg$, where f and g are polynomials with integer coefficients. Let ε be a root of Φ_n. We may assume that $f(\varepsilon) = 0$ and f is irreducible. Let p be a prime relatively prime to n. Then ε^p is a root of Φ_n. We want to prove that ε^p is a root of f. Suppose that ε^p is not a root of f. Then we may assume that $\Phi_n = fgh$, where f and g are irreducible monic polynomials, $f(\varepsilon) = 0$ and $g(\varepsilon^p) = 0$.

The polynomial $x^n - 1$ and the irreducible polynomial $f(x)$ have a common root ε, and hence $x^n - 1$ is divisible by $f(x)$. Similarly, $x^n - 1$ is divisible by $g(x)$. Since f and g are relatively prime, it follows that $x^n - 1$ is divisible by their product. Therefore the discriminant D of $x^n - 1$ is divisible by the resultant $R(f, g)$, cf. Theorem 1.3.4 on page 24.

It is easy to verify that $D = \pm n^n$ (see Example 1.3.7 on page 25). To get a contradiction, it suffices to show that $R(f, g)$ is divisible by p. We require the following lemma.

Lemma. *If p is a prime and $f(x)$ is a polynomial with integer coefficients, then $\bigl(f(x)\bigr)^p \equiv f(x^p) \pmod{p}$.*

Proof. Let $f(x) = a_n x^n + \cdots + a_1 x + a_0$. Then

$$\bigl(f(x)\bigr)^p = \sum_{k_0 + \cdots + k_n = p} \frac{p!}{k_0! \cdot \ldots \cdot k_n!} (a_n x^n)^{k_n} \cdot \ldots \cdot (a_0)^{k_0}.$$

The number $\frac{p!}{k_0! \cdot \ldots \cdot k_n!}$ is not divisible by p only if one of the numbers $k_0, \ldots, k_n$ is equal to p. Hence

$$\bigl(f(x)\bigr)^p \equiv (a_n x)^p + \cdots + (a_0)^p \pmod{p}.$$

Since $a^p \equiv a \pmod{p}$ for all a, we get the statement desired. $\square$

Returning to the main proof, we let $y_1 = \varepsilon^p$, y_2, ..., y_k be the roots of g. By the Lemma, $f(\varepsilon^p) \equiv \bigl(f(\varepsilon)\bigr)^p \equiv 0 \pmod{p}$, i.e., $f(y_1) = p\psi(y_1)$, where ψ is a polynomial with integer coefficients. The polynomial $f - p\psi$ and the irreducible polynomial g have a common root y_1. Hence $f - p\psi$ is divisible by g, and therefore $f(y_i) - p\psi(y_i) = 0$ for all i. Therefore

$$R(f, g) = \pm f(y_1) \cdot \ldots \cdot f(y_k) = \pm p^k \psi(y_1) \cdot \ldots \cdot \psi(y_k).$$

The expression $\psi(y_1) \cdot \ldots \cdot \psi(y_k)$ is a symmetric polynomial with integer coefficients in the roots of g. Therefore this expression is an integer, i.e., $R(f, g)$ is divisible by p^k.

Thus, if Φ_n is divisible by the irreducible polynomial f and ε is a root of f, then, for any prime p relatively prime to n, the number ε^p is also a root of f. Now it is easy to demonstrate that all the roots of Φ_n are also the roots of f, i.e., $f = \pm\Phi_n$. Indeed, any root ω of Φ_n is of the form ε^m, where

$(m, n) = 1$. Let us represent m in the form $m = p_1 \cdot \ldots \cdot p_s$, where $p_1, \ldots, p_s$ are primes among which some may coincide. The condition $(m, n) = 1$ implies that $(p_i, n) = 1$ for all i. Therefore $\varepsilon^{p_1}, \varepsilon^{p_1 p_2}, \ldots, \varepsilon^{p_1 \cdots p_s} = \omega$ are the roots of f. $\square$

3.3.4 The expression for Φ_{mn} in terms of Φ_n

In many cases the cyclotomic polynomial $\Phi_{mn}(x)$ can be expressed in terms of $\Phi_n(x)$. We confine ourselves to the case when $m = p$ is a prime.

Theorem 3.3.5. *Let p be a prime. Then*

$$\Phi_{pn}(x) = \begin{cases} \Phi_n(x^p) & \text{if } (n, p) = p; \\ \dfrac{\Phi_n(x^p)}{\Phi_n(x)} & \text{if } (n, p) = 1. \end{cases}$$

Proof. First, consider the case where n is divisible by p. If ε is a primitive root of degree pn, then $\omega = \varepsilon^p$ is a primitive root of degree n. To the root ω, the roots $\varepsilon_1, \ldots, \varepsilon_p$ correspond so that $(x - \varepsilon_1) \cdot \ldots \cdot (x - \varepsilon_p) = x^p - \omega$. Therefore

$$\Phi_{pn}(x) = \prod_\varepsilon (x - \varepsilon) = \prod_{\omega = \varepsilon^p} (x^p - \omega) = \Phi_n(x^p).$$

Now consider the case where n is not divisible by p. In this case the divisors of pn consist of the divisors of n and their products by p. Hence

$$\Phi_{pn}(x) = \prod_{d \mid pn} (x^d - 1)^{\mu(pn/d)} = \prod_{d \mid n} (x^d - 1)^{\mu(pn/d)} \prod_{d \mid n} (x^{dp} - 1)^{\mu(n/d)}.$$

Since $\mu\left(\dfrac{pn}{d}\right) = -\mu\left(\dfrac{n}{d}\right)$, it follows that

$$\Phi_{pn}(x) = \prod_{d \mid n} \frac{(x^{dp} - 1)^{\mu(n/d)}}{(x^d - 1)^{\mu(n/d)}} = \frac{\Phi_n(x^p)}{\Phi_n(x)}. \ \square$$

Using Theorem 3.3.5 we can compute $\Phi_n(\pm 1)$. Let us start with $\Phi_n(1)$. If n is divisible by p, then by Theorem 3.3.5 $\Phi_n(1) = \Phi_{n/p}(1)$. Therefore, if $n = p_1^{\alpha_1} \cdot \ldots \cdot p_k^{\alpha_k}$ and $m = p_1 \cdot \ldots \cdot p_k$, then $\Phi_n(1) = \Phi_m(1)$. It remains to compute $\Phi_m(1)$. If $m = p$ is a prime, then $\Phi_p(1) = p$. If $m = p_1 \cdot \ldots \cdot p_k$, where $k > 1$, we set $p = p_1$ and $n = \dfrac{m}{p}$. By Theorem 3.3.5 we have $\Phi_m(1) = \dfrac{\Phi_n(1)}{\Phi_n(1)} = 1$.

Thus, if $n > 1$,

$$\Phi_n(1) = \begin{cases} p & \text{if } n = p^\lambda; \\ 1 & \text{if } n \neq p^\lambda. \end{cases}$$

Now let us compute $\Phi_n(-1)$. The following cases are possible:
1) $n > 1$ is odd. Then $\Phi_n(-1) = \Phi_{2n}(1) = 1$.

2) $n = 2^k$. Then $\Phi_n(x) = \dfrac{x^n - 1}{x^{n/2} - 1} = (x^{n/2} + 1)$. Hence $\Phi_n(-1) = 0$ for $n = 2$ and $\Phi_n(-1) = 2$ for $n = 2^k$, where $k > 1$.

3) $n = 2m$, where $m > 1$ is odd. In this case $\Phi_n(-1) = \Phi_m(1)$. Therefore $\Phi_n(-1) = p$ if $m = p^\alpha$ and $\Phi_n(-1) = p$ if m has more than one prime divisor.

4) $n = 2^k m$, where $k > 1$ and $m > 1$ is odd. Let $m = p_1^{\alpha_1} \cdots p_t^{\alpha_t}$. Then $\Phi_n(x) = \Phi_{2r}(x^s)$, where $r = p_1 \cdots p_t$ and $s = 2^{k-1} p_1^{\alpha_1 - 1} \cdots p_t^{\alpha_t - 1}$. Hence $\Phi_n(-1) = \Phi_{2r}(1) = 1$.

3.3.5 The discriminant of a cyclotomic polynomial

Let us represent the cyclotomic polynomial $\Phi_n(x)$ in the form

$$\Phi_n(x) = \prod_{d|n} (x^d - 1)^{\mu(n/d)} = (x^n - 1) \prod_{d|n,\, d \neq n} (x^d - 1)^{\mu(n/d)}.$$

If ε is a root of Φ_n, then

$$\Phi_n'(\varepsilon) = n\varepsilon^{n-1} \prod_{d|n,\, d \neq n} (\varepsilon^d - 1)^{\mu(n/d)}.$$

Therefore the absolute value of the discriminant of Φ_n is

$$\prod_\varepsilon |\Phi_n'(\varepsilon)| = n^{\varphi(n)} \prod_{d|n,\, d \neq n} \prod_\varepsilon |1 - \varepsilon^d|^{\mu(n/d)}.$$

Clearly, ε^d is a primitive root of degree $\dfrac{n}{d}$, i.e.,

$$\prod_\varepsilon (1 - \varepsilon^d) = \left(\Phi_{n/d}(1)\right)^{\varphi(n)/\varphi(n/d)},$$

since $\dfrac{\deg \Phi_n}{\deg \Phi_{n/d}} = \dfrac{\varphi(n)}{\varphi\left(\dfrac{n}{d}\right)}$.

The value of $\Phi_{n/d}(x)$ at $x = 1$ can be distinct from 1 only if $\dfrac{n}{d} = p^\lambda$. On the other hand, $\mu\left(\dfrac{n}{d}\right) \neq 0$ only if $\dfrac{n}{d}$ is not divisible by the square of a prime. Hence there remain only the values of d for which $\dfrac{n}{d}$ is a prime. Thus,

$$\prod_\varepsilon |\Phi_n'(\varepsilon)| = \dfrac{n^{\varphi(n)}}{\prod_{p|n} p^{\frac{\varphi(n)}{p-1}}}.$$

It remains to determine the sign of the discriminant of Φ_n. To this end, we use the fact that Φ_n has no real roots and its degree is equal to $\varphi(n)$. By Theorem 1.3.5 on page 24 the sign of a discriminant should then be equal to $(-1)^{\varphi(n)/2}$.

3.3.6 The resultant of a pair of cyclotomic polynomials

We begin by calculating the resultant $R(\Phi_n, x^m - 1)$. The polynomial $x^m - 1$ is divisible by $\Phi_1 = x - 1$, and hence $R(\Phi_1, x^m - 1) = 0$.

Let $n \geq 2$. Let $d = (n, m)$ and $n_1 = \dfrac{n}{d}$. Let $\xi_1, \xi_2, \ldots$ be nth primitive roots of unity, and let $\eta_1, \eta_2, \ldots$ be n_1th primitive roots of unity. Then

$$R(\Phi_n, x^m - 1) = \prod(\xi_i^m - 1) = \prod(1 - \xi_i^m) =$$
$$= \left(\prod(1 - \eta_i)\right)^{\varphi(n)/\varphi(n_1)} = \left(\Phi_{n_1}(1)\right)^{\varphi(n)/\varphi(n_1)}.$$

If $n_1 = 1$, i.e., if m is divisible by n, then $\Phi_{n_1}(1) = 0$, and hence

$$R(\Phi_n, x^m - 1) = 0.$$

If $n_1 \neq 1$, then $\Phi_{n_1}(1) = p$ for $n_1 = p^\lambda$ and $\Phi_{n_1}(1) = 1$ for $n_1 \neq p^\lambda$.

Passing to the calculation of $R(\Phi_n, \Phi_m)$, we observe that it is an integer that divides both $R(\Phi_n, x^m - 1)$ and $R(\Phi_m, x^n - 1)$. Indeed,

$$x^m - 1 = \Phi_m(x) f(x),$$

where $f(x)$ is a polynomial with integer coefficients, and so

$$R(\Phi_n, x^m - 1) = R(\Phi_n, \Phi_m f) = R(\Phi_n, \Phi_m) R(\Phi_n, f).$$

Further, if $n > m > 1$, then

$$R(\Phi_m, \Phi_n) = (-1)^{\varphi(m)\varphi(n)} R(\Phi_n, \Phi_m) = R(\Phi_n, \Phi_m)$$

and $R(\Phi_m, \Phi_n) > 0$. The latter inequality can be proved, for example, as follows. Clearly,

$$0 \neq R(\Phi_n, x^m - 1) = \prod_{d \mid m} R(\Phi_n, \Phi_d),$$

and hence

$$R(\Phi_n, \Phi_m) = \prod_{d \mid m} R(\Phi_n, x^d - 1)^{\mu(m/d)} > 0.$$

If m is not divisible by n and n is not divisible by m, then the numbers $\dfrac{m}{d}$ and $\dfrac{n}{d}$, where $d = (m, n)$, are distinct from 1 and are relatively prime. Therefore $R(\Phi_n, x^m - 1)$ and $R(\Phi_m, x^n - 1)$ are relatively prime, and therefore $R(\Phi_n, \Phi_m) = 1$.

To be definite, let m be divisible by n. If $m = n$, then $R(\Phi_n, \Phi_m) = 0$. If $\dfrac{m}{n} \neq p^\lambda$, then $R(\Phi_m, x^n - 1) = 1$, and hence $R(\Phi_n, \Phi_m) = 1$. It remains to consider the case $\dfrac{m}{n} = p^\lambda$. Clearly,

$$R(\Phi_n, \Phi_m) = \prod_{\delta \mid n} R(\Phi_m, x^\delta - 1)^{\mu(n/\delta)}.$$

On the right-hand side all the factors are equal to 1 except those for which $\frac{m}{\delta} = p^a$.

If $(n, p) = 1$, the factor distinct from 1 only appears for $\delta = n$. In this case

$$R(\Phi_n, \Phi_m) = R(\Phi_m, x^n - 1) = p^{\varphi(m)/\varphi(m/n)} = p^{\varphi(n)}.$$

If n is divisible by p, the factors distinct from 1 only appear for $\delta = n$ and $\delta = \frac{n}{p}$. In this case

$$R(\Phi_n, \Phi_m) = \frac{R(\Phi_m, x^n - 1)}{R(\Phi_m, x^{n/p} - 1)} = p^a,$$

where

$$a = \frac{\varphi(m)}{\varphi\left(\frac{m}{n}\right)} - \frac{\varphi(m)}{\varphi\left(\frac{mp}{n}\right)} = \varphi(m)\left(\frac{1}{p^{\lambda-1}(p-1)} - \frac{1}{p^{\lambda}(p-1)}\right) = \frac{\varphi(m)}{p^{\lambda}} = \varphi(n).$$

Thus, if $m \geq n$,

$$R(\Phi_n, \Phi_m) = \begin{cases} 0 & \text{if } m = n; \\ p^{\varphi(n)} & \text{if } m = np^{\lambda}; \\ 1 & \text{otherwise.} \end{cases}$$

3.3.7 Coefficients of the cyclotomic polynomials

The examples of polynomials $\Phi_n(x)$ for small values of n show that their coefficients are 0 and ± 1. But this is not always the case. In [Su3] it is proved that any integer may serve as a coefficient of a cyclotomic polynomial. This proof is based on the following auxiliary statement.

Lemma. *For any positive integer $t \geq 3$, there exist primes $p_1 < p_2 < \cdots < p_t$ such that $p_1 + p_2 > p_t$.*

Proof. Fix $t \geq 3$. Suppose on the contrary that, for any set of primes $p_1 < p_2 < \cdots < p_t$, the inequality $p_1 + p_2 \leq p_t$ holds. In this case $2p_1 < p_t$, and so between 2^{k-1} and 2^k there are less than t primes. This means that $\pi(2^k) < kt$, where $\pi(s)$ is the number of primes between 1 and s.

By Chebyshev's theorem (see [GNS], [Da2] or [Ch1]) $\pi(x) > \frac{cx}{\ln x}$, where c is a positive constant. Hence $\frac{c2^k}{\ln 2^k} < kt$, i.e., $c2^k < k^2 t \ln 2$. For a sufficiently large k, this inequality will be violated. $\square$

Let $t \geq 3$ be an odd integer. Select primes $p_1 < p_2 < \cdots < p_t$ so that $p_1 + p_2 > p_t$. Set $p = p_t$, consider the polynomial $\Phi_n(x)$ modulo x^{p+1}. For t odd, we have

$$\Phi_{p_1 \cdots p_t} = \frac{(x^{p_1} - 1) \cdot \ldots \cdot (x^{p_t} - 1)}{x - 1} \cdot \frac{\prod(x^{p_i p_j p_k} - 1)}{\prod(x^{p_i p_j} - 1)} \cdot \ldots$$

But $x^{p_i p_j} \equiv 0 \pmod{x^{p+1}}$, $x^{p_i p_j p_k} \equiv 0 \pmod{x^{p+1}}$, and so on. Hence

$$\Phi_{p_1 \cdots p_t} \equiv \pm \frac{(1 - x^{p_1}) \cdot \ldots \cdot (1 - x^{p_t})}{1 - x} \pmod{x^{p+1}}.$$

We can select the plus sign here since $\Phi_{p_1 \cdots p_t}(0) = 1$. The inequalities

$$p_i + p_j \geq p_1 + p_2 > p_t = p$$

imply that

$$(1 - x^{p_1}) \cdot \ldots \cdot (1 - x^{p_t}) \equiv (1 - x^{p_1} - \ldots - x^{p_t}) \pmod{x^{p+1}}.$$

It is also clear that $(1 - x)^{-1} \equiv (1 + x + \cdots + x^p) \pmod{x^{p+1}}$. Hence

$$\Phi_{p_1 \cdots p_t} \equiv (1 + x + \cdots + x^p)(1 - x^{p_1} - \cdots - x^{p_t}) \pmod{x^{p+1}}.$$

Among the monomials x^{p_i}, x^{p_i+1}, ..., x^{p_i+p}, the monomial $x^p = x^{p_t}$ occurs for all i whereas the monomials x^{p-1} and x^{p-2} occur for all $i \neq t$. Therefore the coefficient of x^p in $\Phi_{p_1 \cdots p_t}$ is $-t + 1$ and the coefficient of x^{p-2} is

$$-(t - 1) + 1 = -t + 2.$$

As t runs over all the odd numbers starting with 3, the numbers $-t + 1$ and $-t + 2$ run over all the negative integers.

To see that all positive integers can be coefficients of cyclotomic polynomials, consider the polynomial $\Phi_{2p_1 \cdots p_t}$, where $p_1 \geq 3$ and $p_1 + p_2 > p_t$. The number $n = p_1 \cdot \ldots \cdot p_t$ is odd, and hence $\Phi_{2n}(x) = \Phi_n(-x)$. This means that the coefficients of x^p, as well as those of x^{p-2}, in Φ_{2n} and Φ_n differ by a sign, i.e., the coefficients of x^p and x^{p-2} in $\Phi_{2p_1 \cdots p_t}$ are $t - 1$ and $t - 2$, respectively.

3.3.8 Wedderburn's theorem

One of the most interesting applications of cyclotomic polynomials is the proof of Wedderburn's theorem on the commutativity of finite skew fields. This proof is due to Witt [Wi2].

A *skew field* is a ring in which the equations $ax = b$ and $xa = b$ are uniquely solvable for all $a \neq 0$.

Theorem 3.3.6 (Wedderburn). *Any finite associative skew field R is commutative, i.e., it is a field.*

Proof. Let e_1 and e_2 be solutions of the equations $ax = a$ and $xa = a$ respectively. Then $ae_1 a = a^2 = ae_2 a$, and hence $ae_1 = ae_2$ and $e_1 = e_2 = e$.

Let us show that $be = b$ for any b. Indeed, let $xa = b$. Then $be = xae = xa = b$. Similarly, $eb = b$.

Therefore the skew field R contains the identity 1. Consider the field F_p generated by $1 \in R$. The skew field R is a linear space over F_p. Let r be the dimension of this space. Then R consists of p^r elements. Let Z be the center of R, i.e., the set of elements of R that commute with all the elements of R. Clearly, Z is a field containing F_p. Therefore Z consists of $q = p^s$ elements. The skew field is also a linear space over Z. If the dimension of R over Z is equal to t, then R consists of q^t elements. Therefore $p^r = q^t = p^{st}$. We wish to show that $R = Z$, i.e., $t = 1$.

For any element $x \in R$, consider its normalizer $N_x = \{y \in R \mid xy = yx\}$. Clearly, N_x is a skew subfield of R containing Z. On the one hand, the skew field N_x is a linear space over Z, and hence consists of q^d elements. On the other hand, R is a linear space (module) over N_x, and hence $q^t = (q^d)^k = q^{dk}$, i.e., $d \mid t$.

In the multiplicative group R^*, we consider for every element x the orbit

$$O_x = \{yxy^{-1} \mid y \in R^*\}.$$

Clearly, O_x consists of

$$|O_x| = \frac{|R^*|}{|N_x^*|} = \frac{q^t - 1}{q^d - 1}$$

elements. The orbits of distinct elements either coincide or do not intersect. Hence, R^* splits into the disjoint union of orbits and the orbit of every element from Z^* consists of a single element. Therefore

$$q^t - 1 = (q - 1) + \sum \frac{q^t - 1}{q^d - 1}, \tag{1}$$

where the sum runs over the divisors d of t such that $d < t$ (the equality $d = t$ corresponds to the case $x \in Z^*$; such elements are separated and correspond to the summand $q - 1$).

With the help of the cyclotomic polynomial $\Phi_t(x)$ we will show that (1) is only possible for $t = 1$. Indeed, the polynomial $x^t - 1$ is divisible by $\Phi_t(x)$. Moreover, if $d \mid t$ and $d < t$, the polynomial $\dfrac{x^t - 1}{x^d - 1}$ is also divisible by $\Phi_t(x)$ since in this case the polynomials $x^d - 1$ and $\Phi_t(x)$ have no common roots. In particular, the numbers $q^t - 1$ and $\dfrac{q^t - 1}{q^d - 1}$ are divisible by $\Phi_t(q)$. Relation (1) then shows that $q - 1$ is divisible by $\Phi_t(q)$. On the other hand,

$$|\Phi_t(q)| = \prod |q - \varepsilon_i| > q - 1$$

since $|\varepsilon_i| = 1$ and $\varepsilon_i \neq 1$. $\square$

3.3.9 Polynomials irreducible modulo p

With the help of Möbius's inversion formula we can obtain an expression for the number of irreducible monic polynomials of degree n over $\mathbb{F}_p$. Let us prove first the following statement.

Theorem 3.3.7. *Let $F_d(x)$ be the product of all irreducible monic polynomials of degree d over $\mathbb{F}_p$. Then*

$$x^{p^n} - x = \prod_{d|n} F_d(x).$$

Proof. The polynomial $x^{p^n} - x$ is relatively prime to its derivative $p^n x^{p^n-1} - 1 = -1$, and hence it has no multiple roots. Therefore it suffices to prove that if $f(x)$ is an irreducible monic polynomial of degree d, then $f(x)$ divides $x^{p^n} - x$ if and only if d divides n.

Let α be a root of f and $K = \mathbb{F}_p(\alpha)$ an extension of degree d of $\mathbb{F}_p$. This extension consists of p^d elements and all its elements satisfy the equation $x^{p^d} - x = 0$. Indeed, the multiplicative group of $\mathbb{F}_p$ is of order $p^d - 1$, and so any nonzero element $x \in \mathbb{F}_p$ satisfies $x^{p^d-1} = 1$.

Lemma. a) *Over an arbitrary field, the polynomial $x^n - 1$ divides $x^m - 1$ if and only if n divides m.*

b) *If $a \geq 2$ is a positive integer, then $a^n - 1$ divides $a^m - 1$ if and only if n divides m.*

Proof. a) Let $m = qn + r$, where $0 \leq r < n$. Then

$$\frac{x^m - 1}{x^n - 1} = x^r \frac{x^{qn} - 1}{x^n - 1} + \frac{x^r - 1}{x^n - 1}.$$

The polynomial $x^{qn} - 1$ is divisible by $x^n - 1$. Hence $x^m - 1$ is divisible by $x^n - 1$ if and only if $x^r - 1$ is divisible by $x^n - 1$. But $r < n$ and so $x^r - 1$ is divisible by $x^n - 1$ only for $r = 0$.

b) is similarly proved. □

Let us continue with the proof of the theorem. First, suppose that d divides n. Then $p^d - 1$ divides $p^n - 1$, and hence $x^{p^d} - x$ divides $x^{p^n} - x$. A root α of the irreducible polynomial $f(x)$ is also a root of the equation $x^{p^d} = x$, and hence $f(x)$ divides $x^{p^d} - x$.

Suppose now that $f(x)$ divides $x^{p^n} - x$. Then $\alpha^{p^n} - \alpha = 0$. For an arbitrary $b_1\alpha^{d-1} + b_2\alpha^{d-2} + \cdots + b_d \in K$, we have

$$(b_1\alpha^{d-1} + \cdots + b_d)^{p^n} = b_1(\alpha^{p^n})^{d-1} + \cdots + b_d = b_1\alpha^{d-1} + \cdots + b_d.$$

Thus any element of K satisfies the equation $x^{p^n} = x$, i.e., $x^{p^n} - x$ is divisible by $x^{p^d} - x$. Hence n is divisible by d. □

Let N_d be the number of irreducible monic polynomials of degree d over $\mathbb{F}_p$. Then the degree of F_d is dN_d, and hence

$$p^n = \sum_{d|n} dN_d.$$

Applying Möbius inversion formula we obtain a formula for N_n:

$$N_n = \frac{1}{n} \sum_{d|n} \mu\left(\frac{n}{d}\right) p^d.$$

In particular, $N_n \neq 0$ since the sum $\sum_{d|n} \mu\left(\frac{n}{d}\right) p^d$ is of the form

$$p^{d_1} \pm p^{d_2} \pm \cdots \pm p^{d_k},$$

where the numbers d_i are distinct.

3.4 Chebyshev polynomials

3.4.1 Definition and main properties of Chebyshev polynomials

Chebyshev polynomials $T_n(x)$ constitute one of the most remarkable families of polynomials. They often appear in various branches of mathematics — from approximation theory to number theory to topology of three-dimensional manifolds. We will discuss several simpler but rather important properties of Chebyshev polynomials.

We use the definition of Chebyshev polynomials based on the fact that $\cos n\varphi$ is polynomially expressed in terms of $\cos\varphi$, i.e., there exists a polynomial $T_n(x)$ such that $T_n(x) = \cos n\varphi$ for $x = \cos\varphi$. Indeed, the formula

$$\cos(n+1)\varphi + \cos(n-1)\varphi = 2\cos\varphi\cos n\varphi$$

shows that the polynomials $T_n(x)$ recursively defined by the relation

$$T_{n+1}(x) = 2xT_n(x) - T_{n-1}(x)$$

with the initial values $T_0(x) = 1$ and $T_1(x) = x$, possess the property required. These polynomials $T_n(x)$ are called the *Chebyshev polynomials*.

The fact that $T_n(x) = \cos n\varphi$ for $x = \cos\varphi$ directly implies that $|T_n(x)| \leq 1$ for $x \leq 1$. The above recurrence implies that

$$T_n(x) = 2^{n-1}x^n + a_1 x^{n-1} + \cdots + a_n, \tag{$*$}$$

where $a_1, \ldots, a_n$ are integers.

The most important property of Chebyshev polynomials was discovered by Chebyshev himself. It consists of the following.

Theorem 3.4.1. *Let $P_n(x) = x^n + \cdots$ be a monic polynomial of degree n such that $|P_n(x)| \le \dfrac{1}{2^{n-1}}$ for $|x| \le 1$. Then $P_n(x) = \dfrac{T_n(x)}{2^{n-1}}$. In other words, the polynomial $\dfrac{T_n(x)}{2^{n-1}}$ is the monic polynomial of degree n that has the least deviation from zero on segment $[-1, 1]$.*

Proof. We use only one property of the polynomial $(*)$, namely, the fact that

$$T_n\left(\cos\left(\frac{k\pi}{n}\right)\right) = \cos k\pi = (-1)^k \text{ for } k = 0, 1, \ldots, n.$$

Consider the polynomial

$$Q(x) = \frac{1}{2^{n-1}}T_n(x) - P_n(x).$$

Its degree does not exceed $n - 1$ since the leading terms of $\frac{1}{2^{n-1}}T_n(x)$ and $P_n(x)$ are equal. Since $|P_n(x)| \le \frac{1}{2^{n-1}}$ for $|x| \le 1$, it follows that at the point $x_k = \cos\left(\frac{k\pi}{n}\right)$ the sign of $Q(x_k)$ coincides with the sign of $T_n(x_k)$. Therefore, at the end points of each segment $[x_{k+1}, x_k]$, the polynomial $Q(x)$ takes values of opposite signs, and hence $Q(x)$ has a root on each of these segments.

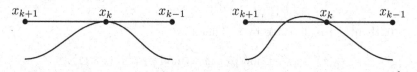

$$x_{k+1} \qquad x_k \qquad x_{k-1} \qquad\qquad x_{k+1} \qquad x_k \qquad x_{k-1}$$

FIGURE 3.1

If $Q(x_k) = 0$ we need a slightly more accurate arguments. In this case either x_k is a double root or within one of the segments $[x_{k+1}, x_k]$ and $[x_k, x_{k-1}]$ there is one more root. This follows from the fact that at x_{k+1} and x_{k-1} the values of $Q(x)$ have the same sign (Fig. 3.1)

The number of segments $[x_{k+1}, x_k]$ is equal to n, and hence the polynomial $Q(x)$ has at least n roots. For a polynomial of degree not greater than $n - 1$, this means that it is identically zero, i.e., $P_n(x) = \frac{1}{2^{n-1}}T_n(x)$. $\square$

If $z = \cos\varphi + i\sin\varphi$, then $z + z^{-1} = 2\cos\varphi$ and $z^n + z^{-n} = 2\cos n\varphi$. Therefore

$$T_n\left(\frac{z + z^{-1}}{2}\right) = \frac{z^n + z^{-n}}{2}.$$

Using this property we can prove the following statement.

Theorem 3.4.2. *Let* $m = \left[\frac{n}{2}\right]$. *Then*

$$T_n(x) = \sum_{j=0}^{m} \binom{n}{2j} x^{n-2j} (x^2 - 1)^j.$$

Proof. Let $x = \frac{1}{2}(z + z^{-1})$ and $y = \frac{1}{2}(z - z^{-1})$. Then $y^2 = x^2 - 1$ and

$$z^n + z^{-n} = (x+y)^n + (x-y)^n = \sum_{i=0}^{n} \binom{n}{i} \left(1 + (-1)^i\right) x^{n-i} y^i =$$

$$= 2 \sum_{j=0}^{m} \binom{n}{2j} x^{n-2j} y^{2j} = 2 \sum_{j=0}^{m} \binom{n}{2j} x^{n-2j} (x^2 - 1)^j.$$

It remains to observe that $T_n(x) = \frac{1}{2}(z^n + z^{-n})$. $\square$

Corollary. *Let p be an odd prime. Then*

$$T_p(x) \equiv T_1(x) \pmod{p}.$$

Proof. We write $p = 2m + 1$. Then

$$T_p(x) = \sum_{j=0}^{m} \binom{p}{2j} x^{p-2j} (x^2 - 1)^j.$$

If $j > 0$, then $\binom{p}{2j}$ is divisible by p. Therefore

$$T_p(x) \equiv x^p \pmod{p} \equiv x \pmod{p} = T_1(x). \ \square$$

For any pair of polynomials P and Q, define their *composition* naturally, by setting

$$P \circ Q(x) = P(Q(x)).$$

The polynomials P and Q are said to *commute* if $P \circ Q = Q \circ P$, i.e., if

$$P(Q(x)) = Q(P(x)).$$

Theorem 3.4.3. *The polynomials $T_n(x)$ and $T_m(x)$ commute.*

Proof. Let $x = \cos \varphi$. Then $T_n(x) = \cos(n\varphi) = y$ and $T_m(y) = \cos m(n\varphi)$, and hence $T_m(T_n(x)) = \cos mn\varphi$. Similarly, $T_n(T_m(x)) = \cos mn\varphi$. Hence the identity $T_n(T_m(x)) = T_m(T_n(x))$ holds for $|x| < 1$, and therefore it holds for all x. $\square$

Chebyshev polynomials are the only non-trivial example of commuting polynomials. Indeed, the following classification theorem for pairs of commuting polynomials holds. Let $l(x) = ax + b$, where $a, b \in \mathbb{C}$ and $a \neq 0$. We will say that the pair of polynomials $l \circ f \circ l^{-1}$ and $l \circ g \circ l^{-1}$ is *equivalent* to the pair of polynomials f and g.

Theorem 3.4.4 (Ritt). *Let f and g be commuting polynomials. Then the pair (f, g) is equivalent to one of the following pairs:*

1) x^m *and* εx^n, *where* $\varepsilon^{m-1} = 1$.

2) $\pm T_m(x)$ *and* $\pm T_n(x)$, *where* T_m *and* T_n *are Chebyshev polynomials;*

3) $\varepsilon_1 Q^{(k)}(x)$ *and* $\varepsilon_2 Q^{(l)}(x)$, *where* $\varepsilon_1^q = \varepsilon_2^q = 1$, *and* $Q(x) = xP(x^q)$, *and where* $Q^{(1)} = Q$, $Q^{(2)} = Q \circ Q$, $Q^{(3)} = Q \circ Q \circ Q$, *and so on.*

This theorem was proved in 1922 by the American mathematician Ritt; all the known proofs of it are rather complicated. A modern exposition of the proof of Ritt's theorem is given in [Pr4].

Sometimes instead of $T_n(x)$ it is convenient to consider the monic polynomial $P_n(x) = 2T_n\left(\dfrac{x}{2}\right)$. The polynomials $P_n(x)$ satisfy the recurrence relation

$$P_{n+1}(x) = xP_n(x) - P_{n-1}(x).$$

Hence $P_n(x)$ is a polynomial with integer coefficients.

If $z = \cos\varphi + i\sin\varphi = e^{i\varphi}$, then $z + z^{-1} = 2\cos\varphi$ and $z^n + z^{-n} = 2\cos n\varphi$. Therefore

$$P_n(z + z^{-1}) = 2T_n(\cos\varphi) = 2\cos n\varphi = z^n + z^{-n},$$

i.e., the polynomial $P_n(x)$ polynomially expresses $z^n + z^{-n}$ via $z + z^{-1}$.

With the help of the polynomials P_n we can prove the next theorem.

Theorem 3.4.5. *If both α and $\cos(\alpha\pi)$ are rational, then $2\cos(\alpha\pi)$ is an integer, i.e., $\cos(\alpha\pi) = 0$, $\pm\frac{1}{2}$ or ± 1.*

Proof. Let $\alpha = \dfrac{m}{n}$ be an irreducible fraction. Set $x_0 = 2\cos t$, where $t = \alpha\pi$. Then

$$P_n(x_0) = 2\cos(nt) = 2\cos(n\alpha\pi) = 2\cos(m\pi) = \pm 2.$$

Hence x_0 is a root of the polynomial with integer coefficients

$$P_n(x) \mp 2 = x^n + b_1 x^{n-1} + \cdots + b_n.$$

Let $x_0 = 2\cos(\alpha\pi) = \dfrac{p}{q}$ be an irreducible fraction. Then

$$p^n + b_1 p^{n-1}q + \cdots + b_n q^n = 0,$$

and hence p^n is divisible by q. But p and q are relatively prime, and so $q = \pm 1$, i.e., $2\cos(\alpha\pi)$ is an integer. $\square$

It is convenient to compute the derivatives of Chebyshev polynomials starting directly from the relation $T_n(x) = \cos n\varphi$, where $x = \cos\varphi$. For example,

$$T'_n(x) = \frac{\dfrac{d\cos n\varphi}{d\varphi}}{\dfrac{d\cos\varphi}{d\varphi}} = \frac{n\sin n\varphi}{\sin\varphi},$$

$$T''_n(x) = \frac{d}{d\varphi}\left(\frac{n\sin n\varphi}{\sin\varphi}\right)\frac{-1}{\sin\varphi} = \frac{n\cos\varphi\sin n\varphi - n^2\cos n\varphi\sin\varphi}{\sin^3\varphi}.$$

These formulas imply that

$$(1-x^2)T'_n(x) = n\left(T_{n-1}(x) - xT_n(x)\right),$$

$$(1-x^2)\left(T'_n(x)\right)^2 = n^2\left(1 - T_n(x)\right)^2,$$

$$(1-x^2)T''_n(x) - xT'_n(x) + n^2 T_n(x) = 0.$$

The identity

$$(1-x^2)\left(T'_n(x)\right)^2 = n^2\left(1 - T_n(x)\right)^2$$

can be rewritten in the form

$$1 = T_n^2(x) - (1-x^2)U_n^2(x), \qquad (1)$$

where $U_n(x) = \dfrac{\sin n\varphi}{\sin\varphi}$ and $x = \cos\varphi$. It is easy to verify that U_n is a polynomial with integer coefficients. Indeed, induction on n shows that

$$\sin nx = p_n(\cos x)\sin x, \quad \cos nx = q_n(\cos x),$$

where p_n and q_n are polynomials with integer coefficients.

Identity (1) can be used to solve *Pell's equation*

$$x^2 - dy^2 = 1.$$

Indeed, if (x_1, y_1) is a positive integer solution of this equation, then

$$1 = T_n^2(x_1) - \frac{(1-x_1^2)}{y_1^2}\left(y_1 U_n(x_1)\right)^2$$

$$= T_n^2(x_1) - d\left(y_1 U_n(x_1)\right)^2,$$

so that $(T_n(x_1), y_1 U_n(x_1))$ is also a positive integer solution of this equation.

Remark. One can prove that if (x_1, y_1) is the least positive integer solution of Pell's equation, then all its positive integer solutions are of the form $(T_n(x_1), y_1 U_n(x_1))$.

3.4.2 Orthogonal polynomials

The polynomials $f_k(x)$, $k = 0, 1, \ldots$ are said to be *orthogonal* polynomials on the interval $[a, b]$ with weight function $w(x) \geq 0$ if $\deg f_k = k$ and

$$\int_a^b f_m(x) f_n(x) w(x) \, dx = 0$$

for $m \neq n$.

In the space V^{n+1} of polynomials of degree $\leq n$, we define the inner product by

$$(f, g) = \int_a^b f(x) g(x) w(x) \, dx.$$

The orthogonal polynomials f_0, f_1, $\ldots$, f_n form an orthogonal basis in the space V^{n+1} with this inner product.

Given an interval $[a, b]$ and a weight function $w(x)$, the orthogonal polynomials are uniquely determined up to proportionality. Indeed, they are obtained by orthogonalization of the basis $1, x, x^2, \ldots$

The best-known ones[1] are the following orthogonal polynomials:

a	b	$w(x)$	Name of the polynomial
-1	1	1	Legendre
-1	1	$(1 - x^2)^{\lambda - 1/2}$	Gegenbauer
-1	1	$(1 - x)^\alpha (1 + x)^\beta$	Jacobi
$-\infty$	∞	$\exp(-x^2)$	Hermite
0	∞	$x^\alpha e^{-x}$	Laguerre

Theorem 3.4.6. *Chebyshev polynomials form an orthogonal system on the interval $[-1, 1]$ with weight function* $w(x) = \dfrac{1}{\sqrt{1 - x^2}}$.

Proof. Making a change of variable $x = \cos \varphi$, we have

$$\int_{-1}^1 T_n(x) T_m(x) \frac{dx}{\sqrt{1 - x^2}} = \int_0^\pi \cos n\varphi \cos m\varphi \, d\varphi =$$

$$= \int_0^\pi \frac{\cos(m + n)\varphi + \cos(m - n)\varphi}{2} d\varphi.$$

---

[1] Bochner showed that up to a complex linear change of variable, the only polynomial solutions that arise as the eigenfunctions of the hypergeometric equation are the Jacobi, Laguerre, Hermite and Bessel polynomials, of which Legendre, Gegenbauer and other famous polynomials, e.g., Tchebyshev ones, are particular cases, see [BCS].

It remains to observe that

$$\int_0^\pi \cos k\varphi \, d\varphi = 0 \qquad \text{if } k \neq 0. \ \square$$

Corollary. *If $P_n(x)$ is a polynomial of degree n and*

$$\int_{-1}^1 P_n(x) \frac{x^k dx}{\sqrt{1-x^2}} = 0$$

for $k = 0, 1, \ldots, n-1$, then $P_n(x) = \lambda T_n'(x)$, where λ is independent of x.

Proof. In the space V^{n+1} with inner product

$$(f, g) = \int_{-1}^1 f(x)g(x) \frac{dx}{\sqrt{1-x^2}},$$

the orthogonal complement[1] of the space generated by the polynomials $1, x, x^2, \ldots, x^{n-1}$ is spanned by the Chebyshev polynomial $T_n(x)$. $\square$

The corollary of Theorem 3.4.6 is often convenient in proving that a given polynomial is indeed a Chebyshev polynomial. For example, using the Corollary we prove the following statement.

Theorem 3.4.7. *Chebyshev polynomials can be defined by the formula*

$$T_n(x) = \frac{(-1)^n \sqrt{1-x^2}}{1 \cdot 3 \cdot 5 \cdots (2n-1)} \frac{d^n}{dx^n} (1-x^2)^{n-1/2}.$$

Proof. By induction on m we easily prove that for $m \leq n$

$$\frac{d^m}{dx^m} (1-x^2)^{n-1/2} = P_m(x)(1-x^2)^{n-m-1/2},$$

where $P_m(x)$ is a polynomial of degree m such that

$$P_0(x) = 1,$$
$$P_1(x) = -(2n-1)x,$$
$$\cdots\cdots\cdots\cdots\cdots\cdots\cdots\cdots\cdots\cdots\cdots\cdots$$
$$P_{m+1} = 1 - x^2 - (2n - 2m - 1)xP_m(x) \text{ for } m \geq 1.$$

Hence

$$\sqrt{1-x^2} \frac{d^n}{dx^n}(1-x^2)^{n-1/2} = P_n(x)$$

is a polynomial of degree n.

---

[1] Recall that the *orthogonal complement* of the subspace $V \subset W$ consists of all vectors in W which are orthogonal to V.

Let us verify that $P_n(x) = \lambda T_n(x)$, i.e.,

$$\int\limits_{-1}^{1} x^k \frac{d^n}{dx^n}(1-x^2)^{n-1/2}dx = 0$$

for $k = 0, 1, \ldots, n-1$. Integrating by parts we obtain

$$\int\limits_{-1}^{1} x^k \frac{d^n}{dx^n}(1-x^2)^{n-1/2}dx =$$

$$= x^k P_{n-1}(x)(1-x^2)^{1/2}\Big|_{-1}^{1} - \int\limits_{-1}^{1} kx^{k-1}\frac{d^{n-1}}{dx^{n-1}}(1-x^2)^{n-1/2}dx.$$

The first term on the right vanishes since $1 - x^2 = 0$ at $x = \pm 1$. Then we integrate by parts the integral term and repeat the process. In order to obtain 0 at the end, we have to integrate by parts $k+1$ times. At the last step we get the derivative $\dfrac{d^{n-k-1}}{dx^{n-k-1}}$. This means that $n - k - 1$ should be non-negative, i.e., $k \leq n - 1$.

It remains to verify that $\lambda = (-1)^n 1 \cdot 3 \cdot 5 \cdot \ldots \cdot (2n-1)$. For this, one can compute $P_n(1)$. Indeed, for $x = 1$, the recurrence

$$P_{m+1}(x) = 1 - x^2 - (2n - 2m - 1)xP_m(x)$$

takes the form

$$P_{m+1}(1) = -(2n - 2m - 1)P_m(1).$$

Therefore $P_n(1) = (-1)^n 1 \cdot 3 \cdot 5 \cdot \ldots \cdot (2n-1)$. It is also clear that $T_n(1) = 1$. $\square$

3.4.3 Inequalities for Chebyshev polynomials

We have shown that Chebyshev polynomials only slightly deviate from zero on the interval $[-1, 1]$. This is compensated by the fact that these polynomials and their derivatives grow rapidly outside this segment. More precisely, the following statement holds.

Theorem 3.4.8. [Ro1] *Let the polynomial* $p(x) = a_0 + a_1x + \cdots + a_nx^n$, *where* $a_i \in \mathbb{C}$, *be such that* $|p(x)| \leq 1$ *for* $-1 \leq x \leq 1$. *Then* $|p^{(k)}(x)| \leq |T_n^{(k)}(x)|$ *for* $|x| \geq 1$, $x \in \mathbb{R}$.

Proof. We use only the fact that $|p(x_i)| \leq 1$ for $x_i = \cos\frac{(n-i)\pi}{n}$, where $i = 0, 1, \ldots, n$. The polynomial $p(x)$ is completely determined by these values $p(x_i)$. Indeed,

$$p(x) = \sum_{i=0}^{n} \frac{p(x_i)}{g_i(x_i)}g_i(x), \tag{1}$$

where $g_i(x) = \prod_{j \neq i}(x - x_j)$. By differentiating (1) k times we obtain

$$p^{(k)}(x) = \sum_{i=0}^{n} \frac{p(x_i)}{g_i(x_i)} g_i^{(k)}(x).$$

Since $|p(x_i)| \leq 1$, it follows that

$$|p^{(k)}(x)| \leq \sum_{i=0}^{n} \left| \frac{g_i^{(k)}(x)}{g_i(x_i)} \right|. \tag{2}$$

The value of $T_n(x)$ at x_i is $\cos(n - i)\pi = (-1)^{n-i}$. Hence

$$|T_n^{(k)}(x)| = \left| \sum_{i=0}^{n} \frac{(-1)^{n-i}}{g_i(x_i)} g_i^{(k)}(x) \right|.$$

It is also clear that $\operatorname{sgn} g_i(x_i) = (-1)^{n-i}$. Further, for $|x| \geq 1$, the sign of $g_i^{(k)}(x)$ does not depend on i. Indeed, all roots of $g_i(x)$ belong to $[-1, 1]$, and hence all the roots of $g_i^{(k)}(x)$ also belong to this interval. Therefore

$$\operatorname{sgn} g_i^{(k)}(x) = \begin{cases} 1 & \text{for } x \geq 1 \\ (-1)^{n-k} & \text{for } x \leq 1. \end{cases}$$

As a result for $|x| \geq 1$ we obtain

$$|T_n^{(k)}(x)| = \left| \sum_{i=0}^{n} \frac{g_i^{(k)}(x)}{g_i(x_i)} \right|.$$

In this case inequality (2) implies that $|p^{(k)}(x)| \leq |T_n^{(k)}(x)|$. $\square$

Theorem 3.4.8 yields several useful corollaries. We formulate them as separate theorems.

Theorem 3.4.9. *Let $p(x) = a_0 + a_1 x + \cdots + a_n x^n$, where $a_i \in \mathbb{C}$, be such that $|p(x)| \leq 1$ for $-1 \leq x \leq 1$. Then $|a_n| \leq 2^{n-1}$.*

Proof. Recall that $T_n(x) = b_0 + b_1 x + \cdots + b_n x^n$, where $b_n = 2^{n-1}$. Therefore, applying Theorem 3.4.8 for $k = n$, we deduce that $|a_n| \leq |b_n| = 2^{n-1}$. $\square$

Theorem 3.4.10. *For $x \leq -1$ and $x \geq 1$, we have*

$$|T_{n-1}^{(k)}(x)| \leq |T_n^{(k)}(x)|.$$

Proof. For the polynomial $p(x) = T_{n-1}(x)$, the conditions of Theorem 3.4.8 hold, and hence $|T_{n-1}^{(k)}(x)| = |p(x)| \leq |T_n^{(k)}(x)|$. $\square$

Theorem 3.4.11 ([As]). *For $x, y \geq 1$, we have*

$$T_n(xy) \leq T_n(x)T_n(y).$$

Proof. Fix $y \geq 1$ and consider the polynomial $p(x) = \dfrac{T_n(xy)}{T_n(y)}$. Let us verify that this polynomial satisfies the condition of Theorem 3.4.8, i.e., $|p(x)| = \dfrac{|T_n(xy)|}{T_n(y)} \leq 1$ for $|x| \leq 1$. For real s, the function $|T_n(s)|$ only depends on $|s|$. Moreover, if $|s| \geq 1$, then $|T_n(s)|$ monotonically increases with $|s|$. Clearly, $|T_n(s)| \leq 1 \leq T_n(y)$ for $|s| \leq 1$. Therefore, if $y \geq 1$ and $|x| \leq 1$, we have $|T_n(xy)| \leq T_n(y)$.

By Theorem 3.4.8 for $x \geq 1$, we have $|p(x)| \leq T_n(x)$, i.e., $T_n(xy) \leq T_n(x)T_n(y)$. □

3.4.4 Generating functions

For a sequence of functions $a_n(x)$ one can consider the series

$$F(x, z) = \sum_{n=0}^{\infty} a_n(x)z^n.$$

If the radius of convergence of this series is positive, the function $F(x, z)$ is called the *generating function* of the sequence $a_n(x)$.

Theorem 3.4.12. *For $-1 < x < 1$ and $|z| < 1$, we have*

$$(a) \qquad 2\sum_{n=1}^{\infty} \frac{T_n(x)}{n} z^n = -\ln(1 - 2xz + z^2);$$

$$(b) \qquad 1 + 2\sum_{n=1}^{\infty} T_n(x)z^n = \frac{1 - z^2}{1 - 2xz + z^2}.$$

Proof. a) Let $x = \cos\varphi$. Then

$$1 - 2xz + z^2 = (1 - e^{i\varphi}z)(1 - e^{-i\varphi}z).$$

Hence $\ln(1 - 2xz + z^2) = \ln(1 - e^{i\varphi}z) + \ln(1 - e^{-i\varphi}z)$. It is also clear that

$$-\ln(1 - e^{\pm i\varphi}z) = \sum_{n=1}^{\infty} \frac{e^{\pm in\varphi}}{n} z^n$$

for $|z| < 1$. Hence

$$-\ln(1 - 2xz + z^2) = \sum_{n=1}^{\infty} \frac{2\cos n\varphi}{n} z^n = 2\sum_{n=1}^{\infty} \frac{T_n(x)}{n} z^n.$$

b) By differentiating both parts of (a) with respect to z we get

$$2 \sum_{n=1}^{\infty} T_n(x) z^{n-1} = \frac{2x - 2z}{1 - 2xz + z^2}.$$

Therefore

$$1 + 2 \sum_{n=1}^{\infty} T_n(x) z^n = 1 + \frac{z(2x - 2z)}{1 - 2xz + z^2} = \frac{1 - z^2}{1 - 2xz + z^2}. \quad \square$$

With the help of Theorem 3.4.12 we can obtain the following explicit expression for Chebyshev polynomials.

Theorem 3.4.13. *Let $n \geq 1$ and $m = \left[\frac{n}{2}\right]$. Then*

$$T_n(x) = \frac{1}{2} \sum_{k=0}^{m} (-1)^k \frac{n}{n-k} \binom{n-k}{k} (2x)^{n-2k}.$$

Proof. By Theorem 3.4.12 (a)

$$2 \sum_{n=1}^{\infty} \frac{T_n(x)}{n} z^n = -\ln(1 - 2xz + z^2) = \sum_{p=1}^{\infty} \frac{(2xz - z^2)^p}{p} =$$

$$= \sum_{p=1}^{\infty} \sum_{k=0}^{p} (-1)^k \frac{1}{p} \binom{p}{k} z^{p+k} (2x)^{p-k}.$$

Hence

$$T_n(x) = \frac{1}{2} \sum_{p+k=n} (-1)^k \frac{n}{p} \binom{p}{k} (2x)^{p-k} =$$

$$= \frac{1}{2} \sum_{k=0}^{M} (-1)^k \frac{n}{n-k} \binom{n-k}{k} (2x)^{n-2k}.$$

The summation is performed until $n - 2k \geq 0$, and hence $M = \left[\frac{n}{2}\right] = m$. $\square$

For the polynomial $P_n(x) = 2T_n\left(\frac{x}{2}\right)$, we get a neater explicit formula, namely:

$$P_n(x) = \sum_{k=0}^{m} (-1)^k \frac{n}{n-k} \binom{n-k}{k} x^{n-2k}, \tag{1}$$

where $m = \left[\frac{n}{2}\right]$.

Recall that the polynomial $P_n(x)$ corresponds to the polynomial expression of $z^n + z^{-n}$ in terms of $z + z^{-1}$ (this follows from the fact that, for $z = e^{i\varphi}$, we have $z^n + z^{-n} = 2 \cos n\varphi$ and $z + z^{-1} = 2 \cos \varphi$).

It is easy to verify that for $n = 2m + 1$

$$(z + z^{-1})^n = \sum_{k=0}^{m} \binom{n}{k} (z^{n-2k} + z^{2k-n}),$$

and for $n = 2m$

$$(z + z^{-1})^n = \sum_{k=0}^{m-1} \binom{n}{k} (z^{n-2k} + z^{2k-n}) + \binom{n}{m}.$$

Therefore, if $P_0(x) = 1$ and the polynomials $P_n(x)$ for $n \geq 1$ are given by (1), then

$$x^n = \sum_{k=0}^{m} \binom{n}{k} P_{n-2k}(x), \tag{2}$$

where $m = \left[\dfrac{n}{2}\right]$.

Relations (1), (2) can be expressed as follows. Let $a_n = x^n$ and $b_n = P_n(x)$, where x is fixed. Then

$$a_n = \sum_{k=0}^{m} \binom{n}{k} b_{n-2k}, \quad b_n = \sum_{k=0}^{m} (-1)^k \frac{n}{n-k} \binom{n-k}{k} a_{n-2k} \tag{3}$$

(for $n = 0$ the second relation takes the form $b_0 = a_0$). Let us prove that the relations (3) are equivalent not only for the indicated sequences but also for arbitrary sequences.

First of all, observe that the first relation is of the form

$$a_n = b_n + \sum \beta_{n-i} b_{n-i}$$

and the second relation is of the form

$$b_n = a_n + \sum \alpha_{n-i} a_{n-i}.$$

Hence each relation uniquely determines both the sequence a_n in terms of the sequence b_n and vice versa. It is also clear that for the sequences

$$a_n = \sum \lambda_i x_i^n \text{ and } b_n = \sum \lambda_i P_n(x_i), \text{ where the } \lambda_i \text{ and } x_i \text{ are fixed,}$$

the relations (3) are equivalent because they are equivalent for the sequences $a_n = x_i^n$ and $b_n = P_n(x_i)$.

It remains to verify that for any sequence $a_0, a_1, \ldots, a_n$ we can select numbers $\lambda_0, \ldots, \lambda_n$ and $x_0, \ldots, x_n$ so that

$$a_l = \sum_{i=0}^{n} \lambda_i x_i^l \text{ for } l = 0, 1, \ldots, n. \tag{$*$}$$

Select arbitrary distinct numbers $x_0, \ldots, x_n$. Then we obtain a system of linear equations for the numbers $\lambda_0, \ldots, \lambda_n$ with the determinant

$$
\begin{vmatrix}
1 & \cdots & 1 \\
x_0 & \cdots & x_n \\
\cdots & \cdots & \cdots \\
x_0^n & \cdots & x_n^n
\end{vmatrix} \neq 0,
$$

hence the system $(*)$ has solutions for any $a_0, \ldots, a_n$.

Relations (3) enable us to obtain nontrivial identities involving binomial coefficients. Let, for instance, $b_n = 1$ for all n. Then

$$
a_{2m+1} = \sum_{k=0}^{m} \binom{2m}{k} = 2^{2m},
$$

$$
a_{2m} = \sum_{k=0}^{m} \binom{2m+1}{k} = \frac{1}{2}\left(2^{2m} + \binom{2m}{m}\right);
$$

These identities are easy to derive from the expansions of $(1+1)^{2m+1}$ and $(1+1)^{2m}$ via the binomial formulas. In this case the relation

$$
b_n = \sum_{k=0}^{m} (-1)^k \frac{n}{n-k} \binom{n-k}{k} a_{n-2k}
$$

takes the form

$$
1 = \sum_{k=0}^{m} (-1)^k \frac{2m+1}{2m+1-k} \binom{2m+1-k}{k} 2^{2m},
$$

$$
2 = \sum_{k=0}^{m} (-1)^k \frac{2m}{2m-k} \binom{2m-k}{k} \frac{1}{2}\left(2^{2m-2k} + \binom{2m-2k}{m-k}\right).
$$

3.5 Bernoulli polynomials

3.5.1 Definition of Bernoulli polynomials

Consider the function

$$
g(z, t) = \frac{te^{tz}}{e^t - 1}.
$$

For $t = 2k\pi i$ the denominator vanishes, so that at such values of t the function $g(z, t)$ may have singularities. But g is regular at $t = 0$, and therefore we can expand $g(z, t)$ into the series

$$
g(z, t) = \sum_{n=0}^{\infty} \frac{t^n}{n!} B_n(z),
$$

which converges for $|t| < 2\pi$.

As we will see shortly, $B_n(z)$ is a polynomial of degree n. The $B_n(z)$ are called *Bernoulli polynomials* and the numbers $B_n = B_n(0)$ are called *Bernoulli numbers.*

The series for $g(z, t)$ is the product of the series

$$g(0, t) = \sum_{n=0}^{\infty} \frac{t^n}{n!} B_n \quad \text{and} \quad e^{tz} = \sum_{n=0}^{\infty} \frac{t^n z^n}{n!}.$$

Hence

$$\frac{B_n(z)}{n!} = \sum_{k=0}^{\infty} \frac{B_{n-k} z^k}{k! \, (n-k)!},$$

i.e.,

$$B_n(z) = \sum_{k=0}^{\infty} \binom{n}{k} B_{n-k} z^k.$$

Formally, this identity can be expressed as $B_n(z) = (B + z)^n$, where by definition $B^{n-k} = B_{n-k}$.

One of the most important properties of Bernoulli polynomials is as follows:

$$B_n(z + 1) - B_n(z) = n z^{n-1}. \tag{1}$$

To prove (1), it suffices to observe that

$$\sum_{n=0}^{\infty} (B_n(z + 1) - B_n(z)) \frac{t^n}{n!} = g(t, z + 1) - g(t, z) =$$

$$= \frac{t e^{t(z+1)}}{e^t - 1} - \frac{t e^{tz}}{e^t - 1} = t e^{tz} = \sum_{n=0}^{\infty} \frac{t^{n+1} z^n}{n!}.$$

Let us sum the identities (1) for $z = 0, 1, \ldots, m - 1$. As a result we obtain

$$\sum_{k=0}^{m-1} k^{n-1} = \frac{1}{n} \left(B_n(m) - B_n(0) \right). \tag{2}$$

This means, in particular, that the sum

$$1 + 2^{n-1} + \cdots + (m - 1)^{n-1}$$

is a polynomial of degree n in m. This is precisely the property that J. Bernoulli discovered [Be6].

In 1738, Euler suggested the generating function

$$\frac{t e^{tz}}{e^t - 1} = \sum_{n=0}^{\infty} \frac{t^n}{n!} B_n(z).$$

It is convenient to compute Bernoulli polynomials from the recurrence formula

$$\sum_{r=0}^{n-1} \binom{n}{r} B_r(z) = nz^{n-1}, \quad n \geq 2. \tag{3}$$

This formula can be proved as follows:

$$\sum_{n=0}^{\infty} \frac{t^n}{n!} B_n(z+1) = \frac{te^{tz}}{e^t - 1} e^z = \left(\sum_{r=0}^{\infty} \frac{t^r}{r!} B_r(z) \right) \left(\sum_{s=0}^{\infty} \frac{z^s}{s!} \right),$$

and hence

$$B_n(z+1) = \sum_{r=0}^{n} \binom{n}{r} B_r(z) = B_n(z) + \sum_{r=0}^{n-1} \binom{n}{r} B_r(z).$$

It remains to use (1).

When $z = 0$ the relations (3) become a recurrence formula for the Bernoulli numbers

$$\sum_{r=0}^{n-1} \binom{n}{r} B_r = 0, \quad n \geq 2. \tag{4}$$

It is easy to verify that $B_0 = 1$. Therefore from (4) we recursively obtain

$$B_1 = -\frac{1}{2}, \quad B_2 = \frac{1}{6}, \quad B_3 = 0, \quad B_4 = -\frac{1}{30}, \quad B_5 = 0, \quad \ldots$$

It is not difficult to show that $B_{2k+1} = 0$ for $k \geq 1$. Indeed,

$$\frac{t}{e^t - 1} = 1 - \frac{t}{2} + \sum_{n=2}^{\infty} \frac{t^n}{n!} B_n.$$

Hence, it suffices to verify that the function

$$\frac{t}{e^t - 1} + \frac{t}{2} = t\frac{e^t + 1}{e^t - 1}$$

is even, and this is evident.

In 1832, Appel showed that the Bernoulli polynomials satisfy the relation

$$B'_{n+1}(z) = (n+1)B_n(z).$$

To prove this, we differentiate the identity

$$\sum_{n=0}^{\infty} \frac{t^n}{n!} B_n(z) = \frac{te^{tz}}{e^t - 1}$$

with respect to z. We obtain

$$\sum_{n=0}^{\infty} \frac{t^n}{n!} B'_n(z) = \frac{t^2 e^{tz}}{e^t - 1} = \sum_{n=0}^{\infty} \frac{t^{n+1}}{n!} B_n(z).$$

Equating the coefficients of t^{n+1} we get Appel's relation.

3.5.2 Theorems of complement, addition of arguments and multiplication

Bernoulli polynomials possess the following properties:

$$B_n(1 - z) = (-1)^n B_n(z) \qquad (\textit{Theorem of complement});$$

$$B_n(x + y) = \sum_{s=0}^{n} \binom{n}{s} B_s(x) y^{n-s} \qquad (\textit{Theorem of addition of arguments});$$

$$\frac{1}{m} \sum_{k=0}^{m-1} B_n\left(z + \frac{k}{m}\right) = m^{-n} B_n(mz) \qquad (\textit{Theorem of multiplication}).$$

All these theorems are easy to deduce from the relation

$$\sum_{n=0}^{\infty} \frac{t^n}{n!} B_n(z) = \frac{t e^{tz}}{e^t - 1}.$$

To prove the first property, we observe that

$$\sum_{n=0}^{\infty} \frac{t^n}{n!} B_n(1 - z) = \frac{t e^{t(1-z)}}{e^t - 1} = \frac{-t e^{-tz}}{e^{-t} - 1} = \sum_{n=0}^{\infty} \frac{(-t)^n}{n!} B_n(z).$$

The second property, on the addition of arguments, is proved as follows:

$$\sum_{n=0}^{\infty} \frac{t^n}{n!} B_n(x + y) = \frac{t e^{tx} e^{ty}}{e^t - 1} = \left(\sum_{s=0}^{\infty} \frac{t^s}{s!} B_s(x)\right) \left(\sum_{r=0}^{\infty} \frac{t^r y^r}{r!}\right) =$$

$$= \sum_{r,\,s=0}^{\infty} \frac{t^{r+s}}{r! s!} B_s(x) y^r = \sum_{n=0}^{\infty} \sum_{s=0}^{n} \frac{t^n}{n!} \binom{n}{s} B_s(x) y^{n-s}.$$

To prove the multiplication property, we use the identity

$$\frac{1}{e^t - 1} = \frac{1 + e^t + \cdots + e^{(m-1)t}}{e^{mt} - 1}.$$

This gives

$$\sum_{n=0}^{\infty} \frac{t^n}{n!} B_n(mz) = \frac{t e^{mtz}}{e^t - 1} = \frac{1}{m} \frac{e^{mtz} mt(1 + e^t + \cdots + e^{(m-1)t})}{e^{mt} - 1} =$$

$$= \frac{1}{m} \sum_{k=0}^{m-1} \frac{e^{(z+k/m)mt} mt}{e^{mt} - 1} = \frac{1}{m} \sum_{k=0}^{m-1} \sum_{n=0}^{\infty} \frac{m^n t^n}{n!} B_n\left(z + \frac{k}{m}\right).$$

The multiplication property shows that $B_n(x)$ is a solution of the functional equation

$$\frac{1}{m} \sum_{k=0}^{m-1} f\left(x + \frac{k}{m}\right) = m^{-n} f(mx). \tag{1}$$

Theorem 3.5.1 ([Le1]). *For fixed $m, n > 1$, there exists only one monic polynomial of degree n satisfying* (1).

Proof. The existence of the polynomial required can be proved directly, but instead we give a proof which uses the fact already known to us, namely, that Bernoulli polynomials satisfy the functional equation (1).

Let $p(x) = x^n + \cdots$ and $q(x) = x^n + \cdots$ be two distinct polynomials satisfying (1). Their difference $\Delta(x) = a_0 x^d + \cdots$, where $a_0 \neq 0$ and $d < n$, also satisfies (1). Comparing the coefficients of x^d on both sides of the equation

$$\frac{1}{m} \sum_{k=0}^{m-1} \Delta\left(x + \frac{k}{m}\right) = m^{-n}\Delta(mx)$$

we see that $a_0 = m^{d-n}a_0$. This contradicts the condition that $m > 1$ and the assumption that $a_0 \neq 0$ and $d < n$. $\square$

Making a change of variables we can reduce (1) to the form

$$f(x) = m^{s-1} \sum_{k=0}^{m-1} f\left(\frac{x+k}{m}\right), \tag{2}$$

where $s = n$. Kubert [Ku] studied continuous functions $f : (0,1) \to \mathbb{C}$ satisfying (1) with s a positive integer. In the more general case $s \in \mathbb{C}$, the space of such functions is two-dimensional and one can select a basis f_{even}, f_{odd} in it so that

$$f_{\text{even}}(x) = f_{\text{even}}(1-x) \quad \text{and} \quad f_{\text{odd}}(x) = -f_{\text{odd}}(1-x).$$

John Milnor wrote a long and interesting paper [Mi3] on various properties of solutions of (2).

3.5.3 Euler's formula

Let $s \in \mathbb{C}$ and $\operatorname{Re} s > 1$. Then the series $\sum_{n=1}^{\infty} \frac{1}{n^s}$ converges. The function

$$\zeta(s) = \sum_{n=1}^{\infty} \frac{1}{n^s}$$

has an analytic continuation to the whole complex plane $\mathbb{C}$. This continuation has a simple pole at $s = 1$ and is regular elsewhere.

It was already Euler who considered the series $\sum_{n=1}^{\infty} \frac{1}{n^s}$ at integer points s. But the first to consider $\zeta(s)$ as a function of a complex variable was Riemann, and it was Riemann who discovered the most profound properties and most important applications of $\zeta(s)$. This is why $\zeta(s)$ is called the *Riemann zeta-function*.

Theorem 3.5.2 (Euler). *If k is a positive integer, then*

$$\zeta(2k) = \frac{(-1)^{k+1} B_{2k} 2^{2k-1} \pi^{2k}}{(2k)!}.$$

Proof. We use the factorization of $\sin z$ into an infinite product

$$\sin z = z \prod_{n=1}^{\infty} \left(1 - \frac{z^2}{n^2 \pi^2} \right).$$

Differentiating the logarithms of both sides we obtain

$$\frac{\cos z}{\sin z} = \frac{1}{z} + \sum_{n=1}^{\infty} \frac{2z}{z^2 - n^2 \pi^2},$$

i.e.,

$$z \frac{\cos z}{\sin z} = 1 - 2 \sum_{n=1}^{\infty} \sum_{k=1}^{\infty} \left(\frac{z}{n\pi} \right)^{2k}. \tag{1}$$

On the other hand, substituting $t = 2iz$ into the identity

$$\frac{t}{e^t - 1} = 1 - \frac{t}{2} + \sum_{m=2}^{\infty} B_m \frac{t^m}{m!},$$

we obtain

$$z \frac{\cos z}{\sin z} = iz \frac{e^{iz} + e^{-iz}}{e^{iz} - e^{-iz}} = 1 + \sum_{m=2}^{\infty} B_m \frac{(2iz)^m}{m!}. \tag{2}$$

Comparison of the coefficients of z^{2k} in (1) and (2) yields the identity desired. $\square$

Euler's formula for $\zeta(2k)$ makes it obvious that $\zeta(2k)$ is a transcendental number since π is transcendental. There is no similarly convenient formula for $\zeta(2k+1)$. It was only in 1978 that R. Apéry proved the irrationality of $\zeta(3)$. The simplest (known to me) proof of irrationality of $\zeta(3)$ is given in [Be7].

3.5.4 The Faulhaber-Jacobi theorem

Mathematicians were interested in the summation of the series of powers

$$1^m + 2^m + 3^m + \cdots$$

long before Bernoulli. In 1617, the German mathematician Johann Faulhaber (1580–1635) published a book in which he gave sums of such series for $m \leq 11$. In 1631, in the book [Fa], he extended his calculations up to $m = 17$.

Pierre Fermat also studied the summation of such series. In 1636 he wrote to Mersenne:

"...We do not want to dwell on this here, let us just mention that we have obtained a solution, perhaps the simplest in the whole arithmetics of the problem thanks to which we can not only *find the sum of squares or cubes of any progression*, but *in general the sum of all powers up to infinity thanks to the most general method*; squares of squares, squares of cubes, and so on."

Many historians of mathematics are inclined to believe that Fermat really obtained the solution of this problem almost a century before Bernoulli.

Faulhaber, in his book [Fa], observed that all the sums $\sum n^k$ can be polynomially expressed in terms of the first two sums $\sum n$ and $\sum n^2$. Two hundred years later, in 1834, Jacobi rediscovered Faulhaber's theorem. It is known that Jacobi possessed Faulhaber's book but it is not known whether he read it or not.

For convenience, in the proof we introduce the polynomials

$$S_{n-1}(m) = \frac{1}{n}\left(B_n(m) - B_n(0)\right).$$

Formula (2) on page 113 shows that

$$S_n(m) = 1^n + 2^n + \cdots + (m-1)^n.$$

Theorem 3.5.3 (Faulhaber-Jacobi). *Let $U = S_1(x)$ and $V = S_2(x)$. For $k \geq 1$, there exist polynomials P_k and Q_k with rational coefficients, such that $S_{2k+1}(x) = U^2 P_k(U)$ and $S_{2k}(x) = V Q_k(U)$.*

Proof. To obtain the expression for S_{2k+1}, we use the identity

$$(n(n-1))^r = \sum_{m=1}^{n-1}(m^r(m+1)^r - m^r(m-1)^r) =$$
$$= 2\left(\binom{r}{1}\sum m^{2r-1} + \binom{r}{3}\sum m^{2r-3} + \binom{r}{5}\sum m^{2r-5} + \cdots\right), \tag{1}$$

i.e.,

$$(n(n-1))^{i+1} = \sum \binom{i+1}{2(i-j)+1} S_{2j+1}(n).$$

These identities can be expressed in the matrix form

$$\begin{pmatrix} n^2(n-1)^2 \\ n^3(n-1)^3 \\ n^4(n-1)^4 \\ \vdots \end{pmatrix} = 2 \begin{pmatrix} 2\,0\,0\,\ldots \\ 1\,3\,0\,\ldots \\ 0\,4\,4\,\ldots \\ \vdots\,\vdots\,\vdots\,\ddots \end{pmatrix} \begin{pmatrix} S_3(n) \\ S_5(n) \\ S_7(n) \\ \vdots \end{pmatrix}.$$

In the infinite matrix $\begin{pmatrix} 2\,0\,0\,\ldots \\ 1\,3\,0\,\ldots \\ 0\,4\,4\,\ldots \\ \vdots\,\vdots\,\vdots\,\ddots \end{pmatrix}$ all the principal minors of finite order are invertible, and hence the matrix itself is invertible and we can write

$$\begin{pmatrix} S_3(n) \\ S_5(n) \\ S_7(n) \\ \vdots \end{pmatrix} = \frac{1}{2} \frac{1}{\|a_{ij}\|} \begin{pmatrix} n^2(n-1)^2 \\ n^3(n-1)^3 \\ n^4(n-1)^4 \\ \vdots \end{pmatrix}, \quad \text{where} \quad a_{ij} = \begin{pmatrix} i+1 \\ 2(i-j)+1 \end{pmatrix}.$$

This formula shows that $S_{2k+1}(n)$ can be expressed in terms of $n(n-1) = 2U(n)$ and is divisible by $(n(n-1))^2$.

To obtain an expression for S_{2k}, we use the identities

$$n^{r+1}(n-1)^r = \sum_{m=1}^{n-1}(m^r(m+1)^{r+1} - (m-1)^r m^{r+1}) =$$
$$= \sum m^{2r}\left(\binom{r+1}{1} + \binom{r}{1}\right) + \sum m^{2r-1}\left(\binom{r+1}{2} - \binom{r}{2}\right) +$$
$$+ \sum m^{2r-2}\left(\binom{r+1}{3} + \binom{r}{3}\right) + \sum m^{2r-3}\left(\binom{r+1}{4} - \binom{r}{4}\right) + \cdots =$$
$$= \left(\binom{r+1}{1} + \binom{r}{1}\right)\sum m^{2r} + \left(\binom{r+1}{3} + \binom{r}{3}\right)\sum m^{2r-2} + \cdots$$
$$\cdots + \binom{r}{1}\sum m^{2r-1} + \binom{r}{3}\sum m^{2r-3} + \cdots$$

The sums of odd powers can be eliminated with the help of (1). As a result we obtain

$$n^{r+1}(n-1)^r = \frac{n^r(n-1)^r}{2} + \left(\binom{r+1}{1} + \binom{r}{1}\right)\sum m^{2r} +$$
$$+ \left(\binom{r+1}{3} + \binom{r}{3}\right)\sum m^{2r-2} + \cdots,$$

i.e.,

$$n^i(n-1)^i\left(\frac{2n-1}{2}\right) = \sum\left(\binom{i+1}{2(i-j)+1} + \binom{i}{2(i-j)+1}\right)S_{2j}(n).$$

Now, similarly to the above, we obtain

$$\begin{pmatrix} S_2(n) \\ S_4(n) \\ S_6(n) \\ \vdots \end{pmatrix} = \frac{2n-1}{2} \frac{1}{\|b_{ij}\|} \begin{pmatrix} n(n-1) \\ n^2(n-1)^2 \\ n^3(n-1)^3 \\ \vdots \end{pmatrix},$$

where $b_{ij} = \binom{i+1}{2(i-j)+1} + \binom{i}{2(i-j)+1}$. Simple calculations show that

$$S_2(n) = \frac{2n-1}{2} \cdot \frac{n(n-1)}{3}.$$

Therefore the polynomials $S_4(n)$, $S_6(n)$, ... are divisible by $S_2(n)$, and $\dfrac{S_{2k}(n)}{S_2(n)}$ is a polynomial in $n(n-1) = 2U(n)$. $\square$

3.5.5 Arithmetic properties of Bernoulli numbers and Bernoulli polynomials

In this section we prove several theorems on the denominators of the values of Bernoulli polynomials at rational points. The most interesting are the values of Bernoulli polynomials at 0, i.e., Bernoulli numbers.

When we formulate statements on denominators of rational numbers it is convenient to use the notion of p-integers. If p is a prime, the rational number r is said to be a p-*integer* if p does not enter the denominator of r, i.e., if the denominator t of the irreducible fraction $\dfrac{s}{t} = r$ is not divisible by p.

For a rational r, the expression $r \equiv 0 \pmod p$ will mean that the numerator s of the irreducible fraction $\dfrac{s}{t} = r$ is divisible by p. It is easy to verify that if $r_1 \equiv r_2 \equiv 0 \pmod p$, then $r_1 r_2 \equiv 0 \pmod p$ and $r_1 \pm r_2 \equiv 0 \pmod p$. Moreover, if r_1 is a p-integer and $r_2 \equiv 0 \pmod p$, then $r_1 r_2 \equiv 0 \pmod p$.

Theorem 3.5.4 (Kummer). *Let p be a prime. If the positive integer n is not divisible by $p - 1$, then $\dfrac{B_n}{n}$ is a p-integer and*

$$\frac{B_{n+p-1}}{n+p-1} - \frac{B_n}{n} \equiv 0 \pmod p.$$

Proof. The multiplicative group of the field $\mathbb{F}_p$ is cyclic, and hence has a generator. This means that there exists a positive integer a lying between 1 and p for which $a^k \not\equiv 1 \pmod p$ for $k = 1, \ldots, p - 2$. Consider the function

$$A(t) = \frac{at}{e^{at} - 1} - \frac{t}{e^t - 1} = \sum_{k=1}^{\infty} (a^k - 1) B_k \frac{t^k}{k!} =$$

$$= t \sum_{k=1}^{\infty} (a^k - 1) \frac{B_k}{k} \frac{t^{k-1}}{(k-1)!} = t \sum_{k=1}^{\infty} A_k \frac{t^{k-1}}{(k-1)!},$$

where $A_{k-1} = (a^k - 1) \frac{B_k}{k}$.

It suffices to prove that all the numbers A_k are p-integers and

$$A_{k+p-1} - A_k \equiv 0 \pmod p.$$

Indeed, if n is not divisible by $p - 1$, then $a^n \not\equiv 1 \pmod p$ and so the equation $\dfrac{B_n}{n} = \dfrac{A_{n-1}}{a^n - 1}$ shows that $\dfrac{B_n}{n}$ is also a p-integer. But since $a^{p-1} \equiv 1 \pmod p$ we have

$$A_{k+p-1} - A_k = (a^{n+p-1} - 1) \frac{B_{n+p-1}}{n+p-1} - (a^n - 1) \frac{B_n}{n} \equiv$$

$$\equiv \left(\frac{B_{n+p-1}}{n+p-1} - \frac{B_n}{n} \right) (a^n - 1) \pmod p.$$

Now we use the identity

$$\sum_{k=1}^{\infty} A_k \frac{t^{k-1}}{(k-1)!} = \frac{a}{e^{at} - 1} - \frac{1}{e^t - 1}.$$

Set $u = e^t - 1$. Then

$$\frac{a}{e^{at} - 1} - \frac{1}{e^t - 1} = \frac{a}{(1+u)^a - 1} - \frac{1}{u} = \frac{a}{au + \sum b_s u^s} - \frac{1}{u} =$$

$$= \frac{1}{u}\left(\frac{1}{1 + \sum \frac{b_s}{a} u^{s-1}} - 1\right) = \sum_{r=0}^{\infty} c_r u^r.$$

All the numbers c_r are p-integers since this is true of the numbers $\frac{b_s}{a}$. Therefore

$$\sum_{k=1}^{\infty} A_k \frac{t^{k-1}}{(k-1)!} = \sum_{r=0}^{\infty} c_r(e^t - 1)^r,$$

where all the coefficients c_r are p-integers.

The function $(e^t - 1)^r$ can be represented as a linear combination with integer coefficients of the functions e^{mt}, where $m = 0, 1, \ldots, r$. In turn,

$$e^{mt} = \sum_{l=0}^{\infty} m^l \frac{t^l}{l!}.$$

Therefore A_k can be represented as a linear combination of the numbers m^{k-1} with p-integer coefficients. The equality $c_r(e^t - 1)^r = c_r t^r + \cdots$ shows that this linear combination contains finitely many summands.

Since the sum and the product of p-integers is a p-integer, it follows that A_k is also a p-integer. It is also clear that

$$m^{(k-1)+(p-1)} - m^{k-1} = m^{k-1}(m^{p-1} - 1) \equiv 0 \pmod{p}.$$

Therefore $A_{k+p-1} - A_k$ is a linear combination with p-integer coefficients of rational numbers whose numerators are divisible by p. This means that

$$A_{k+p-1} - A_k \equiv 0 \pmod{p}. \quad \square$$

Theorem 3.5.5 (von Staudt). *Let n be even and p a prime. If n is not divisible by $p - 1$, then B_n is a p-integer and, if n is divisible by $p - 1$, then $pB_n \equiv -1 \pmod{p}$.*

Proof. By the preceding theorem if p is prime and n is not divisible by $p - 1$, then the denominator of B_n is not divisible by p. Thus it remains to consider the case where n is divisible by $p - 1$.

By multiplying the identities $\sum_{k=0}^{\infty} B_k \frac{t^k}{k!} = \frac{t}{e^t - 1}$ and $\sum_{n=0}^{\infty} \frac{p^n t^{n-1}}{n!} = \frac{e^{pt} - 1}{t}$

we obtain

$$\sum_{n,k=0}^{\infty} \frac{p^n B_k t^{n+k-1}}{n! \, k!} = \sum_{r=0}^{p-1} e^{rt} = \sum_{r=0}^{p-1} \sum_{s=0}^{\infty} \frac{r^s t^s}{s!}.$$

Comparison of the coefficients of t^n on each side shows that

$$\sum_{k=0}^{n+1} \frac{p^{n+1-k} B_k}{(n+1-k)! \, k!} = \sum_{r=1}^{p-1} \frac{r^n}{n!}.$$

Since $B_{n+1} = B_{n-1} = 0$ we obtain that

$$pB_n = -\sum_{k=0}^{n-2} pB_k \frac{p^{n-k}}{n+1-k} + \sum_{r=1}^{p-1} r^n.$$

For $n - k \geq 2$, the number $\dfrac{p^{n-k}}{n-k+1}$ is, clearly, a p-integer. Induction on n shows that the numbers $pB_2, \ldots, pB_n$ are p-integers. (The start of the induction: $pB_0 = p$, $pB_1 = -\frac{1}{2}p$.)

For $n - k \geq 2$, the number $\dfrac{p^{n-k}}{n-k+1}$ is not only p-integer but also its numerator after all possible simplifications is divisible by p, i.e.,

$$\frac{p^{n-k}}{n-k+1} \equiv 0 \pmod{p}.$$

Therefore

$$pB_n \equiv \sum_{r=1}^{p-1} r^n \pmod{p}.$$

In the case considered, n is divisible by $p - 1$, and hence $r^n \equiv 1 \pmod{p}$ for $r = 1, 2, \ldots, p - 1$. As a result, we obtain

$$pB_n \equiv -1 \pmod{p}. \quad \square$$

We now define $\widetilde{B}_n(t) = B_n(t) - B_n(0)$, and we recall that

$$\widetilde{B}_n(m) = n \left(1^{n-1} + 2^{n-1} + \cdots + (m-1)^{n-1} \right)$$

for all positive integers m.

Theorem 3.5.6 (Almkvist-Meurman). *For all positive integers h, k, and n, the number $k^n \widetilde{B}_n \left(\dfrac{h}{k} \right)$ is an integer.*

Proof. [Su2] The theorem on the addition of arguments

$$B_n(x+y) = \sum_{s=0}^{n} B_s(x)y^{n-s}$$

can be expressed in the form

$$\widetilde{B}_n(x+y) = \sum_{s=0}^{n} \widetilde{B}_s(x)y^{n-s} + \widetilde{B}_n(y).$$

Therefore it suffices to prove the statement required for $h = 1$.

For brevity, set

$$a_n = k^n \widetilde{B}_n \left(\frac{h}{k}\right).$$

Clearly, $B_0(z) = B_0$. Therefore

$$\sum_{n=1}^{\infty} \frac{t^n}{n!}\widetilde{B}_n(z) = \sum_{n=0}^{\infty} \frac{t^n}{n!}B_n(z) - \sum_{n=0}^{\infty} \frac{t^n}{n!}B_n(0) = \frac{te^{tz}}{e^t - 1} - \frac{t}{e^t - 1} = \frac{t(e^{tz} - 1)}{e^t - 1}.$$

Set $z = \dfrac{1}{k}$ and make the change $x = kt$. This gives

$$\sum_{n=1}^{\infty} \frac{a_n x^n}{n!} = \frac{kx(e^x - 1)}{e^{kx} - 1}.$$

Let us express this relation in two forms:

$$\sum_{n=1}^{\infty} \frac{a_n x^n}{n!}(e^{kx} - 1) = kx(e^x - 1)$$

and

$$\sum_{n=1}^{\infty} \frac{a_n x^n}{n!}(1 + e^x + \cdots + e^{(k-1)x}) = kx$$

and use the expansions

$$e^{kx} - 1 = \sum_{r=1}^{\infty} \frac{k^r x^r}{r!} \quad \text{and} \quad e^x - 1 = \sum_{s=1}^{\infty} \frac{x^s}{s!}.$$

Comparing the coefficients of x^n on the two sides in the first form we obtain for all $n \geq 1$

$$\sum_{l=1}^{n-1} \binom{n+1}{l} a_r k^{n-l} = (n+1)(1 - a_n). \tag{2}$$

In the second form we get $a_1 = 1$ and for $n \geq 2$ we get

$$\sum_{l=1}^{n-1} \binom{n}{l} a_r s_{n-l} = -k a_n,\tag{3}$$

where $s_0 = k$ and $s_m = 1^m + 2^m + \cdots + (k-1)^m$.

We now prove by induction that a_n are integers, using (2) and (3). To do this, we need the following Lemma.

Lemma. *Let p be a prime, $2 \leq l \leq r$ and $(p, s) = 1$. Then $\binom{sp^r}{l}$ is divisible by p^{r-l+1}.*

Proof. Let us write $l = tp^a$, where $(t, p) = 1$. Clearly, $a \leq l - 1$ (the equality is only possible for $l = p = 2$). It is easy to verify that $\binom{m}{n} = \frac{m}{n}\binom{m-1}{n-1}$. Therefore

$$\binom{sp^r}{l} = \binom{sp^r}{tp^a} = \frac{s}{t} p^{r-a} \binom{sp^r - 1}{tp^a - 1} = \frac{s}{t} p^{r-a} N,$$

where N is an integer and s and t are relatively prime to p. Hence $\binom{sp^r}{l}$ is divisible by p^{r-a}. In turn, p^{r-a} is divisible by p^{r-l+1} because $a \leq l - 1$. □

Since we have not specify k for some time, recall that

$$a_n = k^n \widetilde{B}_n \left(\frac{h}{k} \right).$$

First consider the case where k is a prime. Suppose that $a_1, \ldots, a_{n-1}$ are integers. Then (2) and (3) imply that $(n+1)a_n$ and ka_n are integers. If $n+1$ is not divisible by k, then a_n is an integer. Now let $n + 1 = sk^r$, where $r \geq 1$ and $(k, s) = 1$. To establish that a_n is an integer, it suffices to prove that $(n+1)a_n = sk^r a_n$ is divisible by k^r. Formula (2) shows that, in turn, it suffices to prove that the numbers $\binom{n+1}{l} k^{n-l}$ are divisible by k^r for $l = 1, 2, \ldots, sk^r - 2$. For $l \leq n - r = sk^r - 1 - r$, this is obvious.

If $sk^r - r \leq l \leq sk^r - 2$, consider the number $l' = sk^r - l$ and apply the Lemma to it. As a result, we see that $\binom{sk^r}{l} = \binom{sk^r}{l'}$ is divisible by $k^{r-l'+1}$, and hence $\binom{n+1}{l} k^{n-l}$ is divisible by $k^{n-l+r-l'+1} = k^r$ as was required.

The case $k = p_1^{a_1} \cdot \ldots \cdot p_m^{a_m}$, where $p_1, \ldots, p_m$ are distinct primes, does not involve any essentially new ideas. Suppose that $a_1, \ldots, a_{n-1}$ are integers. Then (2) and (3) imply that $(n+1)a_n$ and ka_n are integers. Let us express $n+1$ in the form

$$n + 1 = p_1^{b_1} \cdot \ldots \cdot p_m^{b_m} s, \quad \text{where } (s, p_i) = 1.$$

If $b_i \geq 1$, the same arguments as in the preceding case show that $(n+1)a_n$ is divisible by p^{b_i}. Set

$$s_i = \frac{n+1}{p_i^{b_i}}.$$

Then $(s_i, p_i) = 1$ and all the numbers $s_i a_n$ are integers. This means that the denominator of the rational number a_n is not divisible by primes $p_1, \ldots, p_m$. On the other hand, the number ka_n is integer, and hence the denominator of a_n can only contain the prime factors of k. $\square$

3.6 Problems to Chapter 3

3.6.1 Symmetric polynomials

3.1 Let $\sigma_1, \ldots, \sigma_n$ be elementary symmetric polynomials and let the numbers $a_1, \ldots, a_n \in \mathbb{C}$ satisfy the system of equations

$$\sigma_k(a_1, \ldots, a_n) = \sigma_k(a_k, \ldots, a_k) \text{ for } k = 1, \ldots, n.$$

Then $a_1 = \cdots = a_n$.

In problems 3.2 — 3.4 we assume that $x = (x_1, \ldots, x_n)$ and $y = (y_1, \ldots, y_n)$, where all the numbers $x_1, \ldots, x_n, y_1, \ldots, y_n$ are positive. The sum $x + y$ is defined component-wise.

3.2 [Ma5] For $r = 2, \ldots, n$, the elementary symmetric polynomials σ_i satisfy the inequality

$$\frac{\sigma_r(x+y)}{\sigma_{r-1}(x+y)} \geq \frac{\sigma_r(x)}{\sigma_{r-1}(x)} + \frac{\sigma_r(y)}{\sigma_{r-1}(y)}.$$

3.3 [Ma5] For $r = 1, 2, \ldots, n$, we have

$$\sqrt[r]{\sigma_r(x+y)} \geq \sqrt[r]{\sigma_r(x)} + \sqrt[r]{\sigma_r(y)}.$$

3.4 [Wh] Fix k and define the functions $T_r(x)$ by the relation

$$\sum_{r=0}^{\infty} T_r(x)t^r = \begin{cases} \prod_{i=0}^{n}(1 + x_i t)^k & \text{for } k > 0, \\ \prod_{i=0}^{n}(1 - x_i t)^k & \text{for } k < 0. \end{cases}$$

In particular, if $k = 1$, then $T_r(x) = \sigma_r(x)$ is the elementary symmetric polynomial and if $k = -1$, then $T_r(x) = p_r(x)$ is a complete homogeneous polynomial.

a) Prove that if $k > 0$, then $\sqrt[r]{T_r(x+y)} \geq \sqrt[r]{T_r(x)} + \sqrt[r]{T_r(y)}$.

b) Prove that if $k < 0$, then $\sqrt[r]{T_r(x+y)} \leq \sqrt[r]{T_r(x)} + \sqrt[r]{T_r(y)}$.

3.6.2 Integer-valued polynomials

3.5 If a polynomial $f(x)$ of degree n takes integer values at $x = 0, 1, 4, 9,...,n^2$, then it takes integer values at all $x = m^2$, where $m \in \mathbb{N}$.

3.6 Let m and n be positive integers. Prove that the following conditions are equivalent.

a) There exist integers $a_0, \ldots, a_n$ such that
$$\text{GCD}(a_0, \ldots, a_n, m) = 1$$
and the values of the polynomial $a_n x^n + a_{n-1} x^{n-1} + \cdots + a_0$ at all $x \in \mathbb{Z}$ are divisible by m.

b) $\dfrac{n!}{m} \in \mathbb{Z}$.

3.6.3 Chebyshev polynomials

3.7 For $n \geq 2$, the discriminant of the Chebyshev polynomial T_n is equal to $2^{(n-1)^2} n^2$.

3.8 a) If $n \geq 3$ is odd, the Chebyshev polynomial T_n is reducible.

b) If $n \neq 2^k$, T_n is reducible.

c) If $n \geq 3$ is odd, the polynomial $T_n(x)/x$ is irreducible if and only if n is a prime.

3.9 Let $u(x)$ and $v(x)$ be polynomials with real coefficients and let $\sqrt{1-u^2} = v\sqrt{1-x^2}$. Prove that

a) $u'(x) = \pm n v(x)$, where $n = \deg u$;

b) $u(x) = \pm T_n(x)$.

3.10 a) Prove that the function $y = T_n(x)$ satisfies the differential equation
$$(1-x^2)y'' - xy' + ny^2 = 0$$
and any polynomial solution of this differential equation is of the form cT_n, where c is a constant.

b) Prove that the function $y = T_n(x)$ satisfies the differential equation
$$(1-x^2)(y')^2 = n^2(1-y^2),$$
and this equation has only two polynomial solutions, namely, $y = \pm T_n(x)$.

3.11 Let $\Delta_n(x)$ be the determinant of the $n \times n$ matrix with the diagonal elements $(x, \ldots, x)$, and elements $\left(1, \frac{1}{2}, \ldots, \frac{1}{2}\right)$ in the first super-diagonal, and elements $\left(\frac{1}{2}, \ldots, \frac{1}{2}\right)$ in the first sub-diagonal, the other elements being zero, i.e., $a_{ij} = 0$ for $|i-j| > 1$. Prove that $T_n(x) = 2^{n-1}\Delta_n(x)$.

3.12 [Da1] Recall that the matrix $\|a_{ij}\|$ is called a *circulant* if $a_{ij} = b_i - b_j$, where $b_k = b_l$ for $k \equiv l \pmod{n}$. Let $\Delta_n(x)$ be the determinant of the circulant matrix with $b_0 = 1$, $b_1 = -2x$, $b_2 = 1$. Prove that
$$\Delta_n(x) = 2\left(1 - T_n(x)\right).$$

3.7 Solution of selected problems

3.2. For $r = 2$, the desired inequality follows from the identity

$$\frac{\sigma_2(x+y)}{\sigma_1(x+y)} - \frac{\sigma_2(x)}{\sigma_1(x)} - \frac{\sigma_2(y)}{\sigma_1(y)} = \frac{\sum\limits_{i=1}^{n}\left(x_i \sum\limits_{j=1}^{n} y_j - y_i \sum\limits_{j=1}^{n} x_j\right)^2}{2\sigma_1(x)\sigma_1(y)\sigma_1(x+y)}.$$

Now suppose that $r > 2$ and the desired inequality is already proved for $r - 1$. Consider the system of numbers $\widehat{x}_i = (x_1, \ldots, x_{i-1}, x_{i+1}, \ldots, x_n)$. It is easy to verify that

$$\sum_{i=1}^{n} x_i \sigma_{r-1}(\widehat{x}_i) = r\sigma_r(x), \tag{3.1}$$

$$x_i \sigma_{r-1}(\widehat{x}_i) + \sigma_r(\widehat{x}_i) = \sigma_r(x). \tag{3.2}$$

The sum of equations (2) for $i = 1, \ldots, n$ gives

$$\sum_{i=1}^{n} x_i \sigma_{r-1}(\widehat{x}_i) + \sum_{i=1}^{n} \sigma_r(\widehat{x}_i) = n\sigma_r(x).$$

Subtracting equation (1) from this equation we obtain

$$\sum_{i=1}^{n} \sigma_r(\widehat{x}_i) = (n - r)\sigma_r(x). \tag{3}$$

It is also clear that

$$\sigma_r(x) - \sigma_r(\widehat{x}_i) = x_i \sigma_{r-1}(\widehat{x}_i) = x_i \sigma_{r-1}(x) - x_i^2 \sigma_{r-2}(\widehat{x}_i).$$

The sum of these equalities for $i = 1, \ldots, n$, with (3) taken into account, gives

$$r\sigma_r(x) = \sum_{i=1}^{n} x_i \sigma_{r-1}(x) - \sum_{i=1}^{n} x_i^2 \sigma_{r-2}(\widehat{x}_i),$$

i.e.,

$$\frac{\sigma_r(x)}{\sigma_{r-1}(x)} = \frac{1}{r}\left(\sum_{i=1}^{n} x_i - \sum_{i=1}^{n} \frac{x_i^2 \sigma_{r-2}(\widehat{x}_i)}{\sigma_{r-1}(x)}\right) =$$

$$= \frac{1}{r}\left(\sum_{i=1}^{n} x_i - \sum_{i=1}^{n} \frac{x_i^2}{x_i + \dfrac{\sigma_{r-1}(\widehat{x}_i)}{\sigma_{r-2}(\widehat{x}_i)}}\right).$$

Let us write similar identities for y and $x + y$. The inequality required follows from the fact that, if x_i, y_i, a_i, b_i, c_i are positive integers such that $c_i \geq a_i + b_i$, then

$$
\frac{x_i^2}{x_i + a_i} + \frac{y_i^2}{y_i + a_i} - \frac{(x_i + y_i)^2}{x_i + y_i + c_i} \geq \frac{x_i^2}{x_i + a_i} + \frac{y_i^2}{y_i + a_i} - \frac{(x_i + y_i)^2}{x_i + y_i + a_i + b_i} =
$$

$$
= \frac{(a_i x_i - b_i y_i)^2}{(x_i + a_i)(y_i + b_i)(x_i + y_i + a_i + b_i)} \geq 0.
$$

3.3. We use Problem 3.2 and the following auxiliary statement.

Lemma. *If $a_1, \ldots, a_r, b_1, \ldots, b_r$ are non-negative numbers then*

$$
\sqrt[r]{(a_1 + b_1) \cdot \ldots \cdot (a_r + b_r)} \geq \sqrt[r]{a_1 \cdot \ldots \cdot a_r} + \sqrt[r]{b_1 \cdot \ldots \cdot b_r}.
$$

Proof. Let

$$
R(z) = \{(z_1, \ldots, z_r) \in \mathbb{R}^r \mid z_1 \cdot \ldots \cdot z_r = 1, \; z_i > 0\}.
$$

The inequality between the arithmetic and geometric means gives

$$
\sqrt[r]{a_1 \cdot \ldots \cdot a_r} = \min_{z \in R(z)} \frac{a_1 z_1 + \ldots + a_r z_r}{r}.
$$

Therefore

$$
\sqrt[r]{(a_1 + b_1) \cdot \ldots \cdot (a_r + b_r)} = \min_{z \in R(z)} \frac{(a_1 + b_1)z_1 + \ldots + (a_r + b_r)z_r}{r} \geq
$$

$$
\geq \min_{z \in R(z)} \frac{a_1 z_1 + \ldots + a_r z_r}{r} + \min_{z \in R(z)} \frac{b_1 z_1 + \ldots + b_r z_r}{r} \geq
$$

$$
\geq \sqrt[r]{a_1 \cdot \ldots \cdot a_r} + \sqrt[r]{b_1 \cdot \ldots \cdot b_r}. \quad \square
$$

Let us express $\sigma_r(x + y)$ as the product

$$
\frac{\sigma_r(x + y)}{\sigma_{r-1}(x + y)} \cdot \frac{\sigma_{r-1}(x + y)}{\sigma_{r-2}(x + y)} \cdot \ldots \cdot \frac{\sigma_1(x + y)}{1}.
$$

By Problem 3.2

$$
\frac{\sigma_k(x + y)}{\sigma_{k-1}(x + y)} \geq \frac{\sigma_k(x)}{\sigma_{k-1}(x)} + \frac{\sigma_k(y)}{\sigma_{k-1}(y)}.
$$

Now use the Lemma to obtain

$$
\sqrt[r]{\sigma_r(x + y)} \geq \sqrt[r]{\frac{\sigma_r(x)}{\sigma_{r-1}(x)} \cdot \frac{\sigma_{r-1}(x)}{\sigma_{r-2}(x)} \cdot \ldots \cdot \frac{\sigma_1(x)}{1}} +
$$

$$
+ \sqrt[r]{\frac{\sigma_r(y)}{\sigma_{r-1}(y)} \cdot \frac{\sigma_{r-1}(y)}{\sigma_{r-2}(y)} \cdot \ldots \cdot \frac{\sigma_1(y)}{1}} = \sqrt[r]{\sigma_r(x)} + \sqrt[r]{\sigma_r(y)}.
$$

3.4. We only consider the case $k < 0$ involving complete homogeneous polynomials. Let $l = -k > 0$. In this case we have an integral representation for the Γ-function:

$$\Gamma(l) = \int_0^\infty e^{-t} t^{l-1} \, dt.$$

For $a > 0$, we can make a change $t = as$ and get

$$\Gamma(l) = a^l \int_0^\infty e^{-as} s^{l-1} \, ds,$$

i.e.,

$$a^k = a^{-l} = \frac{1}{\Gamma(l)} \int_0^\infty e^{-as} s^{l-1} \, ds.$$

Set $a_i = 1 - x_i t$. For small values of $|t|$, the number a_i is positive, and therefore

$$\prod_{i=1}^n (1 - x_i t)^k = \left(\frac{1}{\Gamma(l)}\right)^n \int_0^\infty \cdots \int_0^\infty f(s_1, \ldots, s_n) \, ds_1 \cdots ds_n,$$

where

$$f(s_1, \ldots, s_n) = e^{-s_1 - \cdots - s_n} e^{t(x_1 s_1 + \cdots + x_n s_n)} (s_1 \cdot \ldots \cdot s_n)^{l-1}.$$

Since

$$e^{t(x_1 s_1 + \cdots + x_n s_n)} = \sum_{r=0}^\infty \frac{t^r (x_1 s_1 + \cdots + x_n s_n)^r}{r!},$$

we obtain

$$T_r(x) = \frac{1}{r!} \left(\frac{1}{\Gamma(l)}\right)^n \int_0^\infty \cdots \int_0^\infty (x_1 s_1 + \cdots + x_n s_n)^r \varphi(s_1, \ldots, s_n) \, ds_1 \cdots ds_n,$$

where $\varphi(s_1, \ldots, s_n) = e^{-s_1 - \cdots - s_n} (s_1 \cdot \ldots \cdot s_n)^{l-1}$. The required inequality now follows from the Minkowski inequality

$$\sqrt[r]{\int_a^b (g(x) + h(x))^r \, dx} \leq \sqrt[r]{\int_a^b g^r(x) \, dx} + \sqrt[r]{\int_a^b h^r(x) \, dx},$$

where g and h are non-negative on $[a, b]$.

3.11. Simple calculations show that $T_n(x) = 2^{n-1} \Delta_n(x)$ for $n = 1$ and $n = 2$. For $n \geq 2$, expanding the determinant $\Delta_n(x)$ with respect to the last row, we obtain

$$\Delta_{n+1}(x) = x\Delta_n(x) - \frac{1}{4}\Delta_{n-1}(x).$$

This relation corresponds to the recurrence formula

$$T_{n+1}(x) = 2xT_n(x) - T_{n-1}(x).$$

3.12. Let $f(t) = c_0 + c_1 t + \cdots + c_n t^n$ and let $\varepsilon_1, \ldots, \varepsilon_n$ be distinct nth roots of unity. The determinant of the circulant matrix with elements $a_{ij} = c_{i-j}$ is equal to $f(\varepsilon_1) \cdot f(\varepsilon_2) \cdot \ldots \cdot f(\varepsilon_n)$. Indeed, e.g., for $n = 3$, we obtain

$$\begin{pmatrix} 1 & 1 & 1 \\ 1 & \varepsilon_1 & \varepsilon_1^2 \\ 1 & \varepsilon_2 & \varepsilon_2^2 \end{pmatrix} \begin{pmatrix} c_0 & c_2 & c_1 \\ c_1 & c_0 & c_2 \\ c_2 & c_1 & c_0 \end{pmatrix} = \begin{pmatrix} f(1) & f(1) & f(1) \\ f(1) & \varepsilon_1 f(\varepsilon_1) & \varepsilon_1^2 f(\varepsilon_1) \\ f(1) & \varepsilon_2 f(\varepsilon_2) & \varepsilon_2^2 f(\varepsilon_2) \end{pmatrix} =$$

$$= f(\varepsilon_1) \cdot f(\varepsilon_2) \cdot \ldots \cdot f(\varepsilon_n) \begin{pmatrix} 1 & 1 & 1 \\ 1 & \varepsilon_1 & \varepsilon_1^2 \\ 1 & \varepsilon_2 & \varepsilon_2^2 \end{pmatrix}.$$

Since the determinant of the matrix $\begin{pmatrix} 1 & 1 & 1 \\ 1 & \varepsilon_1 & \varepsilon_1^2 \\ 1 & \varepsilon_2 & \varepsilon_2^2 \end{pmatrix}$ does not vanish, we can divide by it. As a result we obtain the identity required. For $n > 3$, the arguments are similar.

In our case, $f(t) = 1 - 2xt + t^2$, and therefore

$$\Delta_n(x) = \prod_{k=1}^{n} (1 - 2x\varepsilon_k + \varepsilon_k^2).$$

In other words, we have to prove that

$$2(1 - \cos n\varphi) = \prod_{k=1}^{n} (1 - 2\varepsilon_k \cos \varphi + \varepsilon_k^2).$$

First, we prove that

$$2(1 - \cos n\varphi) = 2^n \prod_{k=1}^{n} \left(1 - \cos\left(\varphi + \frac{2k\pi}{n}\right)\right).$$

This identity follows from the fact that

$$x^{2n} - 2x^n \cos n\varphi + 1 = (x^n - \exp(in\varphi))(x^n - \exp(-in\varphi)) =$$

$$= \prod_{k=1}^{n} \left(x - \exp i\left(\varphi + \frac{2k\pi}{n}\right)\right) \prod_{k=1}^{n} \left(x - \exp i\left(-\varphi - \frac{2k\pi}{n}\right)\right) =$$

$$= \prod_{k=1}^{n} \left(x^2 - 2x \cos\left(\varphi + \frac{2k\pi}{n}\right) + 1\right).$$

Then, for $x = 1$, we get the identity required.

Let us prove now that

$$2^n \prod_{k=1}^{n} \left(1 - \cos\left(\varphi + \frac{2k\pi}{n}\right)\right) = \prod_{k=1}^{n}(1 - 2\cos\varphi\,\varepsilon_k + \varepsilon_k^2),$$

where $\varepsilon_k = \exp\left(\dfrac{2k\pi i}{n}\right)$. Clearly,

$$(x - \varepsilon_k)(x - \varepsilon_{-k}) = x^2 - 2x\cos\left(\frac{2k\pi}{n}\right) + 1.$$

Hence $\varepsilon_k^2 + 1 = 2\varepsilon_k \cos\left(\dfrac{2k\pi}{n}\right)$, and therefore

$$\prod_{k=1}^{n}(1 - 2\cos\varphi\,\varepsilon_k + \varepsilon_k^2) = \prod_{k=1}^{n} 2\varepsilon_k \left(\cos\left(\frac{2k\pi}{n}\right) - \cos\varphi\right) =$$

$$= 2^{2n} \left(\prod_{k=1}^{n} \varepsilon_k\right) \prod_{k=1}^{n} \sin\left(\frac{\varphi}{2} + \frac{k\pi}{n}\right) \sin\left(\frac{\varphi}{2} - \frac{k\pi}{n}\right).$$

It remains to observe that the last expression is equal to

$$2^{2n} \prod_{k=1}^{n} \sin^2\left(\frac{\varphi}{2} + \frac{k\pi}{n}\right) = 2^n \prod_{k=1}^{n}\left(1 - \cos\left(\varphi + \frac{2k\pi}{n}\right)\right).$$

Indeed, $\prod \varepsilon_k = (-1)^{n+1}$ since $x^n - 1 = \prod(x - \varepsilon_k)$. It is also clear that

$$\sin\left(\frac{\varphi}{2} + \frac{k\pi}{n}\right) = -\sin\left(\frac{\varphi}{2} - \frac{(n-k)\pi}{n}\right) \quad \text{for } k = 1,\ldots,n-1$$

and

$$\sin\left(\frac{\varphi}{2} + \frac{k\pi}{n}\right) = \sin\left(\frac{\varphi}{2} - \frac{k\pi}{n}\right) \quad \text{for } k = n.$$

4

Certain Properties of Polynomials

4.1 Polynomials with prescribed values

4.1.1 Lagrange's interpolation polynomial

Let $x_1, \ldots, x_{n+1}$ be distinct points in the complex plane $\mathbb{C}$. Then there exists precisely one polynomial $P(x)$ of degree not greater than n which takes a prescribed value a_i at x_i. Indeed, the uniqueness of P follows from the fact that the difference of two such polynomials vanishes at points $x_1, \ldots, x_{n+1}$ and at the same time has degree not greater than n. The following polynomial clearly possesses all the necessary properties:

$$P(x) = \sum_{k=1}^{n+1} a_k \frac{(x - x_1) \cdot \ldots \cdot (x - x_{k-1}) \cdot (x - x_{k+1}) \cdot \ldots \cdot (x - x_{n+1})}{(x_k - x_1) \cdot \ldots \cdot (x_k - x_{k-1}) \cdot (x_k - x_{k+1}) \cdot \ldots \cdot (x_k - x_{n+1})} =$$

$$= \sum_{k=1}^{n+1} a_k \frac{\omega(x)}{(x - x_k)\omega'(x_k)},$$

where

$$\omega(x) = (x - x_1) \cdot \ldots \cdot (x - x_{n+1}).$$

The polynomial $P(x)$ is called *Lagrange's interpolation polynomial* and the points $x_1, \ldots, x_{n+1}$ are called the *interpolation nodes*.

If $a_k = f(x_k)$, where f is a given function, then P is called *Lagrange's interpolation polynomial for f*.

Theorem 4.1.1. *Let $f \in C^{n+1}[a, b]$ and P Lagrange's interpolation polynomial for f with nodes $x_1, \ldots, x_{n+1} \in [a, b]$. Then*

$$\max_{a \le x \le b} |P(x) - f(x)| \le \frac{M}{(n+1)!} \max_{a \le x \le b} |\omega(x)|,$$

where $M = \max_{a \le x \le b} |f^{(n+1)}(x)|$.

V.V. Prasolov, *Polynomials*, Algorithms and Computation in Mathematics 11, DOI 10.1007/978-3-642-03980-5_4, © Springer-Verlag Berlin Heidelberg 2004

Proof. It suffices to verify that, for any point $x_0 \in [a,b]$, there exists a point $\xi \in [a,b]$ such that

$$f(x_0) - P(x_0) = \frac{f^{(n+1)}(\xi)}{(n+1)!}\omega(x_0).$$

For $x_0 = x_i$, where $1 \le i \le n$, this equality is obvious. We therefore assume that $x_0 \ne x_i$. Consider the function

$$u(x) = f(x) - P(x) - \lambda\omega(x),$$

where λ is a constant. Since $\omega(x_0) \ne 0$ this constant can be selected so that $u(x_0) = 0$. It is also clear that $u(x_1) = \cdots = u(x_{n+1}) = 0$. The function $u(x)$ has at least $n+2$ zeros on the interval $[a,b]$. Hence $u'(x)$ has at least $n+1$ zeros on this interval and $u^{(k)}(x)$ has at least $n+2-k$ zeros. For $k = n+1$, we see that

$$u^{(n+1)}(x) = f^{(n+1)}(x) - (n+1)!\,\lambda$$

vanishes at a point $\xi \in [a,b]$. This means that $\lambda = \dfrac{f^{(n+1)}(\xi)}{(n+1)!}$, i.e.,

$$f(x_0) - P(x_0) = \frac{f^{(n+1)}(\xi)}{(n+1)!}\omega(x_0). \quad \square$$

For a fixed interval $[a,b]$ and a fixed degree n, the estimate given by Theorem 4.1.1 is optimal if $\omega(x)$ is a monic polynomial of degree n with the least deviation from zero on $[a,b]$. Since ω is fixed, this is a condition on its roots, i.e., the interpolation nodes.

For example, if $[a,b] = [-1,1]$, then we should have $\omega(x) = \frac{1}{2^n}T_{n+1}(x)$, where $T_{n+1}(x)$ is a Chebyshev polynomial. Recall that

$$T_{n+1}(x) = \cos((n+1)\arccos x) \quad \text{for } x \le 1.$$

The roots of T_{n+1} are

$$x_k = \cos\frac{(2k-1)\pi}{2(n+1)}, \quad k = 1,\ldots,n+1.$$

For such nodes, the interpolation polynomial is

$$P(x) = \frac{1}{n+1}\sum_{k=1}^{n+1} f(x_k)(-1)^{k-1}\sqrt{1-x_k^2}\,\frac{T_{n+1}(x)}{x-x_k}.$$

Indeed, if $\omega(x) = \frac{1}{2^n}T_{n+1}(x)$, then

$$\frac{\omega(x)}{(x-x_k)\omega'(x)} = \frac{T_{n+1}(x)}{(x-x_k)T'_{n+1}(x)},$$

and therefore we have to prove that

$$T'_{n+1}(x_k) = \frac{(n+1)(-1)^{k-1}}{\sqrt{1-x_k^2}}.$$

Since $T_{n+1}(x) = \cos(n+1)\varphi$, where $x = \cos\varphi$, we obtain

$$T'_{n+1}(x_k) = \frac{(n+1)\sin(n+1)\varphi}{\sin\varphi}.$$

If $\cos\varphi = x_k$, then $\sin\varphi = \sqrt{1-x_k^2}$ and $\sin(n+1)\varphi = (-1)^{k-1}$.

In addition to Chebyshev interpolation nodes, one also makes use of the nodes uniformly distributed on a segment of the circle. For the nodes

$$x_k = \exp\left(\frac{2\pi i k}{n+1}\right), \quad \text{where } k = 1, \ldots, n+1,$$

the interpolation polynomial is

$$P(x) = \frac{1}{n+1} \sum_{k=1}^{n+1} x_k f(x_k) \frac{x^n - 1}{x - x_k}.$$

To prove this formula, it suffices to observe that

$$\frac{d}{dx}(x^{n+1} - 1)\bigg|_{x=x_k} = (n+1)x_k^n = (n+1)x_k^{-1}.$$

The interpolation polynomial for the nodes $x_k = a + (k-1)h$, where $k = 1, \ldots, n+1$, can be expressed in the form

$$P(x) = f(a) + \frac{\Delta f(a)}{h}(x - a) + \frac{\Delta^2 f(a)}{h^2} \frac{(x-a)(x-a-h)}{2!} + \cdots$$

$$\cdots + \frac{\Delta^n f(a)}{h^n} \frac{(x-a)(x-a-h)\cdot\ldots\cdot(x-a-(n-1)h)}{n!},$$

where

$$\Delta f(x) = f(x+h) - f(x),$$

$$\cdots\cdots\cdots\cdots\cdots\cdots\cdots\cdots\cdots\cdots\cdots\cdots\cdots\cdots$$

$$\Delta^{k+1} f(x) = \Delta(\Delta^k f(x)) = \sum_{j=0}^{k}(-1)^{k-j}\binom{k}{j}f(x+jh).$$

This polynomial $P(x)$ is called *Newton's interpolation polynomial*. It is easy to verify that $P(x_k) = f(x_k)$. Indeed,

$$P(a) = f(a),$$

$$P(a+h) = f(a) + \Delta f(a),$$

$$P(a+2h) = f(a) + 2\Delta f(a) + \Delta^2 f(a),$$

$$\cdots\cdots\cdots\cdots\cdots\cdots\cdots\cdots\cdots\cdots\cdots\cdots$$

$$P(a+mh) = \sum_{j=0}^{m}\binom{m}{j}\Delta^j f(a) = f(a+mh).$$

The last identity follows from the fact that $\Delta^k f(x+h) = \Delta^{k+1} f(x) + \Delta^k f(x)$.

4.1.2 Hermite's interpolation polynomial

Let $x_1, \ldots, x_n$ be distinct points of the complex plane $\mathbb{C}$ and $\alpha_1, \ldots, \alpha_n$ positive integers whose sum is equal to $m + 1$. For each point x_i, let $y_i^{(0)}, y_i^{(1)}, \ldots, y_i^{(\alpha_i - 1)}$ be given numbers. Then there is a unique polynomial $H_m(x)$ of degree not greater than m such that

$$H_m(x_i) = y_i^{(0)}, \quad H_m'(x_i) = y_i^{(1)}, \quad \ldots, \quad H_m^{(\alpha_i - 1)}(x_i) = y_i^{(\alpha_i - 1)},$$

for $i = 1, \ldots, n$. In other words, at x_i the polynomial H_m has prescribed values for its derivatives up to order $\alpha_i - 1$ inclusive. This polynomial H_m is called *Hermite's interpolation polynomial*.

The uniqueness of $H_m(x)$ is quite obvious. Indeed, if $G(x)$ is the difference of two Hermite interpolation polynomials, then $\deg G \leq m$ and $G(x)$ is divisible by

$$\Omega(x) = (x - x_1)^{\alpha_1} \cdot \ldots \cdot (x - x_n)^{\alpha_n}.$$

To construct Hermite's interpolation polynomial, we have only to define polynomials $\varphi_{ik}(x)$ ($i = 1, \ldots, n$ and $k = 0, 1, \ldots, \alpha_i - 1$) such that
1) $\deg \varphi_{ik} \leq m$;

2) $\varphi_{ik}(x)$ is divisible by the polynomial $\dfrac{\Omega(x)}{(x - x_i)^{\alpha_i}}$, i.e., $\varphi_{ik}(x)$ is divisible by $(x - x_j)^{\alpha_j}$ for $j \neq i$;

3) the expansion of $\varphi_{ik}(x)$ as a power series in $x - x_i$ begins with

$$\frac{1}{k!}(x - x_i)^k + (x - x_i)^{\alpha_i}.$$

We then have

$$\varphi_{ik}^{(0)}(x_j) = \cdots = \varphi_{ik}^{(\alpha_i - 1)}(x_j) = 0 \text{ for } j \neq i,$$

$$\varphi_{ik}^{(k)}(x_i) = 1$$

and

$$\varphi_{ik}^{(l)}(x_i) = 0 \text{ for } 0 \leq l \leq \alpha_i - 1, \ l \neq k.$$

Thus we can define

$$H_m(x) = \sum_{i=1}^{n} \sum_{k=0}^{\alpha_i - 1} y_i^{(k)} \varphi_{ik}(x).$$

To define the φ_{ik}, we note that the function $\dfrac{1}{k!} \dfrac{(x - x_i)^{\alpha_i}}{\Omega(x)}$ is regular at x_i, and therefore in a neighborhood of x_i it can be expanded into the Taylor series

$$\frac{1}{k!} \frac{(x - x_i)^{\alpha_i}}{\Omega(x)} = \sum_{s=0}^{\infty} a_{iks}(x - x_i)^s = l_{ik}(x) + \sum_{s=\alpha_i - k}^{\infty} a_{iks}(x - x_i)^s.$$

Here $l_{ik}(x)$ is a polynomial of degree not greater than $\alpha_i - k - 1$ which is the initial part of the Taylor series. It is not difficult to verify that the polynomial

$$\varphi_{ik}(x) = \frac{\Omega}{(x - x_i)^{\alpha_i}} \left(l_{ik}(x)(x - x_i)^k \right)$$

possesses all the properties required. Properties 1) and 2) are obvious, and property 3) is proved as follows:

$$\varphi_{ik}(x) = \frac{l_{ik}(x)(x - x_i)^k}{k! l_{ik}(x) + a(x - x_i)^{\alpha_i - k} + \cdots} = \frac{(x - x_i)^k}{k!} \left(1 + b(x - x_i)^{\alpha_i - k} + \cdots \right).$$

Looking at the explicit form of the initial part of the Taylor series $l_{ik}(x)$, we obtain

$$H_m(x) = \sum_{i=1}^{n} \sum_{k=0}^{\alpha_i - 1} \sum_{s=0}^{\alpha_i - k - 1} y_i^{(k)} \frac{1}{k!} \frac{1}{s!} \left(\frac{(x - x_i)^{\alpha_i}}{\Omega(x)} \right)^{(s)} \bigg|_{x = x_i} \frac{\Omega(x)}{(x - x_i)^{\alpha_i - k - s}}.$$

4.1.3 The polynomial with prescribed values at the zeros of its derivative

In 1956, Andrushkiw [An3] announced a statement that

  for any n complex numbers $a_1, \ldots, a_n$, there exists a monic polynomial P of degree $n + 1$ which takes the values $a_1, \ldots, a_n$ at the zeros of its derivative, P'.

The first published proof of this statement was given, however, only 9 years later by René Thom [Th]. We give the proof due to Yan Mycielski [My].

Theorem 4.1.2. *For any given numbers $a_1, \ldots, a_n \in \mathbb{C}$, there exist numbers $b_1, \ldots, b_n \in \mathbb{C}$ and a polynomial $P(x) = x^{n+1} + p_1 x^n + \cdots + p_n x$ such that $P(b_i) = a_i$ and $P'(b_i) = 0$ for $i = 1, \ldots, n$. Moreover, if a number β occurs k times in the sequence $b_1, \ldots, b_n$, then $P(x) - P(\beta)$ is divisible by $(x - \beta)^{k+1}$.*

Proof. For $b = (b_1, \ldots, b_n) \in \mathbb{C}^n$, we define

$$P_b(x) = (n + 1) \int_0^x \left(\prod_{i=1}^{n} (t - b_i) \right) dt.$$

Clearly, $P_b(0) = 0$ and $P_b(x)$ is a monic polynomial of degree $n + 1$. Further,

$$P_b'(x) = (n + 1)(x - b_1) \cdot \ldots \cdot (x - b_n),$$

and hence $P_b'(b_i) = 0$. If β occurs exactly k times in the sequence $b_1, \ldots, b_n$, then $P_b'(\beta) = \cdots = P_b^{(k)}(\beta) = 0$. Observe that any number β is a root of the polynomial $P_b(x) - P_b(\beta)$ and $\big(P_b(x) - P_b(\beta)\big)' = P_b'(x)$. Hence $P_b(x) - P_b(\beta)$ is divisible by $(x - \beta)^{k+1}$.

It remains to prove that the map

$$\varphi : \mathbb{C}^n \to \mathbb{C}^n, \qquad \varphi(b) = \big(P_b(b_1), \ldots, P_b(b_n)\big)$$

is surjective. First we prove that φ is a local homeomorphism at any point $b = (b_1, \ldots, b_n)$ such that $b_i \neq b_j$ for $i \neq j$ and $b_1 \cdot \ldots \cdot b_n \neq 0$. For this, it suffices to verify that $\det\left(\frac{\partial P_b(b_i)}{\partial b_j}\right) \neq 0$. Suppose on the contrary that $\det\left(\frac{\partial P_b(b_i)}{\partial b_j}\right) = 0$. Then there are numbers $c_1, \ldots, c_n$, not all zero, such that

$$\sum_{j=1}^{n} c_j \frac{\partial P_b(b_i)}{\partial b_j} = 0 \quad \text{for} \quad i = 1, \ldots, n. \tag{1}$$

It is easy to verify that

$$\frac{\partial P_b(b_i)}{\partial b_j} = -(n+1) \int_0^{b_i} \left(\prod_{s \neq j}(t - b_s)\right) dt$$

(for $i = j$, there appears an extra summand $(n+1) \prod_{s=1}^{n}(b_i - b_s)$, but it vanishes). Hence (1) can be expressed in the form

$$F(x) = \int_0^x \left(\sum_{j=1}^{n} c_j \prod_{s \neq j}(t - b_s)\right) dt = 0 \quad \text{for} \quad x = b_1, \ldots, b_n.$$

The integrand is a polynomial in t of degree not greater than n which takes the value $c_j \prod_{s \neq j}(t - b_s)$ at $t = b_j$. By hypothesis, $\prod_{s \neq j}(t - b_s) \neq 0$ and $c_j \neq 0$ for some j. Hence $F(x)$ is a nonzero polynomial of degree not greater than $n+1$. On the other hand, $F(x) = 0$ at $x = 0, b_1, \ldots, b_n$: a contradiction.

The map $\varphi : \mathbb{C}^n \to \mathbb{C}^n$ induces a map $\widetilde{\varphi} : \mathbb{C}P^n \to \mathbb{C}P^n$ given by the formula[1]

$$\widetilde{\varphi}\big((b_0 : \ldots : b_n)\big) = \big(b_0^{n+1} : P_b(b_1) : \ldots : P_b(b_n)\big).$$

The point is that

$$\varphi(\lambda b) = \lambda^{n+1} \varphi(b) \quad \text{and} \quad \varphi^{-1}(0) = 0.$$

The first of these properties is proved by the change of variable $\tau = \lambda t$:

---

[1] The notation $(b_0 : b_1 : \cdots : b_n)$, standard in projective geometry, means the n-tuple $(b_0, b_1, \ldots, b_n)$ defined up to a nonzero factor.

$$\int_0^b \left(\prod_{i=1}^n (t - b_i)\right) dt = \int_0^{\lambda b} \left(\prod_{i=1}^n (\lambda^{-1}\tau - b_i)\lambda^{-1}\right) d\tau =$$

$$\lambda^{-n-1} \int_0^{\lambda b} \left(\prod_{i=1}^n (\tau - \lambda b_i)\right) d\tau.$$

The second property is proved as follows. Let $\varphi(b_1, \ldots, b_n) = (0, \ldots, 0)$. Suppose that the sequence $b_1, \ldots, b_n$ consists of k_1 numbers β_1, k_2 numbers β_2, and so on, k_m numbers β_m (here $\beta_i \neq \beta_j$ for $i \neq j$). Then the polynomial $P_b(x) = P_b(x) - P_b(\beta_i)$ is of degree $n = k_1 + \cdots + k_m$ and is divisible by x and by $(x - \beta_1)^{k_1+1} \cdot \ldots \cdot (x - \beta_m)^{k_m+1}$. This is only possible if $b_1 = \cdots = b_n = 0$.

Let Δ be the set of points $(b_0 : b_1 : \cdots : b_n) \in \mathbb{C}P^n$ whose coordinates satisfy one of the following equations:

$$b_i = 0 \ (i = 0, \ldots, n); \qquad b_i = b_j \ (1 \leq i < j \leq n).$$

The restriction of $\widetilde{\varphi}$ to $\mathbb{C}P^n \setminus \Delta$ is a local homeomorphism.

Moreover, $\widetilde{\varphi}(\Delta) \subset \Delta$. Indeed, if $b_0 = 0$, then

$$\widetilde{\varphi}((b_0 : b_1 : \ldots : b_n)) = \left(0 : P_b(b_1) : \ldots : P_b(b_n)\right).$$

If $b_i = 0$ for $i \geq 1$, then $P_b(b_i) = 0$ and if $b_i = b_j$ for $1 \leq i < j \leq n$, then $P_b(b_i) = P_b(b_j)$.

The image of $\mathbb{C}P^n$ under $\widetilde{\varphi}$ is a compact set; in particular, it is closed. The image of $\mathbb{C}P^n \setminus \Delta$ under $\widetilde{\varphi}$ is an open set whose boundary belongs to $\widetilde{\varphi}(\Delta) \subset \Delta$. The set Δ does not divide $\mathbb{C}P^n$ since it is of real codimension 2. Hence, $\widetilde{\varphi}(\mathbb{C}P^n \setminus \Delta) \supset \mathbb{C}P^n \setminus \Delta$ and the closure of $\widetilde{\varphi}(\mathbb{C}P^n \setminus \Delta)$ coincides with $\mathbb{C}P^n$. $\square$

Remark. One should not think that $\widetilde{\varphi}(\mathbb{C}P^n \setminus \Delta) \subset \mathbb{C}P^n \setminus \Delta$. This is false, as the example $\varphi(1, 2, 3) = (-9, -8, -9)$ shows.

4.2 The height of a polynomial and other norms

4.2.1 Gauss's lemma

Let K be a field, and let $x \mapsto |x|_v \in \mathbb{R}$ be a function on K. This function is called an *absolute value* if it satisfies the following conditions:

(1) $|x|_v \geq 0$ and $|x|_v = 0 \Leftrightarrow x = 0$;

(2) $|xy|_v = |x|_v |y|_v$;

(3) $|x + y|_v \leq |x|_v + |y|_v$.

If instead of (3) a stronger condition

(3') $|x + y|_v \leq \max\{|x|_v, |y|_v\}$

holds, then this absolute value is said to be a *non-Archimedean*.

For the ring of rationals $\mathbb{Q}$, there is a natural absolute value, namely $|x|_v = |x|$ (the conventional absolute value of x). This absolute value is Archimedean.

But there are also so-called *p-adic absolute values* defined for every prime p as follows. Let us represent a rational x in the form $x = p^r \frac{m}{n}$, where m and n are integers not divisible by p. Set

$$|x|_p = \frac{1}{p^r}.$$

It is easy to verify that this absolute value is non-Archimedean. Indeed, let $x = p^r \frac{m_1}{n_1}$ and $y = p^s \frac{m_2}{n_2}$ be such that $r \le s$. Then $\max\{|x|_p, |y|_p\} = \frac{1}{p^r}$ and

$$x + y = p^r \frac{m_1 n_2 + p^{s-r} m_2 n_1}{n_1 n_2}.$$

Hence $|x + y|_p \le \frac{1}{p^r}$ (the strict inequality is only possible for $s = r$).

The *height* of the polynomial $f(x) = \sum a_i x^i$ relative to a given absolute value $|\cdot|_v$ is the number $H(f) = \max_i |a_i|_v$. We will be interested in estimates of the heights of the product of polynomials in terms of the heights of the factors. The simplest estimate is obtained if the absolute value is non-Archimedean.

Lemma. *Let $H(f)$ be the height of the polynomial f with respect to a non-Archimedean absolute value $|\cdot|$. Then $H(fg) = H(f)H(g)$.*

Proof. Let $f(x) = a_n x^n + \cdots + a_0$ and $g(x) = b_m x^m + \cdots + b_0$. Among the coefficients $a_n, \ldots, a_0$ consider those with the maximal absolute value (there can be several such coefficients) and select among them the coefficient a_r and with the greatest index r. Similarly select the coefficient b_s of maximal absolute value with the greatest index s.

Clearly,

$$(fg)(x) = f(x)g(x) = c_{n+m} x^{n+m} + \cdots + c_0, \quad \text{where } c_k = \sum_{i+j=k} a_i b_j.$$

Since $|\cdot|$ is a non-Archimedean absolute value, we deduce that

$$|c_k| \le \max_{i+j=k} \{|a_i| \cdot |b_j|\}.$$

Hence,

$$|c_k| < |a_r| \cdot |b_s| \quad \text{if } k > r + s,$$
$$c_{r+s} = a_r b_s (1 + \alpha), \quad \text{where } |\alpha| < 1,$$
$$|c_k| \le |a_r| \cdot |b_s| \quad \text{if } k < r + s$$

For non-Archimedean absolute values, $|\alpha| < 1$ implies that $|1 + \alpha| = 1$. To see this, we note that

$$|1 + \alpha| \le \max\{1, |\alpha|\} = 1 \text{ and } 1 = |1 + \alpha - \alpha| \le \max\{|1 + \alpha|, |\alpha|\},$$

and hence, $1 \le |1 + \alpha|$.

Therefore $|c_{r+s}| = |a_r| \cdot |b_s|$ and the absolute values of the remaining coefficients c_k do not exceed $|a_r| \cdot |b_s|$. Hence $H(fg) = |c_{r+s}| = |a_r| \cdot |b_s| = H(f)H(g)$. $\square$

It goes without saying that Gauss formulated and proved his lemma using simpler language. That is, he proved that

the greatest common divisor of the coefficients of the product of the polynomials f and g is equal to the product of the greatest common divisor of the coefficients of f and the greatest common divisor of the coefficients of g.

We can arrive at Gauss's formulation as follows. For a p-adic absolute value, $H(f) = p^{-r}$, where r is the greatest power of p which divides the coefficients of f. The identity $H(fg) = H(f)H(g)$ means that, if p enters the greatest common divisors of the coefficients of f and g with powers r and s, respectively, then it enters the greatest common divisor of the coefficients of fg with power $r + s$.

For the polynomial $f(x_1, \ldots, x_n) = \sum a_{i_1 \ldots i_n} x_1^{i_1} \cdot \ldots \cdot x_n^{i_n}$ in n variables, the *height* is similarly defined as

$$H(f) = \max |a_{i_1 \ldots i_n}|.$$

To prove Gauss's lemma for polynomials in n variables, we use of the so-called *Kronecker's substitution*. Let $d = \deg f + \deg g + 1$. To the polynomial $h(x_1, \ldots, x_n) = \sum c_{k_1 \ldots k_n} x^{k_1} \cdot \ldots \cdot x^{k_n}$, we assign the polynomial

$$(S_d h)(y) = h\big(y, y^d, \ldots, y^{d^{n-1}}\big) = \sum c_{k_1 \ldots k_n} x^{k_1 + dk_2 + d^2 k_3 + \cdots + d^{n-1} k_n}.$$

If $\deg h < d$, then the nonzero coefficients of h and $S_d h$ are the same, and hence $H(h) = H(S_d h)$. Moreover, $S_d(fg) = S_d(f)S_d(g)$. Hence, Gauss's lemma for polynomials in n variables follows from Gauss's lemma for one variable.

4.2.2 Polynomials in one variable

For Archimedean absolute values, the height

$$H(f) = \max |a_i|$$

does not possess the multiplicativity property $H(fg) = H(f)H(g)$. But it is exactly for the conventional absolute value that the estimate of the height of polynomials is most interesting. Such estimates are needed for the theory of transcendental numbers. The estimate of the height of polynomials was obtained by A. O. Gelfond [Ge1] as a by-product of the solution of *Hilbert's seventh problem*:

"*If $a \neq 0, 1$ is an algebraic number and b is an irrational algebraic number, then a^b is a transcendental number.*"

Later, K. Mahler[Ma2], [Ma3] found a simplified proof of Gelfond's estimates.

To estimate the height of the polynomial $f(x) = a_d(x - \alpha_1) \cdot \ldots \cdot (x - \alpha_d)$, Mahler used the quantity

$$M(f) = |a_d| \prod_{i=1}^{d} \max\{1, |\alpha_i|\},$$

which is now called *Mahler's measure* of f. Clearly, Mahler's measure is multiplicative:

$$M(fg) = M(f)M(g).$$

Therefore the upper and lower bounds for $M(f)$ in terms of $H(f)$ enable us to estimate $H(fg)$ in terms of $H(f)$ and $H(g)$.

Theorem 4.2.1. *Let* $\deg f = d$. *Then*

$$\frac{M(f)}{\sqrt{d+1}} \leq H(f) \leq 2^{d-1} M(f).$$

Proof. Let us start with a simpler inequality, $H(f) \leq 2^{d-1} M(f)$. Clearly,

$$|a_d| \cdot |\alpha_{i_1} \alpha_{i_2} \cdot \ldots \cdot \alpha_{i_k}| \leq \prod_{i=1}^{d} \max\{1, |\alpha_i|\} = M(f).$$

Hence

$$|a_k| \leq \binom{d}{k} M(f). \tag{1}$$

Starting from the formula $\binom{d+1}{k} = \binom{d}{k-1} + \binom{d}{k}$ it is easy to prove by induction on d that $\binom{d}{k} \leq 2^{d-1}$ for $d \geq 1$. Together with (1) this proves the inequality required.

The proof of the inequality $M(f) \leq \sqrt{d+1}\, H(f)$ is based on Jensen's formula in the next lemma.

Lemma. *If a function* $f(z)$ *is holomorphic in the disk* $|z| \leq 1$ *and has zeros* $z_1, \ldots, z_n$ *(multiplicities counted) inside this disk, then*

$$\frac{1}{2\pi} \int_0^{2\pi} \ln|f(e^{i\varphi})|\, d\varphi = \ln|f(0)| - \sum_{k=1}^{n} \ln|z_k|.$$

Proof. Consider the auxiliary function

$$f_1(z) = \frac{f(z)}{w_1(z) \cdot \ldots \cdot w_n(z)},$$

where $w_k(z) = \dfrac{z - z_k}{1 - z\bar{z}_k}$. It is easy to verify that w_k conformally maps the unit disk into itself. Indeed, if $|z| = 1$, then $|w_k(z)|^2 = 1$, and, further, $w_k(z_k) = 0$. The function f_1 has no zeros inside the unit disk since the zeros $z_1, \ldots, z_n$ of the numerator $f(z)$ cancel the zeros of the denominator $w_1(z) \cdot \ldots \cdot w_n(z)$. Therefore, by the mean value theorem for the harmonic function $\ln|f_1(z)| = \operatorname{Re}(\ln f_1(z))$, we obtain

$$\frac{1}{2\pi} \int\limits_0^{2\pi} \ln|f_1(e^{i\varphi})| \, d\varphi = \ln|f_1(0)|.$$

But $|f_1(e^{i\varphi})| = |f(e^{i\varphi})|$ and $\ln|f_1(0)| = \ln|f(0)| - \sum\limits_{k=1}^n \ln|z_k|$. $\square$

Corollary. *Let f be a polynomial. Then*

$$M(f) = \exp \int\limits_0^1 \ln|f(e^{2\pi it})| \, dt. \tag{2}$$

Proof. Both sides of (2) are multiplicative with respect to f. Therefore it suffices to consider the case $f(x) = x - \alpha$. For $|\alpha| \geq 1$, the function has no zeros inside the unit circle and, for $|\alpha| < 1$, it has only the one zero α inside the unit circle. Therefore, by Jensen's formula,

$$\int\limits_0^1 \ln|f(e^{2\pi it})| \, dt = \tfrac{1}{2\pi} \int\limits_0^{2\pi} \ln|f(e^{i\varphi})| \, d\varphi =$$
$$= \ln|f(0)| - \varepsilon \ln|\alpha| = (1 - \varepsilon) \ln|\alpha|,$$

where $\varepsilon = 0$ for $|\alpha| \geq 1$ and $\varepsilon = 1$ for $|\alpha| < 1$.

On the other hand, $M(f) = \max\{1, |\alpha|\} = |\alpha|^{1-\varepsilon}$. $\square$

Armed with the formula (2), we can tackle the proof that $M(f) \leq \sqrt{d+1}\, H(f)$. Clearly,

$$\int\limits_0^1 \left| \sum a_k e^{2k\pi it} \right|^2 dt = \int\limits_0^1 \sum a_k \bar{a}_l e^{2(k-l)\pi it} \, dt = \sum |a_k|^2.$$

Also, the convexity of the function $\exp$ implies that

$$\exp \int\limits_0^1 u(t) \, dt \leq \int\limits_0^1 \exp u(t) \, dt$$

for any function $u(t)$. Taking $u(t) = 2\ln|f(e^{2\pi it})|$, we obtain

$$M(f) = \exp \int_0^1 \frac{u(t)\, dt}{2} = \sqrt{\exp \int_0^1 u(t)\, dt} \le$$

$$\le \sqrt{\int_0^1 \exp u(t)\, dt} = \sqrt{\int_0^1 \left| f(e^{2\pi i t}) \right|^2 dt} =$$

$$= \sqrt{\sum |a_k|^2} \le \sqrt{d+1}\, \max |a_k| = \sqrt{d+1}\, H(f). \;\square$$

Using Theorem 4.2.1 we can obtain the following estimates for $H(fg)$ in terms of $H(f)$ and $H(g)$.

Theorem 4.2.2. *Let $d_1 = \deg f$ and $d_2 = \deg g$ be such that $d_1 \le d_2$. Then*

$$\frac{H(f)H(g)}{2^{d_1+d_2-2}\sqrt{d_1+d_2+1}} \le H(fg) \le (1+d_1)H(f)H(g).$$

Proof. We prove that $H(fg) \le (1+d_1)H(f)H(g)$ directly, without appealing to Jensen's formula. To do this, let

$$f(x) = \sum a_i x^i, \quad g(x) = \sum b_j x^j, \quad fg(x) = \sum c_k x^k.$$

Then

$$|c_k| = |a_0 b_k + a_1 b_{k-1} + \cdots + a_{d_1} b_{k-d_1}| \le$$
$$\le (1+d_1) \max |a_i| \max |b_j| = (1+d_1)H(f)H(g).$$

To prove that

$$H(f)H(g) \le 2^{d_1+d_2-2}\sqrt{d_1+d_2-1}\, H(fg),$$

we use Theorem 4.2.1. This gives

$$H(f) \le 2^{d_1-1}M(f), \quad H(g) \le 2^{d_2-1}M(g) \text{ and } \sqrt{d_1+d_2+1}\, M(fg) \le H(fg).$$

It remains to observe that $M(f)M(g) = M(fg)$. $\square$

With Mahler's measure $M(f)$ we can also estimate the *length* $L(f)$ of a polynomial f, defined by

$$L(f) = \sum_{k=0}^d |a_k|.$$

Indeed, on the one hand, the inequality (1) on page 142 gives

$$L(f) = \sum_{k=0}^d |a_k| \le M(f) \sum_{k=0}^d \binom{d}{k} = 2^d M(f). \tag{3}$$

On the other hand,

$$\left|f(e^{2\pi it})\right| = \left|\sum a_k e^{2\pi it}\right| \leq \sum |a_k| = L(f).$$

Therefore

$$M(f) \leq \exp \int_0^1 \ln L(f)\, dt = L(f). \tag{4}$$

By combining (3) and (4) we obtain

$$L(f)L(g) \leq 2^{d_1} M(f) 2^{d_2} M(g) \leq 2^{d_1+d_2} L(fg).$$

The upper estimate for $L(fg)$ is of the form

$$L(fg) \leq L(f)L(g).$$

This follows immediately from the definition of the length of the polynomial:

$$L(fg) \leq \sum_{i,j} |a_i| \cdot |b_j| = \left(\sum_i |a_i|\right)\left(\sum_j |b_j|\right) = L(f)L(g).$$

4.2.3 The maximum of the absolute value and S. Bernstein's inequality

Initially, to estimate the height of a polynomial, Gelfond made use of $\max_{|z|=1} |f(z)|$. He proved the following statement.

Theorem 4.2.3. Let $d_1 = \deg f$, $d_2 = \deg g$ and $d = d_1 + d_2$. Then

$$\max_{|z|=1} |f(z)| \cdot \max_{|z|=1} |g(z)| < 2^{2d} \max_{|z|=1} |fg(z)|.$$

Proof. Without loss of generality we may assume that

$$\max_{|z|=1} |f(z)| = \max_{|z|=1} |g(z)| = 1.$$

Suppose that $\max_{|z|=1} |fg(z)| \leq \dfrac{1}{2^{2d}}$. Then, for $k = 0, 1, \ldots, d$, one of the numbers $|f(\varepsilon_k)|$ and $|g(\varepsilon_k)|$, where $\varepsilon_k = \exp\left(\frac{2\pi ik}{d+1}\right)$, does not exceed $\dfrac{1}{2^d}$. Therefore either $|f(\varepsilon_k)| \leq \frac{1}{2^d}$ for $d_1 + 1$ values of the index k or $|g(\varepsilon_k)| \leq \frac{1}{2^d}$ for $d_2 + 1$ values of the index k. To be definite, let

$$\{\varepsilon_0, \ldots, \varepsilon_d\} = \{\alpha_0, \ldots, \alpha_{d_1}\} \cup \{\beta_1 \ldots, \beta_{d_2}\}$$

and $|f(\alpha_l)| \leq \frac{1}{2^d}$ for $l = 0, 1, \ldots, d_1$.

By Lagrange's interpolation formula, we have

$$f(z) = \sum_{l=0}^{d_1} f(\alpha_l) \frac{(z - \alpha_0) \cdot \ldots \cdot (z - \alpha_{l-1})(z - \alpha_{l+1}) \cdot \ldots \cdot (z - \alpha_{d_1})}{(\alpha_l - \alpha_0) \cdot \ldots \cdot (\alpha_l - \alpha_{l-1})(\alpha_l - \alpha_{l+1}) \cdot \ldots \cdot (\alpha_l - \alpha_{d_1})}.$$

Let us multiply the numerator and denominator of the l-th summand in this formula by $(\alpha_l - \beta_1) \cdot \ldots \cdot (\alpha_l - \beta_{d_2})$. As a result, the denominator becomes equal to

$$\lim_{x \to \alpha_l} \frac{(x - \varepsilon_0)(x - \varepsilon_1) \cdot \ldots \cdot (x - \varepsilon_d)}{x - \alpha_l} = \lim_{x \to \alpha_l} \frac{x^{d+1} - 1}{x - \alpha_l}.$$

Since α_l is a root of $x^{d+1} - 1$, the last limit is equal to the derivative of $x^{d+1} - 1$ at α_l, i.e., to $(d+1)\alpha_l^d$.

If $|z| = 1$, the numerator obtained consists of d factors and the absolute value of each of them does not exceed 2. Hence, if $|z| = 1$, we have

$$\left| f(z) \right| \le (d_1 + 1)2^{-d} \frac{2^d}{d+1} = \frac{d_1 + 1}{d+1} < 1,$$

which contradicts the hypothesis that $\max_{|z|=1} \left| f(z) \right| = 1$. $\square$

Remark. For polynomials with coefficients in $F = \mathbb{C}$ or in $\mathbb{R}$, sharper estimates can be obtained in the form

$$\prod_{k=1}^{m} \left(\max_{|z|=1} |f_k(z)| \right) \le C_F(m, n) \max_{|z|=1} |f(z)|,$$

where $f = f_1 \cdot \ldots \cdot f_m$, $n = \deg f$ and $C_F(m, n)$ is a constant (see [Bo]) which is defined as follows. Let

$$I(\theta) = \int_0^{\theta} \ln\left(2\cos\left(\frac{t}{2}\right)\right) dt.$$

Then

$$C_{\mathbb{C}}(m, n) = \left(\exp\left(\frac{m}{\pi} I\left(\frac{m}{\pi} \right) \right) \right)^n$$

and $C_{\mathbb{R}}(m, n) = C_{\mathbb{C}}(2, n)$. Both estimates are precise.

The *norm*

$$\|f\| = \max_{|z|=1} |f(z)|$$

of the polynomial f satisfies the following inequality.

Theorem 4.2.4 (S. Bernstein). *Let* $\deg f = n$. *Then* $\|f'\| \le n\|f\|$.

Proof. [O] We need the following Lemma.

Lemma. *Let* $\deg f \le n$ *and let* $z_1, \ldots, z_n$ *be the roots of the polynomial* $z^n + 1$. *Then, for any* $t \in \mathbb{C}$, *we have*

$$tf'(t) = \frac{n}{2}f(t) + \frac{1}{n}\sum_{k=1}^{n} f(tz_k)\frac{2z_k}{(z_k - 1)^2}.$$

Proof. Set $g_t(z) = \dfrac{f(tz) - f(t)}{z - 1}$. It is easy to verify that $g_t(1) = tf'(t)$ and g_t is a polynomial in z of degree not higher than $n-1$. Lagrange's interpolation formula with nodes at $z_1, \ldots, z_n$ shows that

$$g_t(z) = \sum_{k=1}^{n} g_t(z_k)\frac{z^n + 1}{(z - z_k)nz_k^{n-1}} = \frac{1}{n}\sum_{k=1}^{n} g_t(z_k)\frac{z^n + 1}{z_k - z}z_k.$$

(We used the fact that $z_k^{n-1} = -\dfrac{1}{z_k}$.)

For $z = 1$, we obtain

$$tf'(t) = \frac{1}{n}\sum_{k=1}^{n} g_t(z_k)\frac{2z_k}{(z_k - 1)^2} = \frac{1}{n}\sum_{k=1}^{n} \frac{f(tz_k) - f(t)}{(z_k - 1)^2} =$$

$$= \frac{1}{n}\sum_{k=1}^{n} f(tz_k)\frac{2z_k}{(z_k - 1)^2} - \frac{f(t)}{n}\sum_{k=1}^{n} \frac{2z_k}{(z_k - 1)^2}.$$

To calculate the sum $\sum_{k=1}^{n} \frac{2z_k}{(z_k-1)^2}$, we consider the choice $f(t) = t^n$. Then $f(tz_k) = -t_n$, and hence

$$nt^n = -\frac{2t^n}{n}\sum_{k=1}^{n} \frac{2z_k}{(z_k - 1)^2},$$

i.e.,

$$\sum_{k=1}^{n} \frac{2z_k}{(z_k - 1)^2} = -\frac{n^2}{2}. \qquad \square \tag{1}$$

Returning to the proof of Bernstein's theorem, we let $|t| = 1$. Then

$$|f'(t)| \le \left(\frac{n}{2}f(t) + \frac{1}{n}\sum_{k=1}^{n} \left|\frac{2z_k}{(z_k - 1)^2}\right|\right)\|f\|.$$

Let us show that $\dfrac{2z_k}{(z_k - 1)^2}$ is a negative real number. Indeed, $z_k = e^{i\varphi} \ne 1$, and hence

$$\frac{2z_k}{(z_k - 1)^2} = \frac{2e^{i\varphi}}{(e^{i\varphi} - 1)^2} = \frac{2}{e^{i\varphi} - 2 + e^{-i\varphi}} = \frac{1}{\cos\varphi - 1} < 0.$$

Hence (1) implies that

$$\frac{1}{n}\sum_{k=1}^{n}\left|\frac{2z_k}{(z_k-1)^2}\right| = -\frac{1}{n}\sum_{k=1}^{n}\frac{2z_k}{(z_k-1)^2} = \frac{n}{2},$$

and the theorem now follows.□

4.2.4 Polynomials in several variables

Mahler's measure also helps to estimate the height $H(F) = \max|a_{k_1\cdots k_n}|$ of the polynomial $F(x_1,\ldots,x_n) = \sum a_{k_1\cdots k_n}x_1^{k_1}\cdot\ldots\cdot x_n^{k_n}$ in n variables (see [Ma3]).

Mahler's measure of a polynomial in one variable can be obtained by either of the two equivalent formulas:

$$M(f) = |a_d|\prod_{i=1}^{d}\max\{1,|\alpha_i|\},$$

$$M(f) = \exp\int_0^1 \ln|f(e^{2\pi it})|\,dt.$$

For polynomials in n variables, only the second definition is relevant:

$$M(F) = \exp\int_0^1\left(\cdots\int_0^1 \ln\left|F(e^{2\pi it_1},\ldots,e^{2\pi it_n})\right|\,dt_1\cdots\right)dt_n. \qquad (*)$$

We recall that on page 142, we obtained the inequality

$$|a_k| \leq \binom{d}{k}M(f)$$

for polynomials in one variable.

Using this inequality we can prove the following inequality for polynomials in n variables:

$$|a_{k_1\ldots k_n}| \leq \binom{d_1}{k_1}\cdot\ldots\cdot\binom{d_n}{k_n}M(F), \qquad (1)$$

where $d_1,\ldots,d_n$ are the degrees of the polynomial F with respect to $x_1,\ldots,x_n$, respectively. To do this, we express F in the form

$$F(x_1,\ldots,x_n) = \sum_{k_1=1}^{d_1}F_{k_1}(x_2,\ldots,x_n)x_1^{k_1}.$$

For fixed $x_2 = \alpha_2,\ \ldots,\ x_n = \alpha_n$, we have the estimate

$$|F_{k_1}(\alpha_2, \ldots, \alpha_n)| \leq \binom{d_1}{k_1} M(g),$$

where $g(x) = F(x, \alpha_2, \ldots, \alpha_n)$.

Set $x = e^{2\pi i t_1}, \alpha_2 = e^{2\pi i t_2}, \ldots, \alpha_n = e^{2\pi i t_n}$, and then take the logarithms of the inequality obtained:

$$\ln|F_{k_1}(e^{2\pi i t_2}, \ldots, e^{2\pi i t_n})| \leq \ln\binom{d_1}{k_1} + \int_0^1 \ln|F(e^{2\pi i t_1}, \ldots, e^{2\pi i t_n})|\, dt_1.$$

Let us integrate both parts of this inequality over $t_2, \ldots, t_n$ from 0 to 1, and then take the exponent. The definition of Mahler's measure (∗) directly implies that as a result we obtain $M(F_{k_1}) \leq \binom{d_1}{k_1} M(F)$.

Next, we express $F_{k_1}(x_2, \ldots, x_n)$ in the form

$$F_{k_1}(x_2, \ldots, x_n) = \sum_{k_2=1}^{d_2} F_{k_1 k_2}(x_3, \ldots, x_n) x_2^{k_2}.$$

We similarly prove that

$$M(F_{k_1 k_2}) \leq \binom{d_2}{k_2} M(F_{k_1}) \leq \binom{d_1}{k_1}\binom{d_2}{k_2} M(F),$$

and so on. It is also clear that $M(a_{k_1 \ldots k_n}) = a_{k_1 \ldots k_n}$.

As we have already mentioned, it is easy to prove by induction on d that $\binom{d}{k} \leq 2^{d-1}$ for $d \geq 1$. Hence (1) implies that, if $d_1 > 0, \ldots, d_n > 0$, then

$$H(F) \leq 2^{d_1 + d_2 + \cdots + d_n - n} M(F). \tag{2}$$

If, in reality, the polynomial $F(x_1, \ldots, x_n)$ depends only on $\nu(F)$ indeterminates, whereas the remaining $n - \nu(F)$ indeterminates enter with degree 0, then instead of (2) we obtain a rougher estimate

$$H(F) \leq 2^{d_1 + d_2 + \cdots + d_n - \nu(F)} M(F). \tag{3}$$

The estimate of $H(F)$ from below is proved in exactly the same way as on page 143 for polynomials in one variable. This estimate is of the form

$$M(F) \leq \sqrt{d_1 + 1} \cdot \ldots \cdot \sqrt{d_n + 1}\, H(F). \tag{4}$$

Let $F_1, \ldots, F_s$ be polynomials in $x_1, \ldots, x_n$ and let $d_{l1}, \ldots, d_{ln}$ be the degrees of the polynomial $F(l)$ with respect to these variables; let $\nu(F_l)$ be the number of variables on which F_l actually depends (i.e., the number of indices j for which $d_{lj} > 0$). Then by (3)

$$\prod_{l=1}^s H(F_l) \leq \prod_{l=1}^s 2^{d_{l1} + d_{l2} + \cdots + d_{ln} - \nu(F_l)} M(F_l) \leq 2^{d_1 + d_2 + \cdots + d_n - \nu(F)} M(F).$$

Now, using (4), we obtain

$$H(F_1) \cdot \ldots \cdot H(F_s) \leq 2^{d_1+d_2+\cdots+d_n-n} \sqrt{d_1+1} \cdot \ldots \cdot \sqrt{d_n+1} \, H(F),$$

where we assume that $\nu(F) = n$, i.e., F actually depends on all the variables $x_1, \ldots, x_n$. The upper estimate

$$H(F) \leq 2^{d_1+d_2+\cdots+d_n} H(F_1) \cdot \ldots \cdot H(F_s)$$

can be proved without appealing to Mahler's measure. Indeed, the number of nonzero coefficients of F_l does not exceed

$$(1+d_{l1}) \cdot \ldots \cdot (1+d_{ln}) \leq 2^{d_{l1}+d_{l2}+\cdots+d_{ln}}.$$

Therefore any coefficient of F is the sum of not more than $2^{d_1+d_2+\cdots+d_n}$ products of the coefficients of the polynomials $F_1, \ldots, F_s$.

Using Mahler's measure, we can also estimate the *length* of the polynomial

$$L(F) = \sum |a_{k_1 \ldots k_n}|.$$

Summing the inequalities (1) we obtain

$$L(F) \leq 2^{d_1+d_2+\cdots+d_n} M(F). \tag{5}$$

For the polynomial $F(x_1, \ldots, x_n) = (1+x_1)^{d_1} \cdot \ldots \cdot (1+x_n)^{d_n}$, inequality (5) becomes an equality, and so the estimate is precise.

The estimate of $L(F)$ from below is simply obtained: from the obvious inequality

$$\left|F(e^{2\pi i t_1}, \ldots, e^{2\pi i t_n})\right| \leq L(F)$$

we derive

$$M(F) \leq L(F). \tag{6}$$

For the polynomial $F(x_1, \ldots, x_n) = x_1^{d_1} \cdot \ldots \cdot x_n^{d_n}$, inequality (6) becomes an equality, and so the estimate is again precise.

From (5) it follows that

$$\prod_{l=1}^{s} L(F_l) \leq \prod_{l=1}^{s} \left(2^{d_{l1}+d_{l2}+\cdots+d_{ln}} M(F_l)\right) = 2^{d_1+d_2+\cdots+d_n} M(F).$$

Then by (6) we obtain

$$L(F_1) \cdot \ldots \cdot L(F_s) \leq 2^{d_1+d_2+\cdots+d_n} L(F_1 \cdot \ldots \cdot F_s).$$

The estimate from above

$$L(F_1 \cdot \ldots \cdot F_s) \leq L(F_1) \cdot \ldots \cdot L(F_s)$$

is obvious.

4.2.5 An inequality for a pair of relatively prime polynomials

Let $f(x)$ and $g(x)$ be relatively prime polynomials over $\mathbb{C}$. Then

$$m(x) = \max\{|f(x)|, |g(x)|\} > 0$$

and $m(x) \to \infty$ as $x \to \infty$. Therefore the quantity

$$E(f, g) = \min_x m(x)$$

is positive. In the theory of transcendental numbers one sometimes needs to estimate $E(f, g)$ from below. N. I. Feldman suggested a method to obtain a precise estimate. We give an exposition of his method following Mahler's paper [Ma4].

Theorem 4.2.5. *Let* $\alpha_1, \ldots, \alpha_m$ *be the roots of a polynomial* f, *and let* $\beta_1, \ldots, \beta_n$ *be the roots of a polynomial* g. *Then*

$$E(f, g) \geq \min_{i,j} \left\{ \frac{|f(\beta_i)|}{2^m}, \frac{|g(\alpha_j)|}{2^n} \right\}. \tag{1}$$

Proof. Fix an arbitrary number $x \in \mathbb{C}$, and let $\alpha = \min_j |x - \alpha_j|$ and $\beta = \min_i |x - \beta_i|$. Then $\alpha = |x - \alpha_k|$ and $\beta = |x - \beta_l|$ for some k and l. Since f and g are relatively prime, it follows that one of the numbers α and β is positive.

Suppose that $\alpha \leq \beta$. First, consider the case where $\alpha > 0$. We show that in this case

$$|x - \beta_i| \geq \frac{|\alpha_k - \beta_i|}{2} \tag{2}$$

for any i. Indeed, if $|\alpha_k - \beta_i| < 2\alpha$, then

$$|x - \beta_i| \geq \beta \geq \alpha > \frac{|\alpha_k - \beta_i|}{2}.$$

If, however, $|\alpha_k - \beta_i| \geq 2\alpha = 2|x - \alpha_k|$, Then again

$$|x - \beta_i| = \big|(x - \alpha_k) + (\alpha_k - \beta_i)\big| \geq |x - \alpha_k| + |\alpha_k - \beta_i| \geq \frac{|\alpha_k - \beta_i|}{2}.$$

Let $g(x) = b_0(x - \beta_1) \cdot \ldots \cdot (x - \beta_n)$. Inequality (2) implies that

$$|g(x)| = \big|b_0(x - \beta_1) \cdot \ldots \cdot (x - \beta_n)\big| \geq \frac{|b_0 \prod_{i=1}^n (\alpha_k - \beta_i)|}{2^n} = \frac{|g(\alpha_k)|}{2^n}.$$

Second, we deal with $\alpha = 0$, i.e., $x = \alpha_k$. Then the inequality $|g(x)| \geq \frac{|g(\alpha_k)|}{2^n}$ is obviously satisfied.

Thus, if $\alpha \leq \beta$, then $|g(x)| \geq \frac{|g(\alpha_k)|}{2^n}$. We similarly prove that if $\alpha \geq \beta$, then $|f(x)| \geq \frac{|f(\beta_l)|}{2^m}$. Thus in either case

$$m(x) \geq \min_{l,k} \left\{ \frac{|f(\beta_l)|}{2^m}, \frac{|g(\alpha_k)|}{2^n} \right\}. \quad \square$$

Remark. If $f(x) = (x-1)^m$ and $g(x) = (x+1)^n$, inequality (1) becomes an equality. Hence (1) is precise.

4.2.6 Mignotte's inequality

In this chapter we have already considered several inequalities for estimating the coefficients of the factors of a given polynomial. An estimate of this type follows from the next theorem due to Mignotte [Mi1].

Theorem 4.2.6. *Let* $f(x) = a_0 + a_1 x + \cdots + a_m x^m$ *and* $g(x) = b_0 + b_1 x + \cdots + b_n x^n$ *be polynomials with integer coefficients. If* f *is divisible by* g, *then*

$$|b_j| \leq \binom{n-1}{j} \|f\| + \binom{n-1}{j-1} |a_m|,$$

where

$$\|f\| = \sqrt{a_0^2 + \cdots + a_m^2}.$$

Proof. Together with $f(x) = a_m \prod (x - \alpha_i)$, we consider a polynomial

$$\widehat{f}(x) = a_m \prod_{|\alpha_i| \geq 1} (x - \alpha_i) \prod_{|\alpha_i| < 1} (\overline{\alpha}_i x - 1).$$

We prove first that $\|\widehat{f}\| = \|f\|$, where the norm $\|\widehat{f}\|$ of $\widehat{f}$ is defined as for f. This follows immediately from the next lemma.

Lemma 4.2.7. *Let* $h(x) = c_0 + c_1 x + \cdots + c_k x^k$, *let* $h_1(x) = (x - \alpha)h(x)$ *and* $h_2(x) = (\overline{\alpha} x - 1)h(x)$. *Then* $\|h_1\| = \|h_2\|$.

Proof. Clearly,

$$\|h_1\|^2 = \sum |c_{i-1} - \alpha c_i|^2 =$$
$$= \sum \left(|c_{i-1}|^2 + |\alpha c_i|^2 - 2\operatorname{Re}(\alpha c_1 \overline{c_{i-1}}) \right) =$$
$$= \sum \left(|\alpha c_{i-1}|^2 + |c_i|^2 - 2\operatorname{Re}(\alpha c_1 \overline{c_{i-1}}) \right) =$$
$$= \sum |\alpha c_{i-1} - c_i|^2 = \|h_2\|^2. \quad \square$$

The coefficient of the highest term x^m of $\widehat{f}$ is $a_m \prod\limits_{|\alpha_i|<1} \overline{\alpha}_i$, and the coefficient of the lowest term is $\pm a_m \prod\limits_{|\alpha_i|>1} \overline{\alpha}_i$ (we assume that the empty product (without any factors) is equal to 1). Set

$$M(f) = \prod_{|\alpha_i|>1} \alpha_i, \quad m(f) = \prod_{|\alpha_i|<1} \alpha_i.$$

Then

$$\|\widehat{f}\|^2 = \|f\|^2 \geq |a_m|^2 \left(M(f)^2 + m(f)^2\right).$$

Hence

$$M(f) \leq \frac{\|f\|}{|a_m|}. \tag{1}$$

Also

$$|a_j| = |a_m| \cdot \left| \sum \alpha_{i_1} \cdot \ldots \cdot \alpha_{i_{m-j}} \right| \leq |a_m| \sum \beta_{i_1} \cdot \ldots \cdot \beta_{i_{m-j}}, \tag{2}$$

where $\beta_i = \max\{1, |\alpha_i|\}$. Clearly, $\prod \beta_i = M(f)$.

Now we need one more lemma.

Lemma 4.2.8. *Let* $x_1 \geq 1, \ldots, x_m \geq 1$ *and* $x_1 \cdot \ldots \cdot x_m = M$. *Then*

$$\sum_{i_1 < \cdots < i_k} x_{i_1} \cdot \ldots \cdot x_{i_k} \leq \binom{m-1}{k-1} M + \binom{m-1}{k}.$$

Proof. We may assume that $x_1 \leq x_2 \leq \cdots \leq x_m$. Let us replace the pair $\{x_{m-1}, x_m\}$ by $\{1, x_{m-1}x_m\}$. As a result, the sum considered will increase by $\sigma(x_{m-1} - 1)(x_m - 1)$, where σ is the sum of products $x_{i_1} \cdot \ldots \cdot x_{i_{k-1}}$, $1 \leq i_1 < \cdots < i_{k-1} \leq m-2$. Therefore, if $x_{m-1} > 1$, the sum considered strictly increases. Hence it will be minimal when $x_1 = \cdots = x_{m-1} = 1$, $x_m = M$. In this case, the sum consists of $\binom{m-1}{k-1}$ terms equal to M and $\binom{m-1}{k}$ terms equal to 1. $\square$

Applying Lemma 4.2.8 to the collection $\beta_1, \ldots, \beta_m$, we see that

$$\sum \beta_{i_1} \cdot \ldots \cdot \beta_{i_{m-j}} \leq \binom{m-1}{m-j-1} M(f) + \binom{m-1}{m-j}.$$

Now, if we take into account that $\binom{m-1}{m-j-1} = \binom{m-1}{j}$ and $\binom{m-1}{m-j} = \binom{m-1}{j-1}$ we can rewrite (2) in the form

$$|a_j| \leq |a_m| \left(\binom{m-1}{j} M(f) + \binom{m-1}{j-1} \right).$$

We similarly prove that

$$|b_j| \leq |b_n| \left(\binom{n-1}{j} M(g) + \binom{n-1}{j-1} \right). \tag{3}$$

All the roots of g are the roots of f, and hence $M(g) \leq M(f)$. Further, $|b_n| \leq |a_m|$ since by hypothesis g divides f. Using these inequalities and (1), we can reduce (3) to the form required:

$$|b_j| \leq \binom{n-1}{j} \|f\| + \binom{n-1}{j-1} |a_m|. \quad \square$$

Corollary. *If f, g and $\dfrac{f}{g}$ are polynomials with integer coefficients, then*

$$\|g\| \leq \binom{2n}{n}^{1/2} \|f\|, \quad \text{where} \quad n = \deg g.$$

Proof. Clearly, $|a_n| \leq \|f\|$, and hence

$$|b_j| \leq \left(\binom{n-1}{j} + \binom{n-1}{j-1} \right) \|f\| = \binom{n}{j} \|f\|.$$

Therefore $\|g\|^2 \leq \sum\limits_{j=0}^{n} \binom{n}{j}^2 \|f\|^2$. It remains to verify the combinatorial identity

$$\sum_{j=0}^{n} \binom{n}{j}^2 = \binom{2n}{n}.$$

To do this, it suffices to compare the coefficients of t^n in both sides of the identity

$$(1+t)^n (1+t)^n = (1+t)^{2n}. \quad \square$$

4.3 Equations for polynomials

4.3.1 Diophantine equations for polynomials

Mason's theorem and its corollaries

In the proof of the insolvability of various Diophantine equations for polynomials the following statement is rather effective.

Theorem 4.3.1 (Mason). *Let $a(x)$, $b(x)$ and $c(x)$ be pairwise relatively prime polynomials such that $a + b + c = 0$. Then the degree of each of these polynomials does not exceed $n_0(abc) - 1$, where $n_0(P)$ is the number of distinct roots of the polynomial P.*

Proof. [La4] Set $f = \dfrac{a}{c}$ and $g = \dfrac{b}{c}$. Then f and g are rational functions which satisfy $f + g + 1 = 0$. Differentiating this equality we see that $f' = -g'$. Hence,

$$\frac{b}{a} = \frac{g}{f} = -\frac{\frac{f'}{f}}{\frac{g'}{g}}.$$

The rational functions f and g are of a particular form:

$$\prod (x - \rho_i)^{r_i}, \text{ where } r_i \in \mathbb{Z}.$$

For the function $R(x) = \prod (x - \rho_i)^{r_i}$, we have

$$\frac{R'}{R} = \sum \frac{r_i}{x - \rho_i}.$$

Let

$$a(x) = \prod (x - \alpha_i)^{a_i}, \quad b(x) = \prod (x - \beta_j)^{b_j}, \quad c(x) = \prod (x - \gamma_k)^{c_k}.$$

Then

$$\frac{f'}{f} = \sum \frac{a_i}{x - \alpha_i} - \sum \frac{c_k}{x - \gamma_k},$$
$$\frac{g'}{g} = \sum \frac{b_j}{x - \beta_j} - \sum \frac{c_k}{x - \gamma_k}.$$

Therefore, after multiplication by the polynomial

$$N_0 = \prod (x - \alpha_i)(x - \beta_j)(x - \gamma_k)$$

of degree $n_0(abc)$, the rational functions $\dfrac{f'}{f}$ and $\dfrac{g'}{g}$ become polynomials of degree not greater than $n_0(abc) - 1$. Then, since $a(x)$ and $b(x)$ are relatively prime and

$$\frac{b}{a} = -\frac{N_0 \frac{f'}{f}}{N_0 \frac{g'}{g}},$$

the degree of each of the polynomials $a(x)$ and $b(x)$ does not exceed $n_0(abc) - 1$. For $c(x)$, the proof is similar. $\square$

Theorem 4.3.1 has several interesting corollaries which we formulate as the following Theorems 4.3.2 — 4.3.4.

Theorem 4.3.2 (Davenport). *Let f and g be relatively prime polynomials of nonzero degree. Then*

$$\deg(f^3 - g^2) \geq \frac{1}{2} \deg f + 1.$$

Proof. If $\deg f^3 \neq \deg g^2$, then

$$\deg(f^3 - g^2) \geq \deg f^3 = 3 \deg f \geq \frac{1}{2} \deg f + 1.$$

Thus, we may assume that $\deg f^3 = \deg g^2 = 6k$.

Now consider the polynomials $F = f^3$, $G = g^2$ and $H = F - G = f^3 - g^2$. Clearly, $\deg H \leq 6k$. By Theorem 4.3.1

$$\max\{\deg F, \deg G, \deg H\} \leq n_0(FGH) - 1 \leq \deg f + \deg g + \deg H - 1,$$

i.e.,

$$6k \leq 2k + 3k + \deg H - 1.$$

Hence $\deg H \geq k + 1 = \frac{1}{2} \deg f + 1.$ □

Remark. For the polynomials

$$f(t) = t^2 + 2, \quad g(t) = t^3 + 3t,$$

Davenport's inequality becomes an equality.

Theorem 4.3.3. *Let f, g and h be relatively prime polynomials, at least one of them not being a constant. Then the identity*

$$f^n + g^n = h^n$$

cannot hold for $n \geq 3$.

Proof. On the assumption that the identity holds, the degree of each of the polynomials f^n, g^n and h^n does not exceed

$$\deg f + \deg g + \deg h - 1,$$

by Theorem 4.3.1. Adding up these three inequalities we obtain

$$n(\deg f + \deg g + \deg h) \leq 3(\deg f + \deg g + \deg h - 1).$$

Hence $n < 3.$ □

The Diaphantine equation $f^\alpha + g^\beta = h^\gamma$ for polynomials f, g, h has an obvious solution if one of the numbers α, β, γ is equal to 1. Therefore in what follows we assume that $\alpha, \beta, \gamma \geq 2$.

Theorem 4.3.4. *Let α, β, γ be positive integers and $2 \leq \alpha \leq \beta \leq \gamma$. Then the equation*

$$f^\alpha + g^\beta = h^\gamma$$

has relatively prime solutions only for the following collections (α, β, γ):

$$(2, 2, \gamma), \quad (2, 3, 3), \quad (2, 3, 4), \quad (2, 3, 5).$$

Proof. Let a, b and c be the degrees of f, g and h respectively. Then by Theorem 4.3.1

$$\alpha a \leq a + b + c - 1, \tag{4.1}$$
$$\beta b \leq a + b + c - 1, \tag{4.2}$$
$$\gamma c \leq a + b + c - 1. \tag{4.3}$$

Hence

$$\alpha(a + b + c) \leq \alpha a + \beta b + \gamma c \leq 3(a + b + c) - 3,$$

giving $\alpha < 3$. By hypothesis $\alpha \geq 2$, and hence $\alpha = 2$. For $\alpha = 2$, inequality (1) becomes

$$a \leq b + c - 1. \tag{4}$$

Adding together inequalities (4), (2) and (3) we obtain

$$\beta b + \gamma c \leq 3(b + c) + a - 3.$$

Since $\beta \leq \gamma$ and applying (4) once again, we obtain

$$\beta(b + c) \leq 4(b + c) - 4,$$

giving $\beta \leq 4$. Hence $\beta = 2$ or 3.

It remains to prove that if $\beta = 3$, then $\gamma \leq 5$. For $\beta = 3$, inequality (2) becomes

$$2b \leq a + c - 1. \tag{5}$$

Adding together (4) and (5) we obtain

$$b \leq 2c - 2.$$

Then (4) implies that

$$a \leq 3c - 3.$$

The last two inequalities and (3) imply that

$$\gamma c \leq 6c - 6,$$

and so $\gamma \leq 5$.

The polynomials satisfying the relation $f^\alpha + g^\beta = h^\gamma$ are closely related to regular polyhedra. Felix Klein described this relation in detail in the book [KF], where a method of constructing these polynomials is also given. Let us recall the final result.

The case $\alpha = \beta = 2$, $\gamma = n$ is related to a degenerate regular polyhedron, namely the planar n-gon. The relation in question is

$$\left(\frac{x^n + 1}{2}\right)^2 - \left(\frac{x^n - 1}{2}\right)^2 = x^n.$$

The case $\alpha = 2$, $\beta = 3$, $\gamma = 3$ is related to a regular tetrahedron. The relation is

$$12i\sqrt{3}\,(x^5 - x)^2 + (x^4 - 2i\sqrt{3}\,x^2 + 1)^3 = (x^4 + 2i\sqrt{3}\,x^2 + 1)^3.$$

The case $\alpha = 2$, $\beta = 3$, $\gamma = 4$ is related to a cube and a regular octahedron. The relation is

$$(x^{12} - 33x^8 - 33x^4 + 1)^2 + 108(x^5 - x)^4 = (x^8 + 14x^4 + 1)^3.$$

The case $\alpha = 2$, $\beta = 3$, $\gamma = 5$ is related to a dodecahedron and an icosahedron. The relation is

$$T^2 + h^3 = 1728 f^5,$$

where

$$
\begin{aligned}
T &= x^{30} + 1 + 522(x^{25} - x^5) - 10005(x^{20} + x^{10}), \\
H &= -(x^{20} + 1) + 228(x^{15} - x^5) - 494x^{10}, \\
f &= x(x^{10} + 11x^5 - 1). \quad \square
\end{aligned}
$$

Theorem 4.3.4 was proved by H. Schwarz [Sc7]. The solution of Diophantine equations of a more general type

$$f^\alpha + g^\beta = l^\mu h^\gamma$$

is given in [Ev].

Theorem 4.3.5 ([Na]). *Let $x(t)$ and $y(t)$ be rational functions; let $m \geq 2$ and $n \geq 2$. Then the equation*

$$x^n - y^m = 1$$

has solutions only for $m = n = 2$.

Proof. Let us express x and y in the form $x = \dfrac{f}{g}$ and $y = \dfrac{h}{k}$, where f and g are relatively prime polynomials and h and k are also relatively prime polynomials. Then the equation considered takes the form

$$f^m k^n - h^n g^m = g^m k^n. \tag{6}$$

Since f and g are relatively prime, it follows that if $g(\alpha) = 0$, then $f(\alpha) = 0$. In this case (6) implies that $k(\alpha) = 0$. Similarly, if $k(\alpha) = 0$, then $g(\alpha) = 0$. Hence $g(t) = \prod(t - \alpha_i)^{a_i}$ and $k(t) = \prod(t - \alpha_i)^{b_i}$, where $a_i, b_i \geq 1$.

The multiplicity of α_i, as a root of the polynomials $f^m k^n$, $h^n g^m$ and $g^m k^n$, is equal to nb_i, ma_i and $nb_i + ma_i$, respectively. If $nb_i \neq ma_i$, then the multiplicity of the root α_i of the polynomial $f^m k^n - h^n g^m$ is strictly less than $nb_i + ma_i$. Hence $nb_i = ma_i$, i.e., $k^n = g^m$.

After division by $k^n = g^m$, equation (6) becomes

$$f^m - h^n = g^m.$$

By Theorem 4.3.4 only two possibilities can occur: $\{m, n\} = \{2, 2\}$ or $\{2, 3\}$. But in the second case $k^n = g^m = l^6$, where l is a polynomial, and the equation $f^3 - h^2 = l^6$ has no solutions. $\square$

Waring's problem for polynomials

The classical Waring's problem is as follows:

given a positive integer n, find the minimal number $k = k(n)$ for which any positive integer m can be represented in the form $m = m_1^n + \cdots + m_k^n$, where $m_1, \ldots, m_k$ are non-negative integers.

Several generalizations of this problem for polynomials are known. Here by *Waring's problem for polynomials*, we will mean the following problem:

given a positive integer n, find the minimal number $k = k(n)$ for which any polynomial $g \in \mathbb{C}[x]$ can be represented in the form $g = f_1^n + \cdots + f_k^n$, where $f_i \in \mathbb{C}[x]$.

To solve Waring's problem, it suffices to confine ourselves to the case $g(x) = x$. Indeed, if $x = f_1^n(x) + \cdots + f_k^n(x)$ and $h(x)$ is an arbitrary polynomial, then $h(x) = f_1^n(h(x)) + \cdots + f_k^n(h(x))$.

The identity $\left(x + \frac{1}{4}\right)^2 - \left(x - \frac{1}{4}\right)^2 = x$ shows that $k(2) = 2$.

Theorem 4.3.6 ([Ne]). *If $n \geq 3$, then*
   a) $k(n) \geq 3$;
   b) $k(n) \leq n < k^2(n) - k(n)$.

Proof. a) Suppose on the contrary that

$$x = f_1^n(x) + f_2^n(x) = \prod_{r=1}^{n} (f_1 + \varepsilon^r f_2),$$

where ε is a primitive n-th root of unity. All the factors $f_1 + \varepsilon^r f_2$, except one, are constants. For $n \geq 3$, there are at least two such factors. Therefore $f_1 + a f_2 = \alpha$ and $f_1 + b f_2 = \beta$, where $a \neq b$ and $a, b, \alpha, \beta \in \mathbb{C}$. Hence, $f_1, f_2 \in \mathbb{C}$ which is impossible.

   b) For any $f(x)$, set

$$\Delta f(x) = f(x + 1) - f(x) \text{ and } \Delta^p f = \Delta(\Delta^{p-1} f) \text{ for } p \geq 2.$$

It is easy to verify that $\deg(\Delta f) = \deg f - 1$, and so $\deg \Delta^{n-1}(x^n) = 1$, i.e., $\Delta^{n-1}(x^n) = ax + b$. On the other hand, the definition directly implies that

$$\Delta^{n-1}(x^n) = (x+n-1)^n + c_1(x+n-2)^n + \cdots + c_{n-1}x^n.$$

Indeed, for example,

$$\Delta^2(x^3) = \left((x+2)^3 - (x+1)^3\right) - \left((x+1)^3 - x^3\right).$$

Making a change of variables $x_1 = ax + b$ we get a representation

$$x_1 = f_1^n(x_1) + \cdots + f_n^n(x_1)$$

which shows that $k(n) \leq n$.

Next, we prove the inequality $n < k^2(n) - k(n)$. Consider the representation $x = f_1^n(x) + \cdots + f_k^n(x)$, where $f_i \in \mathbb{C}[x]$, with the minimal k.

Recall that the *Wronski determinant*, or *Wronskian*, $W(g_1, \ldots, g_k)$ of the functions $g_1(x), \ldots, g_k(x)$ is the determinant of the matrix

$$\begin{pmatrix} g_1(x) & \cdots & g_k(x) \\ g_1'(x) & \cdots & g_k'(x) \\ \vdots & \ddots & \vdots \\ g_1^{(k-1)}(x) & \cdots & g_k^{(k-1)}(x) \end{pmatrix}.$$

Consider two Wronski determinants,

$$W_1 = W(f_1^n, f_2^n, \ldots, f_k^n) \text{ and } W_2 = W(x, f_2^n, \ldots, f_k^n).$$

By hypothesis, $x = f_1^n + \cdots + f_k^n$, and so the first column of W_2 is obtained from the first column of W_1 by adding a linear combination of the other columns of W_1. Hence $W_1 = W_2$.

If the functions $g_1, g_2, \ldots, g_k$ are linearly dependent, then $W(g_1, \ldots, g_k)$ vanishes identically. The converse statement is false. For example, if $g_1(x) = x^2$ and $g_2(x) = x|x|$, then

$$W(g_1, g_2) = \begin{vmatrix} x^2 & x|x| \\ 2x & 2|x| \end{vmatrix} = 0,$$

but the functions g_1 and g_2 are linearly independent in any interval $(-a, a)$. It is known that if $W(g_1, \ldots, g_k)$ vanishes identically for $x \in (a, b)$, then there exists a subinterval $(\alpha, \beta) \subset (a, b)$ on which the functions $g_1, \ldots, g_k$ are linearly dependent (for a simple proof of this statement, see [Kr2]). In particular, for polynomials $g_1, \ldots, g_k$, the fact that $W(g_1, \ldots, g_k)$ identically vanishes implies that these polynomials are linearly dependent.

Since the representation $x = f_1^n(x) + \cdots + f_k^n(x)$ is minimal, it follows that the functions $f_1^n, \ldots, f_k^n$ are linearly independent, and so $W(f_1^n, f_2^n, \ldots, f_k^n)$ is a nonzero polynomial.

The r-th derivative of f_i^n is divisible by f_i^{n-r}, and so the i-th column of the Wronskian is divisible by f_i^{n-k+1}. Therefore $W(f_1^n, \ldots, f_k^n)$ is divisible by $\prod_{i=1}^{k} f_i^{n-k+1}$. In particular,

$$\deg W(f_1^n, \ldots, f_k^n) \geq (n - k + 1) \sum_{i=1}^{k} \deg f_i. \tag{1}$$

On the other hand, we now prove that

$$\deg W(f_1^n, \ldots, f_k^n) \leq n \sum_{i=2}^{k} \deg f_i - \frac{k(k-1)}{2} + 1. \tag{2}$$

First, we recall that $W(f_1^n, \ldots, f_k^n) = W(x, f_2^n, \ldots, f_k^n)$. If we multiply the j-th row of the determinant[1] $W(x, f_2^n, \ldots, f_k^n)$ by x^{j-1} we obtain a determinant for which all the nonzero elements in the i-th column (for $i \geq 2$) are of degree $n \deg f_i$. Hence

$$\deg W(f_1^n, \ldots, f_k^n) \leq 1 + n \sum_{i=2}^{k} \deg f_i - 1 - 2 - \cdots - (k-1) =$$

$$= n \sum_{i=2}^{k} \deg f_i - \frac{k(k-1)}{2} + 1.$$

Comparing (1) with (2) we see that

$$(n - k + 1) \sum_{i=1}^{k} \deg f_i \leq n \sum_{i=2}^{k} \deg f_i - \frac{k(k-1)}{2} + 1,$$

i.e.,

$$n \deg f_1 \leq (k-1) \sum_{i=1}^{k} \deg f_i - \frac{k(k-1)}{2} + 1.$$

We may assume that f_1 is the polynomial of the highest degree. Then

$$n \deg f_1 \leq k(k-1) \deg f_1 - \frac{k(k-1)}{2} + 1 < k(k-1) \deg f_1.$$

The last inequality follows from the fact that $1 - \frac{1}{2}k(k-1) < 0$ for $k \geq 3$. After division by $\deg f_1$ we obtain $n < k(k-1)$. $\square$

---

[1] By abuse of the language we mean here the matrix whose determinant — a number without rows or columns — we consider.

4.3.2 Functional equations for polynomials

Functional equations that determine polynomials

Every polynomial f of degree $n + 1$ satisfies the identity

$$f(x) = f(y) + (x - y)f'(y) + \cdots + (x - y)^{n+1}\frac{f^{(n+1)}(y)}{(n+1)!}.$$

Here $f^{(n+1)}$ is a constant, hence,

$$(x - y)^{n+1}\frac{f^{(n+1)}(y)}{(n+1)!} =$$
$$= (x - y)^n \left(\frac{-yf^{(n+1)}(y)}{(n+1)!} - c\right) + (x - y)^n \left(\frac{xf^{(n+1)}(x)}{(n+1)!} + c\right).$$

Therefore the functional equation

$$f(x) = \sum_{k=0}^{n}(x - y)^k g_k(y) + (x - y)^n h(x) \tag{1}$$

has a solution of one of the following form:

   a) f is a polynomial of degree not higher than $n + 1$;

   b) $g_k(y) = \dfrac{f^{(k)}(y)}{k!}$ for $k = 0, 1, \ldots, n - 1$;

   c) $g_n(y) = \dfrac{f^{(n)}(y)}{n!} - \dfrac{yf^{(n+1)}(y)}{(n+1)!} - c$;

   d) $h(x) = \dfrac{xf^{(n+1)}(x)}{(n+1)!} + c$.

Theorem 4.3.7 ([Cr]). *Let $f, g_0, \ldots, g_n, h : \mathbb{R} \to \mathbb{R}$ be arbitrary functions which satisfy (1) for any $x, y \in \mathbb{R}$ such that $x \neq y$. Then these functions are of one of the above forms a)—d).*

Proof. For $y = 0$ and $y = 1$, equation (1) takes the form

$$f(x) = \sum_{k=0}^{n} c_k x^k + x^n h(x) \text{ for } x \neq 0, \tag{2}$$

$$f(x) = \sum_{k=0}^{n} d_k x^k + (x - 1)^n h(x) \text{ for } x \neq 1. \tag{3}$$

Hence

$$h(x) = \frac{\sum_{k=0}^{n}(d_k - c_k)x^k}{x^n - (x - 1)^n}.$$

This equality holds for $x \neq 0, 1$ and, if n is even, we should also exclude $x = \frac{1}{2}$.

Fix $y = 2$ and $y = 4$. We similarly obtain the equality

$$h(x) = \frac{\sum_{k=0}^{n}(f_k - e_k)x^k}{(x-2)^n - (x-4)^n},$$

which holds for $x \neq 2, 3, 4$. As a result, we see that $h \in C^\infty(\mathbb{R})$ but then (2) and (3) imply that $f \in C^\infty(\mathbb{R})$.

Differentiating (1) n times with respect to x, we see that

$$f^{(n)}(x) = n!g_n(y) + \frac{d^n}{dx^n}\left((x-y)^n h(x)\right).$$

For a fixed x, this equality implies that $g_n(y)$ is a polynomial. Now we may differentiate (1) with respect to x not n but $n-1$ times and similarly deduce that $g_{n-1}(y)$ is a polynomial, and so on. In particular, $g_0, \ldots, g_n \in C^\infty(\mathbb{R})$.

Next, we differentiate (1) with respect to y and set $y = 0$. As a result we obtain

$$0 = -\sum_{k=0}^{n} kx^{k-1}g_k(0) + \sum_{k=0}^{n} kx^k g_k'(0) - nx^{n-1}h(x).$$

This equality implies that $x^{n-1}h(x)$ is a polynomial of degree not greater than n. If we differentiate (1) n times with respect to y and set $y = 0$ we can show that $h(x)$ is a polynomial (of degree also not higher than n). But if $h(x)$ is a polynomial and $x^{n-1}h(x)$ is a polynomial of degree not higher than n, then $h(x) = ax + b$.

Since $h(x) = ax + b$ and $f \in C^\infty(\mathbb{R})$, it follows that (2) implies that $f(x)$ is a polynomial of degree no higher than $n + 1$. Therefore

$$f(x+y) = \sum_{k=0}^{n+1} \frac{f^{(k)}(y)}{k!}x^k.$$

On the other hand, replacing x by $x + y$ we can rewrite (1) as

$$f(x+y) = \sum_{k=0}^{n} x^k g_k(y) + x^n(ax + ay + b).$$

This means that

$$g_k(y) = \frac{f^{(k)}(y)}{k!} \quad \text{for} \quad k = 0, 1, \ldots, n-1,$$

$$g_n(y) = \frac{f^{(n)}(y)}{n!} - ay - b \quad \text{and} \quad \frac{f^{(n+1)}(y)}{(n+1)!} = a. \ \square$$

Corollary 1. *Let $f \in C^n(\mathbb{R})$ and suppose that*

$$\frac{f(x) - \sum_{k=0}^{n-1} \frac{(x-y)^k}{k!} f^{(k)}(y)}{(x-y)^n} = \frac{f^{(n)}(x) + f^{(n)}(y)}{(n+1)!}$$

for any $x, y \in \mathbb{R}$ such that $x \neq y$. Then f is a polynomial of degree no higher than n.

Corollary 2. ([Ha]) *If*

$$\frac{f(x) - g(y)}{x - y} = \frac{\varphi(x) + \varphi(y)}{2} \tag{4}$$

for any $x, y \in \mathbb{R}$ such that $x \neq y$, then f is a polynomial of degree no higher than 2, and $g = f$ and $\varphi = f'$.

The functional equation

$$\frac{f(x) - g(y)}{x - y} = \varphi\left(\frac{x+y}{2}\right) \tag{5}$$

also reduces to the functional equation (4). Indeed, if (5) holds for any $x, y \in \mathbb{R}$ such that $x \neq y$, then

$$\varphi\left(\frac{x+y}{2}\right) = \frac{\varphi(x) + \varphi(y)}{2}.$$

To prove this, in (5) we replace x by $x + y$ and y by $x - y$. Then we obtain

$$\frac{f(x+y) - g(x-y)}{2y} = \varphi(x)$$

for any $x, y \in \mathbb{R}$ such that $y \neq 0$. Now replacing y by $-y$ we obtain

$$\frac{f(x-y) - g(x+y)}{-2y} = \varphi(x).$$

Hence

$$f(u + v + y) - g(u + v - y) = 2y\varphi(u + v),$$
$$f(u - v + y) - g(u - v - y) = 2y\varphi(u - v)$$

and

$$f(u + v + y) - g(u - v - y) = 2(v + y)\varphi(u),$$
$$f(u - v + y) - g(u + v - y) = -2(v - y)\varphi(u).$$

Therefore

$$\varphi(u + v) + \varphi(u - v) = 2\varphi(u).$$

Setting $u + v = x$ and $u - v = y$ we obtain the identity required

$$\varphi(x) + \varphi(y) = 2\varphi\left(\frac{x+y}{2}\right).$$

Polynomial solutions of the equation $f(\alpha x + \beta) = f(x)$

For $\alpha = \pm 1$, the polynomial solutions of the equation $f(\alpha x + \beta) = f(x)$ are easy to find. If $\alpha = 1$ and $f(x) = a_0 x^n + a_1 x^{n-1} + \cdots + a_n$, where $a_0 \neq 0$, we obtain the identity

$$f(x) = a_0 x^n + a_1 x^{n-1} + \cdots + a_n = a_0(x + \beta)^n + a_1(x + \beta)^{n-1} + \cdots + a_n.$$

This identity is only possible when $a_1 = a_1 + a_0 n \beta$, i.e., $\beta = 0$.

If $\alpha = -1$, then the equation $f(-x + \beta) = f(x)$ reduces under the change $g(x) = f\left(x + \dfrac{\beta}{2}\right)$ to the equation $g(x) = g(-x)$ whose solutions are polynomials of the form

$$a_0 x^{2n} + a_2 x^{2n-2} + \cdots + a_{2n}.$$

For an arbitrary α, comparison of the coefficients of the highest term of $f(x) = a_0 x^n + \cdots$ shows that $\alpha^n = 1$. Therefore, for $\alpha \neq \pm 1$, we see that $n \geq 3$, and we deal with this case in the next theorem.

Theorem 4.3.8 ([Oz]). *Let a polynomial f of degree $n \geq 3$ satisfy the relation $f(\alpha x + \beta) = f(x)$, where $\alpha \neq \pm 1$ and $\alpha^n = 1$. Then*

$$f(x) = a_0 \left(x + \frac{\beta}{\alpha - 1}\right)^n + c.$$

Proof. It suffices to consider polynomials of the form

$$f(x) = x^n + a_1 x^{n-1} + \cdots + a_n.$$

We have to prove that $a_j = \binom{n}{j} \dfrac{\beta^j}{(\alpha - 1)^j}$ for $j = 1, \ldots, n - 1$. We use induction on j.

Comparison of the coefficients of x^{n-j} for $f(x)$ and $f(\alpha x + \beta)$ shows that

$$(1 - \alpha^{n-j}) a_j = \sum_{s=0}^{j-1} a_s \binom{n-s}{j-s} \alpha^{n-j} \beta^{j-s}. \tag{1}$$

For $j = 1$, we obtain $(1 - \alpha^{n-1}) a_1 = \binom{n}{1} \alpha^{n-1} \beta$. By the hypothesis $\alpha^n = 1$, and so

$$a_1 = \binom{n}{1} \frac{\alpha^{-1} \beta}{1 - \alpha^{-1}} = \binom{n}{1} \frac{\beta}{\alpha - 1}.$$

The start of the induction is proved.

To prove the inductive step, i.e., the passage from $j = k$ to $j = k + 1$, we again use (1) and the condition $\alpha^n = 1$. By the induction hypothesis, $a_s = \binom{n}{s} \dfrac{\beta^s}{(\alpha - 1)^s}$ for $s = 1, \ldots, k$. Hence

$$(1 - \alpha^{n-k-1}) a_{k+1} = \binom{n}{k+1} \alpha^{n-k-1} \beta^k + \sum_{s=1}^{k} \binom{n}{s} \binom{n-s}{k+1-s} \frac{\alpha^{n-k+1} \beta^{k+1}}{(\alpha - 1)^s}.$$

It is easy to verify that $\binom{n}{s}\binom{n-s}{k+1-s} = \binom{n}{k+1}\binom{k+1}{s}$. Hence

$$(1 - \alpha^{n-k-1})a_{k+1} =$$
$$= \binom{n}{k+1}\alpha^{n-k-1}\beta^{k+1}\left(1 + \binom{k+1}{1}\frac{1}{\alpha-1} + \cdots + \binom{k+1}{k}\frac{1}{(\alpha-1)^k}\right).$$

It is also clear that the expression in the square brackets is equal to

$$\left(1 + \frac{1}{\alpha-1}\right)^{k+1} - \left(\frac{1}{\alpha-1}\right)^{k+1} = \frac{\alpha^{k+1}-1}{(\alpha-1)^{k+1}}.$$

Therefore

$$a_{k+1} = \frac{1}{1-\alpha^{n-k-1}}\binom{n}{k+1}\beta^{k+1}\frac{\alpha^n - \alpha^{n-k-1}}{(\alpha-1)^{k+1}}.$$

Since $\alpha^n = 1$, we have

$$a_{k+1} = \binom{n}{k+1}\frac{\beta^{k+1}}{(\alpha-1)^{k+1}}. \quad \square$$

4.4 Transformations of polynomials

4.4.1 Tchirnhaus's transformation

In 1683, in the Leipzig journal "Acta eruditorum", E. V. von Tchirnhaus (1651–1708) published a method for the transformation of algebraic equations which, he thought, allowed the solution by radicals of algebraic equations of any degree. Leibniz immediately refuted Tchirnhaus's claim on the omnipotence of his transformations. Moreover, it turned out that to solve 5th degree equations by means of Tchirnhaus's transformations one had to solve an equation of degree 24 and rather complicated at that.

Nevertheless, Tchirnhaus's transformation has important applications. For example, with the help of it any equation of degree 5 without multiple roots can be reduced to the form $y^5 + 5y = a$ solving in the process only equations of degree 2 and 3.

Tchirnhaus's transformation of the equation

$$x^n + c_1 x^{n-1} + \cdots + c_n = 0 \tag{$*$}$$

consists of the following. Let $x_1, \ldots, x_n$ be the roots of this equation. Consider a rational function φ that does not tend to infinity at points $x_1, \ldots, x_n$. Set $y_i = \varphi(x_i)$ and consider the equation

$$R(y) = 0, \text{ where } R(y) = y^n + q_1 y^{n-1} + \cdots + q_n, \tag{$**$}$$

whose roots are $y_1, \ldots, y_n$. We show in what follows that if $(*)$ has no multiple roots, then the x_i can be expressed in terms of the y_i. Selecting an appropriate function φ we can make the coefficients $q_1, \ldots, q_{n-1}$ vanish. Unfortunately, to do this one has to solve an equation of order $(n-1)!$ and this was precisely the circumstance that Leibniz pointed to.

Without loss of generality we can take φ to be a polynomial of degree not higher than $n-1$ due to the following s tatement.

Theorem 4.4.1. *Let* $x_1, \ldots, x_n$ *be the roots of a polynomial* f *of degree* n *and let* $\varphi = \dfrac{P}{Q}$, *where* P *and* Q *are polynomials such that* $Q(x_i) \neq 0$ *for all* $i = 1, \ldots, n$. *Then there exists a polynomial* g *of degree no higher than* $n-1$ *whose values at* $x_1, \ldots, x_n$ *coincide with the values of* φ *at these points.*

Proof. By hypothesis the polynomials f and Q have no common roots, so they are relatively prime. Hence there exist polynomials R and S such that $Rf + SQ = 1$. Since $f(x_i) = 0$, it follows that $S(x_i) = \dfrac{1}{Q(x_i)}$. Hence

$$\varphi(x_i) = \frac{P(x_i)}{Q(x_i)} = P(x_i)\, S(x_i).$$

Thus, for the required polynomial g we may take the remainder after division of PS by f. $\square$

In what follows we assume that to equation $(*)$ .we apply the transformation

$$y = g(x) = p_0 + p_1(x) + \cdots + p_{n-1}x^{n-1}.$$

Let us show in this case how we can calculate the coefficients of the polynomial $R(y)$, see $(**)$, with given roots $y_i = g(x_i)$, $i = 1, \ldots, n$.

To avoid cumbersome notations, we confine ourselves to the case $n = 3$. If $x^3 = -c_1 x^2 - c_2 x - c_3$, then

$$yx = p_0 x + p_1 x^2 + p_2(-c_1 x^2 - c_2 x - c_3) = p_0' + p_1' x + p_2' x^2.$$

Similarly, $yx^2 = p_0'' + p_1'' x + p_2'' x^2$, where the p_i'' are linear functions in the p_i. Therefore, if x_i is a root of f and $y_i = g(x_i)$, the system of equations

$$\begin{cases} (p_0 - y)z_0 + p_1 z_1 + p_2 z & = 0, \\ p_0' z_0 + (p_1' - y)z_1 + p_2' z & = 0, \\ p_0'' z_0 + p_1'' z_1 + (p_2'' - y)z & = 0 \end{cases} \tag{1}$$

has a nonzero solution $(z_0, z_1, z_2) = (1, x_i, x_i^2)$. Set

$$A = \begin{pmatrix} p_0 & p_1 & p_2 \\ p_0' & p_1' & p_2' \\ p_0'' & p_1'' & p_2'' \end{pmatrix}.$$

Then $\det(A - yI) = 0$ for $y_i = g(x_i)$. If the polynomial $\det(A - yI)$ has no multiple roots, it coincides with the polynomial $R(y)$ to be found. Since the elements of A linearly depend on the p_i, the coefficient q_k is a polynomial of degree k in the p_i.

If the polynomial $R(y)$ has no multiple roots, the matrix A has no multiple eigenvalues. Therefore to every eigenvalue of A corresponds a unique, up to a factor, solution of (1). This means that the root y_i of the initial polynomial is uniquely recovered from the root x_i of the transformed polynomial, and each x_i depends rationally on y_i.

Tchirnhaus's transformation helps also to solve 3rd and 4th degree equations by radicals. The cubic equation can be reduced to the form

$$y^3 + q_3 = 0,$$

solving a system consisting of a linear equation $q_1 = 0$ and a second degree equation $q_2 = 0$ depending on parameters p_0, p_1, p_2. For this, one needs to solve a quadratic equation.

An arbitrary 4th degree equation can be reduced to the form

$$y^4 + q_2 y^2 + q_4 = 0.$$

For this, one has to solve a system consisting of linear equation $q_1 = 0$ and a 3rd degree equation $q_3 = 0$ which reduces to solution of a cubic equation.

4.4.2 5th degree equation in Bring's form

Any 5th degree equation can be reduced to the form

$$y^5 + q_4 y + q_5 = 0,$$

solving a system of equations $q_1 = q_2 = q_3 = 0$. For this, one has to solve a 6th degree equation. A sharper analysis performed in 1789 by the Swedish mathematician Bring, shows that in this case, instead of a 6th degree Equation, it suffices, actually, to solve equations of degree 2 and 3. To satisfy the condition $q_1 = 0$, we express one of the parameters $p_0, \dots, p_4$ as a linear function of the remaining parameters. Then the coefficient q_2 represents a quadratic form in four of the parameters p_i. This quadratic form can be reduced to the basic form

$$u_1^2 + u_2^2 - v_1^2 - v_2^2,$$

where u_j and v_j are linear functions in the p_i (for this, we have to take square roots). To satisfy the equation $q_2 = 0$, it suffices to solve the system of linear equations $u_1 = v_1$, $u_2 = v_2$. After that there remain two parameters for which the equation $q_3 = 0$ is of a 3rd degree. As a result, we get an equation of the form

$$y^5 + q_4 y + q_5 = 0.$$

If $q_4 \neq 0$, a linear substitution reduces this equation to the form $y^5 + 5y = a$.

Similarly, the equation

$$x^n + c_1 x^{n-1} + \cdots + c_n = 0, \quad n \geq 5, \tag{1}$$

can be reduced to the form

$$y^n + q_4 y^{n-4} + q_5 y^{n-5} + \cdots + q_n = 0$$

by means of the transformation

$$y = p_0 + p_1 x + p_2 x^2 + p_3 x^3 + p_4 x^4.$$

In the process we only have to solve equations of degree 2 and 3.

Moreover, instead of a system $q_1 = q_2 = q_3 = 0$ we can solve the system $q_1 = q_2 = q_4 = 0$, i.e., at the last step, instead of the cubic equation $q_3 = 0$, we solve the 4th degree equation $q_4 = 0$. Then (1) will be reduced to the form

$$y^n + q_3 y^{n-3} + q_5 y^{n-5} + \cdots + q_n = 0.$$

Making use of these transformations and the change of variable $x \mapsto x^{-1}$ we can reduce the general 5th degree equation to any of the following forms

$$x^5 + px + q = 0,$$
$$x^5 + px^2 + q = 0,$$
$$x^5 + px^3 + q = 0,$$
$$x^5 + px^4 + q = 0.$$

4.4.3 Representation of polynomials as sums of powers of linear functions

The problem of representing polynomials as sums of powers of linear functions is simplest for the quadratic $x^2 + 2ax + b$. This problem is as follows:

Represent the given quadratic in the form

$$\lambda_1 (x + \alpha_1)^2 + \cdots + \lambda_m (x + \alpha_m)^2$$

and investigate what is the minimal number of basic linear functions $x + \alpha_1, \ldots, x + \alpha_m$ necessary to perform this.

There are two versions of this problem:

1) The basic functions are the same for all quadratics.

2) The basic functions depend on the given quadratic.

In the first case the minimal m is 3. For the basic functions we can take any three distinct functions $x + \alpha_1$, $x + \alpha_2$, $x + \alpha_3$. Indeed, the system of equations for $\lambda_1, \lambda_2, \lambda_3$ is

$$\lambda_1 + \lambda_2 + \lambda_3 = 1,$$

$$\alpha_1 \lambda_1 + \alpha_2 \lambda_2 + \alpha_3 \lambda_3 = a,$$

$$\alpha_1^2 \lambda_1 + \alpha_2^2 \lambda_2 + \alpha_3^2 \lambda_3 = b.$$

The system always has a solution since its determinant is a non-vanishing Vandermonde determinant. It is also clear that two functions $x + \alpha_1$ and $x + \alpha_2$ are insufficient: $\lambda_3 = 0$ only if a, b, α_1 and α_2 are constrained by a relation $a(\alpha_1 + \alpha_2) = b + \alpha_1 \alpha_2$.

In the second case, the minimal m is 2. The required representation is, e.g., of the form

$$x^2 + 2ax + b = \frac{1}{2}\left(x + a + \sqrt{b - a^2}\right)^2 + \frac{1}{2}\left(x + a - \sqrt{b - a^2}\right)^2.$$

For polynomials of degree n, the problem of selecting the universal basic functions $x + \alpha_1, \dots, x + \alpha_m$ is solved exactly as for $n = 2$. Let us formulate the answer as a theorem (the proof for $n > 2$ is the same as for $n = 2$).

Theorem 4.4.2. a) *If* $\alpha_1, \dots, \alpha_{n+1}$ *are distinct numbers, then any polynomial of degree* n *can be represented in the form*

$$\lambda_1 (x + \alpha_1)^n + \dots + \lambda_{n+1}(x + \alpha_{n+1})^n.$$

b) *If the numbers* $\alpha_1, \dots, \alpha_m$ *are such that any polynomial of degree* n *can be represented in the form*

$$\lambda_1 (x + \alpha_1)^n + \dots + \lambda_m (x + \alpha_m)^n,$$

then $m \geq n + 1$.

It is not difficult to indicate a collection of universal basic linear forms for polynomials of degree n in m variables as well. For convenience, instead of a polynomial $f(x_1, \dots, x_m)$ of degree n, we will consider the homogeneous polynomial

$$F(x_0, x_1, \dots, x_m) = x_0^n f\left(\frac{x_1}{x_0}, \dots, \frac{x_m}{x_0}\right).$$

Theorem 4.4.3 ([So]). *Let* $\alpha_0, \dots, \alpha_n$ *be distinct numbers, and let*

$$Z_j = x_0 + \alpha_s x_1 + \alpha_t x_2 + \dots + \alpha_u x_m, \text{ where } \alpha_s, \alpha_t, \dots, \alpha_u \in \{\alpha_0, \dots, \alpha_n\}.$$

Then the forms $(Z_j)^n$ *generate the linear space of all homogeneous polynomials of degree* n *in* $m + 1$ *variables.*

Remark. The number of forms $(Z_j)^n$ is equal to $(n+1)^m$ while the dimension of the space of homogeneous polynomials of degree n in $m + 1$ variables is equal to $\binom{m+n}{n}$.

Proof. For simplicity, we consider the case $m = 2$. In this case we have to represent the polynomial

$$p(x, y, z) = \sum_{0 \leq i+j \leq n} a_{ij} x^{n-i-j} y^i z^j$$

in the form

$$p(x, y, z) = \sum_{s,t=1}^{n+1} \lambda_{st}(x + \alpha_s y + \alpha_t z)^n = \sum_{s,t=1}^{n+1} \lambda_{st} \sum_{0 \leq i+j \leq n} c_{nij} \alpha_s^i \alpha_t^j x^{n-i-j} y^i z^j,$$

where $c_{nij} = \dfrac{n!}{i!j!(n-i-j)!}$. We obtain the system of equations

$$a_{ij} = c_{nij} \sum_{s,t=1}^{n+1} \lambda_{st} \alpha_s^i \alpha_t^j, \quad 0 \leq i+j \leq n.$$

Let us complement this system with the equations

$$\sum_{s,t=1}^{n+1} \lambda_{st} \alpha_s^i \alpha_t^j = 0 \quad \text{for} \quad i+j > n, \quad 1 \leq i, j \leq n.$$

We then obtain a system of linear equations with the matrix $V \otimes V$, where $V = \|\alpha_i^j\|_0^n$ is a Vandermonde matrix. For arbitrary square matrices A and B, of sizes $a \times a$ and $b \times b$, respectively, it is easy to prove that

$$\det(A \otimes B) = (\det A)^b (\det B)^a,$$

using their Jordan normal form. Hence $\det(V \otimes V) = (\det V)^{(n+1)^2} \neq 0$.

For arbitrary m, we similarly obtain a system of linear equations whose determinant is $\det(V \otimes \cdots \otimes V) = (\det V)^{(n+1)^m}$. $\square$

If for each polynomial we select its own basic linear functions, the problem becomes much more difficult. Adding the summand $b_k(x + \beta_k)^n$ increases the number of variable parameters by 2. The coincidence of the number of parameters on which the polynomial of degree n depends with the number of parameters in the expression

$$b_1(x + \beta_1)^n + \cdots + b_k(x + \beta_k)^n$$

is only possible when n is odd. Then $k = \frac{1}{2}(n+1)$. It turns out that indeed, a generic polynomial of odd degree n can be represented as the sum of $\frac{1}{2}(n+1)$ summands of the form $b(x+\beta)^n$. But in certain degenerate cases several extra summands might become necessary.

Example. The polynomial $x^3 + x^2$ cannot be represented in the form $b_1(x + \beta_1)^3 + b_2(x + \beta_2)^3$.

Proof. Clearly, $\beta_1 \neq \beta_2$ and $\beta_1 \beta_2 \neq 0$. In this case, the conditions $b_1 \beta_1^2 + b_2 \beta_2^2 = 0$ and $b_1 \beta_1^3 + b_2 \beta_2^3 = 0$ imply that $b_1 = b_2 = 0$: a contradiction. $\square$

Let us clarify, therefore, what the term "generic polynomial" does mean in the situation considered. We keep to the case $n = 5$ (for other odd n the arguments are similar). Let us express the polynomial $f(x)$ of degree 5 in the form

$$a_5 x^5 + 5a_4 x^4 + 10a_3 x^3 + 10a_2 x^2 + 5a_1 x + a_0.$$

Let

$$\begin{vmatrix} 1 & z & z^3 & z^3 \\ a_0 & a_1 & a_2 & a_3 \\ a_1 & a_2 & a_3 & a_4 \\ a_2 & a_3 & a_4 & a_5 \end{vmatrix} = p_3 z^3 + p_2 z^2 + p_1 z + p_0 = p(z).$$

We say that $f(x)$ is a *generic polynomial* if $p(z)$ is a polynomial with precisely three distinct roots (in particular, $p_3 \neq 0$).

Theorem 4.4.4 (Sylvester). *The generic polynomial of odd degree n can be represented as the sum*

$$b_1 (x + \beta_1)^n + \cdots + b_k (x + \beta_k)^n,$$

where $k = \dfrac{n+1}{2}$.

Proof. For $n = 5$, we have to solve the system of equations

$$b_1 \beta_1^r + b_1 \beta_2^r + b_1 \beta_3^r = a_r, \quad r = 0, 1, \dots, 5. \tag{1}$$

For β_1, β_2 and β_3 we take the roots of the equation $p(z) = 0$. By the hypothesis these roots are distinct. Hence, from the system (1) for $r = 0, 1, 2$ we uniquely find b_1, b_2 and b_3. It remains to prove that, for the values b_i and β_i obtained, equations (1) are satisfied for $r = 3, 4, 5$.

From the definition of $p(z)$ it follows that

$$\begin{vmatrix} x_0 & x_1 & x_2 & x_3 \\ a_0 & a_1 & a_2 & a_3 \\ a_1 & a_2 & a_3 & a_4 \\ a_2 & a_3 & a_4 & a_5 \end{vmatrix} = p_3 x_3 + p_2 x_2 + p_1 x_1 + p_0 x_0$$

for any numbers x_0, x_1, x_2, x_3. Further, if we replace the row (x_0, x_1, x_2, x_3) by $(a_i, a_{i+1}, a_{i+2}, a_{i+3})$, where $i = 0, 1, 2$, the determinant vanishes. For $i = 0$, we then have

$$p_3 a_3 + p_2 a_2 + p_1 a_1 + p_0 a_0 = 0,$$

i.e.,

$$-p_3 a_3 = p_0 \sum b_i + p_1 \sum b_i \beta_i + p_2 \sum b_i \beta_i^2 =$$
$$= \sum b_i (p + 0 + p_1 \beta_i + p_2 \beta_i^2) = -\sum b_i p_3 \beta_i^3.$$

Since $p_3 \neq 0$, we obtain (1) for $r = 3$. Taking $i = 1$ and 2 we similarly obtain (1) for $r = 4$ and 5. $\square$

4.5 Algebraic numbers

4.5.1 Definition and main properties of algebraic numbers

The number $\alpha \in \mathbb{C}$ is called *algebraic* if it is a root of an irreducible polynomial with rational coefficients. If the highest coefficient of the polynomial is equal to 1 and the remaining coefficients are integers, then the number α is said to be an *algebraic integer*. To every algebraic number α, there corresponds a unique irreducible monic polynomial f. The roots of this polynomial are called numbers *conjugate* to α.

It is not difficult to show that if α is a root of an arbitrary (i.e., not necessarily irreducible) monic polynomial with integer coefficients, then α is an algebraic integer. In other words,

if a monic polynomial with integer coefficients is represented as the product of two monic polynomials with rational coefficients, then all these rational coefficients are integers.

This statement is one of the possible formulations of Gauss's lemma (Lemma 2.1.1 on page 49).

Theorem 4.5.1. *Let α and β be algebraic numbers, and let $\varphi(x,y)$ be an arbitrary polynomial with rational coefficients. Then $\varphi(\alpha, \beta)$ is an algebraic number.*

Proof. Let $\{\alpha_1, \ldots, \alpha_n\}$ and $\{\beta_1, \ldots, \beta_m\}$ be the sets of numbers conjugate to α and β, respectively. Consider the polynomial

$$F(t) = \prod_{i=1}^{n} \prod_{j=1}^{m} (t - \varphi(\alpha_i, \beta_j)).$$

The coefficients of this polynomial are symmetric functions in $\alpha_1, \ldots, \alpha_n$ and $\beta_1, \ldots, \beta_m$. Hence they are rational numbers. $\square$

Remark. One can similarly prove that, if α and β are algebraic integers and $\varphi(x, y)$ is a polynomial with integer coefficients, then $\varphi(\alpha, \beta)$ is an algebraic integer.

In particular, if α and β are algebraic numbers (integers), then so are $\alpha\beta$ and $\alpha \pm \beta$. Further, if $\alpha \neq 0$ is an algebraic number, then α^{-1} is also an algebraic number. Indeed, if α is a root of the polynomial $\sum a_k x^k$ of degree n, then α^{-1} is a root of the polynomial $\sum a_k x^{n-k}$. But if α is an algebraic *integer*, α^{-1} is not necessarily an algebraic integer.

Therefore the algebraic numbers constitute a field and the algebraic integers constitute a ring.

Theorem 4.5.2. *Let α and β be algebraic numbers constrained by the relation $\varphi(\alpha, \beta) = 0$, where φ is a polynomial with rational coefficients. Then, for any number α_i conjugate to α, there exists a number β_j conjugate to β and such that $\varphi(\alpha_i, \beta_j) = 0$.*

Proof. Let $\{\alpha_1, \ldots, \alpha_n\}$ and $\{\beta_1, \ldots, \beta_m\}$ be the sets of all numbers conjugate to α and β, respectively. Consider the polynomial

$$f(x) = \prod_{j=1}^{m} \varphi(x, \beta_j).$$

The coefficients of this polynomial are rational and $f(\alpha) = 0$. Hence $f(x)$ is divisible by $\prod(x - \alpha_i)$, and therefore $f(\alpha_i) = 0$, i.e., $\varphi(\alpha_i, \beta_j) = 0$ for some j. $\square$

Theorem 4.5.3. *Let α be a root of the polynomial*

$$f(x) = x^n + \beta_{n-1} x^{n-1} + \cdots + \beta_0,$$

where $\beta_0, \ldots, \beta_{n-1}$ are algebraic integers. Then α is an algebraic integer.

Proof. Consider the polynomial

$$F(x) = \prod_{i,\ldots,l} (x^n + \beta_{n-1,i} x^{n-1} + \cdots + \beta_{0,l}),$$

where $\{\beta_{n-1,i}\}, \ldots, \{\beta_{0,l}\}$ are the sets of all the numbers conjugate to β_{n-1}, $\ldots, \beta_0$, respectively. It is easy to verify that the coefficients of F are integers. It is also clear that α is a root of F. $\square$

The algebraic number α is called *totally real* if all its conjugates are real, in other words, if all the roots of the irreducible polynomial with root α are real.

Example. The number $\alpha = 2 \cos\left(\dfrac{k\pi}{n}\right)$ is totally real.

Indeed, $\alpha = \varepsilon + \varepsilon^{-1}$, where $\varepsilon = \exp\left(\dfrac{k\pi}{n}\right)$. Let α_1 be conjugate to α. Theorem 4.5.2 implies that $\alpha_1 = \varepsilon_1 + \varepsilon_1^{-1}$, where ε_1 is conjugate to ε. The number ε satisfies the equation $x^n - 1 = 0$, and hence ε_1 is also its root, and therefore $\varepsilon_1 = \exp\left(\dfrac{l\pi}{n}\right)$. In this case $\alpha_1 = 2 \cos\left(\dfrac{lk\pi}{n}\right) \in \mathbb{R}$.

4.5.2 Kronecker's theorem

In 1857, Kronecker [Kr1] proved the following statement.

Theorem 4.5.4 (Kronecker). a) *Let $\alpha \neq 0$ be an algebraic integer. If α is not a root of unity, then at least one number conjugate to α has absolute value strictly greater than 1.*

b) *Let β be a totally real algebraic integer. If $\beta \neq 2\cos r\pi$, where $r \in \mathbb{Q}$, then at least one number conjugate to β has absolute value strictly greater than 2.*

Proof. a) Let $\{\alpha_1, \ldots, \alpha_n\}$ be the set of all numbers conjugate to α. Suppose on the contrary that $|\alpha_i| \leq 1$, $i = 1, \ldots, n$, and consider the polynomial

$$f_k(x) = (x - \alpha_1^k) \cdot \ldots \cdot (x - \alpha_n^k) = x^n + a_{k,n-1}x^{n-1} + \cdots + a_{k,0}.$$

Since α is an algebraic integer, it follows that $a_{k,n-1}, \ldots, a_{k,0} \in \mathbb{Z}$. The conditions $|\alpha_i| \leq 1$ for $i = 1, \ldots, n$ imply that $|a_{k,s}| \leq \binom{n}{s}$. Therefore the coefficients of the polynomials $f_1, f_2, \ldots$ assume only finitely many values, and hence, among these polynomials, there are only finitely many distinct ones. But then the set of roots of these polynomials is also finite, and all the numbers $\alpha, \alpha^2, \alpha^3, \ldots$ are in this set. Therefore

$$\alpha^p = \alpha^q \text{ for some } p, q \in \mathbb{N} \text{ and } p \neq q.$$

Since $\alpha \neq 0$, it follows that $\alpha^{p-q} = 1$.

b) Let $\{\beta_1, \ldots, \beta_n\}$ be the set of all numbers conjugate to β. By hypothesis all of them are real. Suppose on the contrary that $|\beta_i| \leq 2$, $i = 1, \ldots, n$. Then the absolute values of all the numbers conjugate to

$$\alpha = \frac{\beta}{2} + \sqrt{\frac{\beta^2}{4} - 1}$$

are equal to 1. Indeed, the numbers α and β satisfy $\alpha^2 - \beta\alpha + 1 = 0$. Hence, by Theorem 4.5.2, any number α_j conjugate to α is a root of a polynomial of the form $\alpha_j^2 - \beta_i\alpha_j + 1 = 0$. Since $|\beta_i| \leq 2$, we have $\frac{\beta_i^2}{4} \leq 1$, and therefore

$$|\alpha_j|^2 = \left(\frac{\beta_i^2}{2}\right) + 1 - \left(\frac{\beta_i^2}{2}\right) = 1.$$

By Theorem 4.5.3 the number α is an algebraic integer. Therefore we can apply part a) to it. As a result, we see that $\alpha = e^{r\pi i}$, where $r \in \mathbb{Q}$. Therefore

$$\beta = \alpha + \alpha^{-1} = \alpha + \overline{\alpha} = 2\cos r\pi,$$

as was required. $\square$

Let us give an interesting application of Kronecker's theorem.

Theorem 4.5.5 (Minkowski). *Let A be a square matrix with integer elements. Suppose that all the elements of the matrix $A - I$, where I is the unit matrix, are divisible by an integer $n \geq 2$ but $A \neq I$. Then*

a) *If $n > 2$, then $A^m \neq I$ for all positive integers m;*

b) *If $n = 2$ and $A^2 \neq I$, then $A^m \neq I$ for all positive integers m.*

Proof. By hypothesis $A = I + nB$, where B is a matrix with integer elements. In particular, all the eigenvalues of B are algebraic integers. The eigenvalues α of A and the eigenvalues β of B are related by the equation $\alpha = 1 + n\beta$.

Suppose that $A^m = I$. Then $\alpha^m = 1$, and hence $|\alpha| = 1$. Therefore

$$|\beta| = \frac{|\alpha - 1|}{n} \leq \frac{2}{n} \leq 1. \tag{1}$$

Inequality (1) is strict except for the case when $n = 2$ and $|\alpha - 1| = 2$, i.e., $\alpha = -1$.

For $n > 2$, inequality (1) is strict. In this case the absolute value of the algebraic integer β and of all its conjugates are strictly less than 1. Hence $\beta = 0$, and therefore $\alpha = 1$.

The identity $A^m = I$ can only hold if all the Jordan blocks of A are of size 1×1. If all the eigenvalues of A in this case are equal to 1, then $A = I$.

Now consider the case $n = 2$. In this case $\alpha = \pm 1$. Therefore the Jordan form of A is a diagonal matrix with elements ± 1 on the main diagonal. Hence $A^2 = I$ which contradicts our hypothesis. $\square$

Elementary but rather cumbersome estimates enable us to sharpen Kronecker's theorem as follows.

Theorem 4.5.6 ([ScZ]). a) *Let $\alpha \neq 0$ be an algebraic integer which is not a root of unity, and let $\{\alpha_1, \ldots, \alpha_n\}$ be the set of all the conjugates of α. If $2s$ of the numbers $\alpha_1, \ldots, \alpha_n$ are real, then*

$$\max_{1 \leq i \leq n} |\alpha_i| > 1 + 4^{-s-2}.$$

b) *Let β be a totally real algebraic integer such that $\beta \neq 2\cos r\pi$, $r \in \mathbb{Q}$; let $\{\beta_1, \ldots, \beta_n\}$ be the set of all the conjugates of β. Then*

$$\max_{1 \leq i \leq n} |\beta_i| > 2 + 4^{-2n-3}.$$

4.5.3 Liouville's theorem

Euler conjectured that not all numbers are algebraic but he could not prove this. The first to prove the existence of *transcendental* (i.e., not algebraic) numbers was Liouville in 1844. In 1874, Cantor showed that in a sense there are more transcendental numbers than there are algebraic ones, in the sence that the set of algebraic numbers is countable whereas the set of all real (or complex) numbers is uncountable.

Liouville's proof is based on a relatively simple but important remark: every irrational algebraic number does not have too good an approximation by rationals. More precisely, the following statement holds.

Theorem 4.5.7 (Liouville). *Let α be a root of an irreducible polynomial $f(x) = a_n x^n + a_{n-1} x^{n-1} + \cdots + a_0$, where $n \geq 2$. Then there exists a number $c > 0$ (depending only on α) such that*

$$\left| \alpha - \frac{p}{q} \right| > \frac{c}{q^n} \tag{1}$$

for any integer p and any positive integer q.

Proof. If $\left| \alpha - \frac{p}{q} \right| \geq 1$, then (1) holds for $c = 1$. Hence, we assume that $\left| \alpha - \frac{p}{q} \right| < 1$. Let us express $f(x)$ in the form $f(x) = a_n \prod_{i=1}^{n} (x - \alpha_i)$, where $\alpha_1 = \alpha$. Then

$$\left| f\left(\frac{p}{q}\right) \right| = |a_n| \cdot \left| \alpha - \frac{p}{q} \right| \cdot \prod_{i=2}^{n} \left| \frac{p}{q} - \alpha_i \right| \leq$$

$$\leq |a_n| \cdot \left| \alpha - \frac{p}{q} \right| \cdot \prod_{i=2}^{n} (|\alpha| + 1 + |\alpha_i|) = c_1 \left| \alpha - \frac{p}{q} \right|,$$

where c_1 is a positive number that depends only on $|a_n|$ and α.

Let us assume that $a_0, \ldots, a_n$ are integers relatively prime to each other. Then the number $|a_n|$ is completely determined by α. Moreover, the number

$$q^n f\left(\frac{p}{q}\right) = a_n p^n + a_{n-1} p^{n-1} + \cdots + a_0 q^n$$

is an integer, and hence $\left| q^n f\left(\frac{p}{q}\right) \right| \geq 1$. Therefore

$$\left| \alpha - \frac{p}{q} \right| \geq \frac{1}{c_1} \left| f\left(\frac{p}{q}\right) \right| \geq \frac{1}{c_1 q^n} = \frac{c}{q^n},$$

where $c = c_1^{-1}$. $\square$

Theorem 4.5.8 (Liouville). *The number $\alpha = \sum_{k=0}^{\infty} 2^{-k!}$ is transcendental.*

Proof. For any integer N, consider the number $\alpha = \sum_{k=0}^{N} 2^{-k!} = \frac{p}{q}$, where p is an integer and $q = 2^{N!}$. We have

$$\left| \alpha - \frac{p}{q} \right| = \frac{1}{2^{(N+1)!}} \left(1 + \frac{1}{2^{N+2}} + \frac{1}{2^{(N+2)(N+3)}} + \cdots \right) < \frac{2}{2^{(N+1)!}} = \frac{2}{q^{N+1}}.$$

Suppose that α is an algebraic number of degree n, i.e., a root of a polynomial of degree n with rational coefficients. Then by Theorem 4.5.7

$$\left| \alpha - \frac{p}{q} \right| \geq \frac{c}{q^n},$$

and therefore $2q^{-N-1} > cq^{-n}$, i.e., $c < 2q^{n-N-1} = 2 \cdot 2^{N!(n-N-1)}$. But $\lim_{N \to \infty} 2^{N!(n-N-1)} = 0$ which is clearly a contradiction when N is large enough. $\square$

Inequality (1) can be expressed as

$$|q\alpha - p| > \frac{c}{q^{N-1}}.$$

Set $P(x) = qx - p$. Then

$$|P(\alpha)| > \frac{c}{H^{n-1}}, \tag{2}$$

where $H = \max\{|p|, |q|\}$ is the height of P.

An inequality similar to (2) holds also for polynomials P of arbitrary degree.

Theorem 4.5.9. *Let α be an algebraic number of degree n. Then there exists a number $c > 0$ (depending only on α) such that, for any polynomial P of degree k with integer coefficients, either $P(\alpha) = 0$ or*

$$|P(\alpha)| > \frac{c^k}{H^{n-1}},$$

where H is the height of P (i.e., the greatest of the absolute values of the coefficients of P).

Proof. Let $P(x) = a_k x^k + \cdots + a_1 x + a_0$, where $a_i \in \mathbb{Z}$, and $P(\alpha) \neq 0$. For a positive integer r, the number $\beta = r\alpha$ is an algebraic integer. Define the polynomial Q by the relation

$$Q(rx) = r^k P(x).$$

Clearly

$$Q(y) = r^k P\left(\frac{y}{r}\right) = a_k y^k + r a_{k-1} y^{k-1} + \cdots + r^k a_0$$

is a polynomial with integer coefficients. Therefore the product $Q(\beta_1) \cdot \ldots \cdot Q(\beta_n)$, where $\beta_1, \ldots, \beta_n$ are all the numbers conjugate to β, is a nonzero integer. Hence $|Q(\beta)| \cdot |Q(\beta_2) \cdot \ldots \cdot Q(\beta_n)| \geq 1$, i.e.,

$$r^{kn} |P(\alpha)| \cdot |P(\alpha_2) \cdot \ldots \cdot P(\alpha_n)| \geq 1. \tag{3}$$

On the other hand,

$$|P(\alpha_i)| \le H \left(1 + |\alpha_i| + \cdots + |\alpha_i|^k\right) \le H \left(1 + |\alpha_i|\right)^k. \tag{4}$$

Let $h(\alpha) = \max \{|\alpha_1|, \ldots, |\alpha_n|\}$. Then (3) and (4) imply that

$$|P(\alpha)| \cdot H^{n-1} \left(r^n \left(1 + h(\alpha)\right)^{n-1}\right)^k \ge 1.$$

Choosing $c = \frac{(1+h(\alpha))^{1-n}}{r^n}$ we get the inequality desired. $\square$

For $n = 2$, Liouville's theorem cannot be improved in the sense that the inequality

$$\left|\alpha - \frac{p}{q}\right| < \frac{1}{q^2}$$

has infinitely many solutions. For $n > 3$, however, the estimate (1) was consecutively sharpened by Thue, Siegel, Dyson, Gelfond, Schneider, Roth and others. For example, Roth [Ro4] proved the following statement.

Theorem 4.5.10 (Roth). *Let α be an irrational algebraic number and let δ be a positive number however small. Then the inequality*

$$|\alpha - \frac{p}{q}| < \frac{1}{q^{2+\delta}}$$

holds only for finitely many pairs p and q, where q (> 0) and p are integers.

For the proof of Roth's theorem, see, e.g., the book [Ca4].

4.6 Problems to Chapter 4

4.1 Let $z_1, \ldots, z_n$ be the vertices of a regular n-gon, z_0 its center. Prove that, if P is a polynomial of degree no higher than $n-1$, then $\frac{1}{n} \sum_{k=1}^{n} P(z_k) = P(z_0)$.

4.2 Let $P(x, y)$ be a polynomial such that

$$P(x, y) = P(x+1, y+1).$$

Prove that $P(x, y) = \sum a_k (x - y)^k$.

4.3 Let $f(x)$ be a polynomial of degree n with only simple roots $x_1, \ldots, x_n$. Prove that

a) $\sum_{i=1}^{n} \dfrac{x_i^k}{f'(x_i)} = 0$ for $k = 0, 1, \ldots, n - 2$;

b) $\sum_{i=1}^{n} \dfrac{x_i^{n-1}}{f'(x_i)} = 1$.

4.4 Let $P(z)$ be a polynomial of degree n and $\max\limits_{|z|=1}|P(z)| \le 1$. Prove that, if $P(\alpha) = 0$, then

$$\max_{|z|=1}\left|\frac{P(z)}{z-\alpha}\right| \le \frac{n+1}{2} \quad \text{and} \quad \max_{|z|\le 1}\left|\frac{P(z)}{z-\alpha}\right| \le \frac{n}{1+|\alpha|}.$$

4.5 Let $d = x^2 + ax + b \in \mathbb{Z}[x]$.

a) Prove that the equation $p^2 - dq^2 = 1$ has non-trivial solutions $p, q \in \mathbb{Z}[x]$ in exactly the following cases:

1) a is odd and $4b = a^2 - 1$;

2) a is even and $b = \left(\dfrac{a}{2}\right)^2 \pm 1$ or $b = \left(\dfrac{a}{2}\right)^2 \pm 2$.

b) Prove that the equation $p^2 - dq^2 = -1$ has non-trivial solutions $p, q \in \mathbb{Z}[x]$ if and only if a is even and $b = \left(\dfrac{a}{2}\right)^2 + 1$.

4.6 Prove that there exists a unique, up to multiplication by -1, polynomial $f(x)$ of degree n for which the function $(x+1)(f(x))^2 - 1$ is odd.

4.7 Let $P_n(x)$ be a polynomial of degree n over $\mathbb{C}$. Prove that for $n = 4, 6$ and 8 almost all polynomials $P_n(x)$ can be represented in the following form:

$$P_4 = u^4 + v^4 + \lambda u^2 v^2;$$
$$P_6 = u^6 + v^6 + w^6 + \lambda uvw(u - v)(v - w)(w - u);$$
$$P_8 = u^8 + v^8 + w^8 + z^8 \lambda u^2 v^2 w^2 z^2,$$

where u, v, w, z are linear functions and λ is a number.

4.8 Let the numbers $\alpha_1, \ldots, \alpha_n \in \mathbb{C}$ be such that $\sum \alpha_i^m$ is an integer for any integer m. Prove that all the coefficients of the polynomial $\prod(x - \alpha_i)$ are integers.

5

Galois Theory

5.1 Lagrange's theorem and the Galois resolvent

5.1.1 Lagrange's theorem

Let K be a field of characteristic 0 and φ a rational function in variables $x_1, \ldots, x_n$ over K. Let S_n We denote the permutation group of n elements. We can assign to φ the stabilizer of φ, i.e., the group

$$G_\varphi = \{\sigma \in S_n \mid \varphi(x_{\sigma(1)}, \ldots, x_{\sigma(n)}) = \varphi(x_1, \ldots, x_n)\}.$$

For example, if φ is a symmetric function, then $G_\varphi = S_n$, whereas if $\varphi = \sum a_i x_i$, where the numbers $a_1, \ldots, a_n$ are distinct, then G_φ contains only the identity permutation.

Theorem 5.1.1 (Lagrange). *Let $\varphi, \psi \in K(x_1, \ldots, x_n)$ and $G_\varphi \subset G_\psi$. Then $\psi = R(\varphi)$, where R is a rational function whose coefficients are symmetric functions in $x_1, \ldots, x_n$.*

First proof. Let us split G_ψ into non-intersecting cosets $h_1 G_\varphi = G_\varphi$, $h_2 G_\varphi, \ldots, h_k G_\varphi$. To every coset $h_i G_\varphi$ there corresponds a function φ_i, the image of φ under the action of this coset; clearly, $\varphi_i \neq \varphi_j$ for $i \neq j$. The function ψ is G_φ-invariant, and hence, a function ψ_i uniquely corresponds to the coset $h_i G_\varphi$; if $G_\varphi \neq G_\psi$, then among these functions some will coincide.

The function $\sum_{i=1}^{k} \dfrac{\psi_i}{t - \varphi_i}$ is invariant with respect to the action of all permutations from S_n. Hence

$$\sum_{i=1}^{k} \frac{\psi_i}{t - \varphi_i} = \frac{F(t)}{\Omega(t)},$$

where

$$\Omega(t) = (t - \varphi_1) \cdot \ldots \cdot (t - \varphi_k)$$

and $F(t)$ is a polynomial in t whose coefficients are symmetric functions in $x_1, \ldots, x_n$. Since $\varphi_i \neq \varphi_j$ for $i \neq j$, it follows that $\Omega'(\varphi) \neq 0$.

Clearly,

$$\lim_{t \to \varphi_i} \frac{\Omega(t)}{t - \varphi_i} = \lim_{t \to \varphi_i} \frac{\Omega(t) - \Omega(\varphi_i)}{t - \varphi_i} = \Omega'(\varphi_i).$$

Therefore

$$\lim_{t \to \varphi_i} \frac{\Omega(t)}{\Omega'(t)(t - \varphi_i)} = \begin{cases} 0 & \text{for } \varphi_i \neq \varphi; \\ 1 & \text{for } \varphi_i = \varphi. \end{cases}$$

Hence

$$\frac{F(\varphi)}{\Omega'(\varphi)} = \sum_{i=1}^{k} \psi_i \frac{\Omega(\varphi)}{\Omega'(\varphi)(\varphi - \varphi_i)} = \psi.$$

$\square$

Second proof. Let us construct functions $\varphi_1 = \varphi, \ldots, \varphi_k$ and $\psi_1 = \psi, \ldots, \psi_k$ as in the first proof. Clearly,

$$\sum \varphi_i^s \psi_i = T_s \tag{1}$$

is a symmetric function in $x_1, \ldots, x_n$. We consider the equalities (1) for $s = 0, \ldots, k-1$ as a system of linear equations for $\psi_1, \ldots, \psi_k$. Solving this system we obtain $\psi_1 = \dfrac{D_1}{\Delta}$, where

$$\Delta = \begin{vmatrix} 1 & \cdots & 1 \\ \varphi_1 & \cdots & \varphi_k \\ \vdots & \ddots & \vdots \\ \varphi_1^{k-1} & \cdots & \varphi_k^{k-1} \end{vmatrix} \quad \text{and} \quad D_1 = \begin{vmatrix} T_0 & 1 & \cdots & 1 \\ T_1 & \varphi_2 & \cdots & \varphi_k \\ \vdots & \vdots & \ddots & \vdots \\ T_{k-1} & \varphi_2^{k-1} & \cdots & \varphi_k^{k-1} \end{vmatrix}.$$

Let us express the identity obtained in the form $\psi_1 = \dfrac{D_1 \Delta}{\Delta^2}$. Clearly, Δ^2 is a symmetric function. Under the transposition of any pair of functions $\varphi_2, \ldots, \varphi_k$, both determinants D_1 and Δ change sign, and so $D_1 \Delta$ is a symmetric function in $\varphi_2, \ldots, \varphi_k$. Hence

$$D_1 \Delta = S_0 + \varphi_1 S_1 + \cdots + \varphi_1^{k-1} S_{k-1},$$

where $S_0, \ldots, S_{k-1}$ are symmetric polynomials in $\varphi_2, \ldots, \varphi_k$. Therefore $S_0, \ldots, S_{k-1}$ are expressed in terms of

$$\tilde{\sigma}_1 = \varphi_2 + \cdots + \varphi_k,$$
$$\tilde{\sigma}_2 = \varphi_2 \varphi_3 + \cdots,$$
$$\cdots\cdots\cdots\cdots\cdots\cdots$$
$$\tilde{\sigma}_{k-1} = \varphi_2 \cdot \ldots \cdot \varphi_k.$$

But, as is easy to verify,

$$\tilde{\sigma}_1 = \sigma_1 - \varphi_1,$$
$$\tilde{\sigma}_2 = \sigma_2 - \varphi_1\sigma_1 + \varphi_1^2,$$
$$\tilde{\sigma}_3 = \sigma_3 - \varphi_1\sigma_2 + \varphi_1^2\sigma_1 - \varphi_1^3,$$

....

where σ_1, σ_2, ... are elementary symmetric functions in φ_1, ..., φ_k. Hence $\tilde{\sigma}_1$, ..., $\tilde{\sigma}_{k-1}$ are expressed in terms of σ_1, ..., σ_{k-1} and φ_1. Hence $D_1\Delta$ is a polynomial in $\varphi_1 = \varphi$ whose coefficients are symmetric functions in x_1, ..., x_n. $\square$

Lagrange's theorem has numerous corollaries. Let φ and ψ be rational functions in $x_1, \ldots, x_n$. If $\psi = R(\varphi)$, where R is a rational function whose coefficients are symmetric functions in $x_1, \ldots, x_n$, we will briefly say that ψ is *rationally expressed* in terms of φ.

Corollary 1 *Any rational function in* $x_1, \ldots, x_n$ *is rationally expressed in terms of* $a_1 x_1 + \cdots + a_n x_n$, *where* $a_1, \ldots, a_n$ *are distinct numbers.*

Corollary 2 *If* $G_\varphi = G_\psi$, *then the functions* φ *and* ψ *are rationally expressed in terms of each other.*

Corollary 3 *If the rational function* r *is invariant with respect to all the permutations that preserve the functions* $r_1, \ldots, r_n$, *then* r *is rationally expressed in terms of* $r_1, \ldots, r_n$.

Proof. We may assume that the functions $r_1, \ldots, r_n$ are linearly independent. Let $\varphi = a_1 r_1 + \cdots + a_n r_n$, where $a_1, \ldots, a_n$ are distinct numbers. Then any permutation that preserves φ should also preserve all the functions $r_1, \ldots, r_n$. Therefore r is preserved under all the permutations that preserve φ. Hence, r is rationally expressed in terms of $\varphi = a_1 r_1 + \cdots + a_n r_n$. $\square$

Corollary 4. *Let the polynomial* $f(x_1, \ldots, x_n)$ *take only two distinct values under all possible permutations of its variables. Then*

$$f = S_1 + \Delta S_2,$$

where S_1 *and* S_2 *are symmetric functions and* $\Delta = \prod_{i<j}(x_i - x_j)$.

Proof. If f is not a symmetric function, then the substitutions that preserve f form a subgroup of S_n of index 2. Therefore it suffices to prove that S_n has only one subgroup of index 2, namely, the alternating group A_n. Let $G \subset S_n$ be a subgroup of index 2 and $h \in S_n \setminus G$. Then S_n splits into the non-intersecting subsets G and $hG = Gh$. Therefore $hGh^{-1} = G$ and, if

$h_1, h_2 \in S_n \setminus G$, then $h_1 h_2 \in G$. If G had contained a transposition (ij), then it would have contained any other transposition (pq) as well. Indeed, (pq) is obtained from (ij) by conjugation with any permutation that sends i to p and j to q. Therefore all the transpositions lie in $S_n \setminus G$, and hence their products lie in G. Thus, all the products of any even number of transpositions lie in G, and therefore $G \supset A_n$. But $|G| = |A_n|$, and hence $G = A_n$. □

In Lagrange's theorem we deal with the rational functions in $x_1, \ldots, x_n$. The passage from algebraically independent variables $x_1, \ldots, x_n$ to the concrete values of these variables requires certain caution. The point is that the permutations which preserve the value of $\varphi(x_1, \ldots, x_n)$ for given $x_1, \ldots, x_n$ may not form a group. For example, let

$$x_k = \exp\left(\frac{2\pi i k}{7}\right) \text{ for } k = 1, \ldots, 6.$$

Consider the function $f(x_1, \ldots, x_6) = x_1 x_6$. Let

$$\sigma = (12)(56) \text{ and } \tau = (16)(23).$$

The permutations σ and τ send f into $\sigma f = x_2 x_5 = 1$ and $\tau f = x_6 x_1 = 1$ respectively, i.e., both σ and τ preserve f. But τ sends $\sigma f = x_2 x_5$ to $\tau \sigma f = x_3 x_5 \neq 1$.

To avoid such nuisances, Galois suggested considering not all the permutations of the roots of the equation but only the ones that preserve all the rational relations between them.

More precisely, let $f = x^n + a_{n-1} x^{n-1} + \cdots + a_0$ be a polynomial with coefficients from a field K and let $\alpha_1, \ldots, \alpha_n$ be the roots of f. Galois suggested considering those permutations σ that for any rational function $r \in K(x_1, \ldots, x_n)$ the identity $r(\alpha_1, \ldots, \alpha_n) = 0$ implies the identity $r(\alpha_{\sigma(1)}, \ldots, \alpha_{\sigma(n)}) = 0$.

In modern terms this means that the permutation σ corresponds to an automorphism of the field $K(\alpha_1, \ldots, \alpha_n)$ which preserves the ground field K. The group of all such permutations is called the *Galois group* of the polynomial f (this group depends of course on the field K).

In the example considered above, the permutations σ and τ do not enter the Galois group of the polynomial $x^6 + x^5 + \cdots + x + 1$ whose roots are $x_1, \ldots, x_6$. Indeed, the permutations σ and τ do not preserve, for example, the relations $x_2 = x_1^2$ and $x_6 = x_1^6$.

For any rational function in the roots $\alpha_1, \ldots, \alpha_n$ of the polynomial f, we can consider the elements of the Galois group which preserve its value. Clearly, these elements form a group. Indeed, let σ and τ be the elements of the Galois group of f and let $r(x_1, \ldots, x_n)$ be a rational function such that

$$r(\alpha_{\sigma(1)}, \ldots, \alpha_{\sigma(n)}) = r(\alpha_1, \ldots, \alpha_n), \tag{2}$$

$$r(\alpha_{\tau(1)}, \ldots, \alpha_{\tau(n)}) = r(\alpha_1, \ldots, \alpha_n). \tag{3}$$

We can apply σ and τ to any rational relation between the roots $\alpha_1, \ldots, \alpha_n$. Hence, applying τ to the relation (2) we obtain

$$r(\alpha_{\tau\sigma(1)}, \ldots, \alpha_{\tau\sigma(n)}) = r(\alpha_{\tau(1)}, \ldots, \alpha_{\tau(n)}).$$

Relation (3) implies that $\tau\sigma$ also preserves the value of r.

5.1.2 The Galois resolvent

In this section, let

$$f(x) = x^n + a_{n-1}x^{n-1} + \cdots + a_0$$

be a polynomial over a field K of characteristic 0 and let $\alpha_1, \ldots, \alpha_n$ be its roots. Suppose that f has no multiple roots, i.e., the numbers $\alpha_1, \ldots, \alpha_n$ are distinct. Consider a rational function

$$\psi(x_1, \ldots, x_n) = m_1 x_1 + \cdots + m_n x_n,$$

where $m_1, \ldots, m_n$ are integers. Let us show that the numbers $m_1, \ldots, m_n$ can be selected so that all the $n!$ values $\psi_\sigma = \psi(\alpha_{\sigma(1)}, \ldots, \alpha_{\sigma(n)})$ are distinct. Indeed, consider the function

$$D(t_1, \ldots, t_n) = \prod_{\sigma, \tau} \sum_{i=1}^{n} t_i(\alpha_{\sigma(i)} - \alpha_{\tau(i)}),$$

where the product runs over all unordered pairs of distinct permutations σ and τ. The function D is a product of nonzero polynomials in $t_1, \ldots, t_n$, and hence D is a nonzero polynomial in indeterminates $t_1, \ldots, t_n$ over K. But any nonzero polynomial function takes a nonzero value for certain integer values of its arguments $t_1 = m_1, \ldots, t_n = m_n$. These integers $m_1, \ldots, m_n$ are the required ones.

Before we advance further, we prove one auxiliary statement.

Lemma. *Any symmetric polynomial in the roots $\alpha_2, \ldots, \alpha_n$ of f is polynomially expressed in terms of the root α_1 and the coefficients $a_0, \ldots, a_{n-1}$.*

Proof. Any symmetric polynomial in the roots $\alpha_2, \ldots, \alpha_n$ is expressed in terms of the coefficients of the polynomial

$$(x - \alpha_2) \cdot \ldots \cdot (x - \alpha_n) = \frac{f(x)}{x - \alpha_1} = x^{n-1} + b_{n-2}x^{n-2} + \cdots + b_0.$$

Here we have

$$a_{n-1} = b_{n-2} - \alpha_1,$$
$$a_{n-2} = b_{n-3} - b_{n-2}\alpha_1,$$
$$a_{n-3} = b_{n-4} - b_{n-3}\alpha_1,$$
$$\ldots\ldots\ldots\ldots\ldots\ldots$$

i.e.,

$$b_{n-2} = a_{n-1} + \alpha_1,$$
$$b_{n-3} = a_{n-2} + \alpha_1 a_{n-1} + \alpha_1^2,$$
$$b_{n-4} = a_{n-3} + \alpha_1 a_{n-2} + \alpha_1^2 a_{n-1} + \alpha_1^3,$$
$$\dots\dots\dots\dots\dots\dots\dots\dots\dots\dots$$

Thus, the coefficients $b_0, \dots, b_{n-2}$ are polynomially expressed in terms of the root α_1 and the coefficients $a_0, \dots, a_{n-1}$. $\square$

Select the numbers $m_1, \dots, m_n$ so that all the $n!$ values

$$\psi_\sigma = m_1 \alpha_{\sigma(1)} + \cdots + m_n \alpha_{\sigma(n)}$$

are distinct and consider the polynomial

$$F(x) = \prod_{\sigma \in S_n} (x - m_1 \alpha_{\sigma(1)} - \cdots - m_n \alpha_{\sigma(n)}).$$

The coefficients of this polynomial are symmetric polynomials with integer coefficients in the roots of f, and hence they are rationally expressed in terms of the coefficients of f. Therefore, if f is a polynomial over K, then so is F.

Let us factorize F into a product of irreducible over K monic factors. Any such irreducible factor G is called a *Galois resolvent* of f. Clearly, all Galois resolvents are obtained from each other by permutations of the roots. Having numbered the roots we may fix *the* Galois resolvent, the one corresponding to the identity permutation. For definiteness sake, we will assume that G has a root

$$\psi = m_1 \alpha_1 + \cdots + m_n \alpha_n.$$

Theorem 5.1.2 (Galois). *Any root of f is rationally expressed (over K) in terms of one of the roots of G.*

Proof. Consider the polynomial

$$F_1(x) = \prod_{\{\sigma \in S_n | \sigma(1)=1\}} (x - m_1 \alpha_1 - m_2 \alpha_{\sigma(2)} - \cdots - m_n \alpha_{\sigma(n)}).$$

The coefficients of F_1 are symmetric polynomials in $\alpha_2, \dots, \alpha_n$; hence, by the Lemma, they are rationally expressed (over K) in terms of α_1, i.e., $F_1(x) = g(x, \alpha_1)$, where g is a polynomial in two variables over K. Clearly, $g(\psi, \alpha_1) = F_1(\psi) = 0$.

Now consider the polynomial

$$F_2(x) = \prod (x - m_1 \alpha_2 - m_2 \alpha_{\sigma(1)} - \cdots - m_n \alpha_{\sigma(n)}),$$

where the product runs over the permutations $\sigma \in S_n$ such that $\sigma(2) = 2$. From the proof of the Lemma we see that the coefficients of F_2 are the same as for F_1 up to replacement α_1 by α_2, i.e., $F_2(x) = g(x, \alpha_2)$.

By the hypothsis

$$\psi = m_1\alpha_1 + \cdots + m_n\alpha_n \neq m_1\alpha_2 + m_2\alpha_{\sigma(1)} + \cdots + m_n\alpha_{\sigma(n)},$$

i.e., $F_2(\psi) \neq 0$. Therefore α_1 is the only common root of the polynomials $f(x)$ and $g(\psi, x)$. This means that the greatest common divisor of $f(x)$ and $g(\psi, x)$ is $x - \alpha_1$. But the greatest common divisor of two polynomials is found by Euclid's algorithm, and so α_1 is rationally expressed in terms of ψ and the coefficients of f and g, i.e., α_1 is rationally expressed over K in terms of ψ. $\square$

Corollary. *All the roots of the Galois resolvent are rationally expressed in terms of one of its roots.*

Proof. Every root of the Galois resolvent is of the form

$$m_1\alpha_{\sigma(1)} + \cdots + m_n\alpha_{\sigma(n)}.$$

Clearly, they are rationally expressed in terms of $\alpha_1, \ldots, \alpha_n$. In turn, $\alpha_1, \ldots, \alpha_n$ are rationally expressed in terms of $\psi = m_1\alpha_1 + \cdots + m_n\alpha_n$. $\square$

The Galois resolvents are a convenient tool for constructing the splitting field of $K(\alpha_1, \ldots, \alpha_n)$. We recall the definition. Let f be a polynomial over a field k without multiple roots (but not necessarily irreducible), and let $\alpha_1, \ldots, \alpha_n$ be all the roots of f. The field $K = k(\alpha_1, \ldots, \alpha_n)$ is called the *splitting field* of f.

Indeed, $K(\alpha_1, \ldots, \alpha_n) = K(\psi)$, i.e., instead of adjoining to K all the roots of a polynomial, we can adjoin just one root of the Galois resolvent of the polynomial.

Another application of the Galois resolvent is that it can be used to construct the Galois group. (It was precisely with the help of the resolvents that Galois initially constructed the Galois groups of the polynomials.)

Let $\psi_1 (= \psi), \psi_2, \ldots, \psi_r$ be the roots of the Galois resolvent G. As we have shown, all of them can be rationally expressed in terms of ψ, i.e., $\psi_i = R_i(\psi)$, where $R_i \in K(x)$. The relation $\psi_i = R_i(\psi)$ can be considered as a relation between the elements of the field $K(\psi)$. Hence we may assume that R_i is a polynomial of degree not higher than $\deg G - 1 = r - 1$. This polynomial is uniquely determined. The formula $\psi_i = R_i(\psi)$ remains valid if we replace R_i by $R_i + aG$, where a is an arbitrary polynomial.

Consider polynomials $G(x)$ and $G_i(x) = G\big(R_i(x)\big)$. The coefficients of these polynomials lie in K and the polynomials have a common root ψ. Since by the hypothesis the polynomial G is irreducible, it follows that any root ψ_j of G is also a root of G_i, i.e., $R_i(\psi_j) = \psi_p$ for some p. This means that $R_i\big(R_j(\psi)\big) = R_p(\psi)$, i.e., $R_iR_j \equiv R_p \pmod{G}$. Therefore, for any root ψ_s of G, we have $R_i\big(R_j(\psi_s)\big) = R_p(\psi_s)$. In particular, the set of polynomials

$R_1, \ldots, R_r$ and the transformations of the roots $\psi_1 \ldots, \psi_r$ which correspond to them is invariantly defined, i.e., does not depend on the choice of the root ψ_1.

To the number

$$\psi_i = m_1 \alpha_{\sigma_i(1)} + \cdots + m_n \alpha_{\sigma_i(n)},$$

a permutation σ_i uniquely corresponds, and to the permutation, in its turn, there corresponds an automorphism, which we also denote by σ_i, of the field $K(\alpha_1, \ldots, \alpha_n) = K(\psi)$. Thus, to the roots $\psi_1, \ldots, \psi_r$ of the Galois resolvent there correspond permutations $\sigma_1, \ldots, \sigma_r$. To σ_i, we assign the polynomial R_i. Since $\sigma_i(\psi) = \psi_i$, we see that

$$\sigma_i \sigma_j(\psi) = \sigma_i(\psi_j) = \sigma_i R_j(\psi) = R_j(\psi_i) = R_j\big(R_i(\psi)\big),$$

i.e., $\sigma_i \sigma_j \leftrightarrow R_j R_i \bmod G$. Therefore the group of transformation of $R_1, \ldots, R_r$ is anti-isomorphic[1] to the group of permutations of $\sigma_1, \ldots, \sigma_r$. In order to establish a relation with the Galois group of f, it remains to prove the following statement.

Theorem 5.1.3. *The Galois group of the polynomial f consists of the permutations $\sigma_1, \ldots, \sigma_r$.*

Proof. We have to prove that, if $\alpha_1, \ldots, \alpha_n$ are the roots of f and $\tau \in S_n$, then the condition $\tau \in \{\sigma_1, \ldots, \sigma_r\}$ is equivalent to the fact that

$$\varphi(\alpha_1, \ldots, \alpha_n) = 0 \Longleftrightarrow \varphi(\alpha_{\tau(1)}, \ldots, \alpha_{\tau(n)}) = 0$$

for any rational function φ.

Since the roots $\alpha_1, \ldots, \alpha_n$ can be rationally expressed in terms of ψ_1, it follows that

$$\varphi(\alpha_1, \ldots, \alpha_n) = \Phi(\psi_1) = \Phi(m_1 \alpha_1 + \cdots + m_n \alpha_n),$$

where $\Phi \in K(x)$. Therefore

$$\varphi(\alpha_{\tau(1)}, \ldots, \alpha_{\tau(n)}) = \Phi(\widetilde{\psi}_\tau),$$

where $\widetilde{\psi}_\tau = m_1 \alpha_{\tau(1)} + \cdots + m_n \alpha_{\tau(n)}$. The equivalence of the equalities $\Phi(\psi_1) = 0$ and $\Phi(\widetilde{\psi}_\tau) = 0$ for all possible rational functions Φ means that ψ_1 and $\widetilde{\psi}_\tau$ are roots of the same irreducible polynomial, i.e., $\tau \in \{\sigma_1, \ldots, \sigma_r\}$. $\square$

Corollary. *Let f be a polynomial over a field k with distinct roots $\alpha_1, \ldots, \alpha_n$. Then the order $|G|$ of the Galois group G of f is equal to $[K : k]$, where $K = k(\alpha_1, \ldots, \alpha_n)$.*

---

[1] Recall that the map of groups $a : G \longrightarrow H$ is an *anti-isomorphism* if $a(fg) = a(g)a(f)$ for any $f, g \in G$.

Proof. Both $|G|$ and $[K : k]$ are equal to the degree of any Galois resolvent of f. $\square$

The following statement is of considerable importance for Galois theory.

Theorem 5.1.4. *Let $\alpha_1, \ldots, \alpha_n$ be the roots of an irreducible polynomial f over k, and let $\varphi \in k(x_1, \ldots, x_n)$.*

a) *Let $\varphi(\alpha_1, \ldots, \alpha_n) = \varphi(\alpha_{\sigma_i(1)}, \ldots, \alpha_{\sigma_i(n)})$ for any permutation σ_i in the Galois group. Then $\varphi(\alpha_1, \ldots, \alpha_n) \in k$.*

b) *Let $H = \{\sigma_{i_1}, \ldots, \sigma_{i_s}\}$ be a subgroup of the Galois group $\{\sigma_1, \ldots, \sigma_r\}$ such that if $\varphi(\alpha_1, \ldots, \alpha_n) = \varphi(\alpha_{\sigma(1)}, \ldots, \alpha_{\sigma(n)})$ for any permutation $\sigma \in H$, then $\varphi(\alpha_1, \ldots, \alpha_n) \in k$. Then H coincides with the whole Galois group.*

Proof. a) The roots $\alpha_1, \ldots, \alpha_n$ can be rationally expressed in terms of ψ_1, and hence $\varphi(\alpha_1, \ldots, \alpha_n) = \Phi(\psi_1)$, where $\Phi \in k(x)$. Therefore

$$\varphi(\alpha_{\sigma_i(1)}, \ldots, \alpha_{\sigma_i(n)}) = \Phi(\psi_i).$$

Thus, $\Phi(\psi_1) = \cdots = \Phi(\psi_r)$, and hence

$$\Phi(\psi_1) = \frac{1}{r}\big(\Phi(\psi_1) + \cdots + \Phi(\psi_r)\big)$$

is a rational symmetric function in the roots $\psi_1, \ldots, \psi_r$ of the Galois resolvent. Hence $\varphi(\alpha_1, \ldots, \alpha_n) = \Phi(\psi_1) \in k$.

b) Consider the polynomial

$$g(x) = \prod_{\sigma \in H} (x - \sigma(\psi)) = (x - \psi_{i_1}) \cdot \ldots \cdot (x - \psi_{i_s}).$$

Its coefficients are invariant with respect to the H-action, and hence all of them belong to k. Therefore the coefficients of $g(x)$ lie in k and $g(x)$ has common roots with an irreducible over k polynomial

$$G(x) = (x - \psi_1) \cdot \ldots \cdot (x - \psi_r).$$

Hence $g(x) = G(x)$ and $H = \{\sigma_1, \ldots, \sigma_r\}$. $\square$

5.1.3 Theorem on a primitive element

On page 187 we observed that the field $k(\alpha_1, \ldots, \alpha_n)$, where $\alpha_1, \ldots, \alpha_n$ are the roots of a polynomial f, can be generated over k by one element, namely, by a root ψ of a Galois resolvent of f. Usually, to construct the element that generates the field, one uses the following standard construction.

Theorem 5.1.5 (On a primitive element). *Let k be a field of characteristic zero, and let $\alpha_1, \ldots, \alpha_n$ be algebraic elements over k. Then $k(\alpha_1, \ldots, \alpha_n)$ is generated over k by one element.*

Proof. First, consider the case of two algebraic elements, α and β. Let $f(x)$ and $g(x)$ be irreducible over k polynomials with roots α and β, respectively. Let further $\alpha_1 = \alpha, \alpha_2, \ldots, \alpha_r$ be all the roots of f and $\beta_1 = \beta, \beta_2, \ldots, \beta_s$ all the roots of g. Select $c \in k$ so that $\alpha_i + c\beta_j \neq \alpha_1 + c\beta_1$ for $j \neq 1$. Set

$$\theta = \alpha_1 + c\beta_1 = \alpha + c\beta.$$

Clearly, $k(\theta) \subset k(\alpha, \beta)$. It remains to show that $k(\alpha, \beta) \subset k(\theta)$. To do this, it suffices to prove that $\beta \in k(\theta)$. Indeed, in this case $\alpha = \theta - c\beta \in k(\theta)$.

The element β satisfies the equations

$$g(x) = 0 \text{ and } f(\theta - cx) = 0$$

whose coefficients belong to $k(\theta)$. The only common root of the polynomials $g(x)$ and $f(\theta - cx)$ is β since $\theta - c\beta_j \neq \alpha_i$ for $j \neq 1$. The polynomial $g(x)$ has no multiple roots, and so the greatest common divisor of $g(x)$ and $f(\theta - cx)$ is $x - \beta$. The greatest common divisor of two polynomials over the field $k(\theta)$ is a polynomial over $k(\theta)$, and hence $\beta \in k(\theta)$.

The passage from $n = 2$ to an arbitrary n is performed by an obvious induction: if $k(\alpha_1, \ldots, \alpha_{n-1}) = k(\theta)$, then $k(\alpha_1, \ldots, \alpha_n) = k(\theta, \alpha_n) = k(\theta')$ for some θ'. $\square$

With the help of Theorem 1.5 one can prove, for example, the following statement on prime divisors of a set of polynomials.

Theorem 5.1.6 ([Ho]). *Let $f_1(x), \ldots, f_n(x)$ be non-constant integer-valued polynomials, i.e., $f_i(m) \in \mathbb{Z}$ for $m \in \mathbb{Z}$. Further, let M_i be the set of all prime divisors of all the numbers $f_i(m) \in \mathbb{Z}$, where $m \in \mathbb{Z}$ and $f_i(m) \neq 0$. Then the set $M = M_1 \cup \cdots \cup M_n$ is infinite.*

Proof. First, consider the case $n = 1$ (in this case the theorem was proved by Schur [Sc6]). If $f_1(x)$ is an integer-valued polynomial of degree k, then all the coefficients of the polynomial $k! f_1(x)$ are integers by Theorem 3.2.1 on page 85. Passing from f_1 to $k! f_1$ we add to prime divisors of f_1 only prime divisors of the number $k!$, i.e., a certain finite set of divisors. Therefore it suffices to carry out the proof for the case of the polynomial f_1 with integer coefficients.

Now f_1 assumes values 0 and ± 1 at finitely many points only. Therefore the values of f_1 at integer points have at least one prime divisor; in other words, $M_1 \neq \varnothing$. Suppose that $M_1 = \{p_1, \ldots, p_r\}$ is a finite set.

Let $a \in \mathbb{Z}$ and $f_1(a) = b \neq 0$. Let us show that

$$g(x) = \frac{f_1(a + bp_1 \cdot \ldots \cdot p_r x)}{b}.$$

is a polynomial with integer coefficients such that $g(x) \equiv 1 \pmod{p_1 \cdot \ldots \cdot p_r}$ for all $x \in \mathbb{Z}$. It is easy to verify that if $c \in \mathbb{Z}$, then $(a + cx)^l - a^l = ch_l(x)$, where $h_l(x)$ is a polynomial with integer coefficients. Hence

$$f_1(a + bp_1 \cdot \ldots \cdot p_r x) - f_1(a) = bp_1 \cdot \ldots \cdot p_r h(x),$$

where $h(x)$ is a polynomial with integer coefficients. It remains to observe that

$$g(x) = \frac{\left(f_1(a) + bp_1 \cdot \ldots \cdot p_r h(x)\right)}{b} = 1 + p_1 \cdot \ldots \cdot p_r h(x).$$

For a certain $x \in \mathbb{Z}$, the integer $g(x)$ has a prime divisor p. The congruence $g(x) \equiv 1 \pmod{p_1 \cdot \ldots \cdot p_r}$ shows that $p \notin M_1 = \{p_1, \ldots, p_r\}$. On the other hand, the number $f_1(a + bp_1 \cdot \ldots \cdot p_r x) = bg(x)$ is divisible by p, and hence $p \in M_1$. The contradiction obtained means that M_1 is an infinite set.

The passage from $n = 1$ to an arbitrary n is performed with the help of Theorem 1.5. As in the proof for $n = 1$, we assume that $f_1, \ldots, f_n \in \mathbb{Z}[x]$. Let α_i be one of the roots of f_i. By Theorem 1.5, $\mathbb{Q}(\alpha_1, \ldots, \alpha_n) = \mathbb{Q}(\alpha)$, i.e., $\alpha_i = \varphi_i(\alpha)$, where $\varphi_i(t) \in \mathbb{Q}[t]$. Let g be an irreducible polynomial over $\mathbb{Q}$ with root α. If $\varphi_i(0) \neq 0$, then replace φ_i by $\widetilde{\varphi}_i$, where

$$\widetilde{\varphi}_i(t) = \varphi_i(t) - \frac{\varphi_i(0)}{g(0)} g(t).$$

Thus, we may assume that $\varphi_i(0) = 0$. In this case, if a number N is divisible by the denominators of all the coefficients of φ_i, then $\varphi_i(Nt) \in \mathbb{Z}[t]$.

The polynomials $f_1\left(\varphi_1(t)\right), \ldots, f_n\left(\varphi_n(t)\right) \in \mathbb{Q}[t]$ have a common root α, and hence all of them are divisible by $g(t)$. Consider the polynomials

$$F_i(t) = f_i\left(\varphi_i(Nt)\right).$$

Clearly, $F_i(t) \in \mathbb{Z}[x]$ and $F_i(t)$ is divisible over $\mathbb{Q}$ by $g(Nt)$, i.e., $F_i(t) = g(Nt)g_i(t)$, where $g_i(t) \in \mathbb{Q}[t]$. Let us express the polynomials g and g_i in the form

$$g(Nt) = rh(t) \quad \text{and} \quad g_i(t) = s_i h_i(t),$$

where $r, s_i \in \mathbb{Q}$, $h(t), h_i(t) \in \mathbb{Z}[t]$ and $\text{cont}(h) = \text{cont}(h_i) = 1$.

Then $F_i(t) = rs_i h(t) h_i(t)$ and, by Gauss's lemma, $rs_i = \text{cont}(F_i)$ is an integer. This means that $F_i(t)$ is divisible over $\mathbb{Z}$ by $h(t)$. In particular, all the divisors of the values of h at integer points are divisors of the values of F_i which, in turn, are divisors of the values of f_i. Therefore the set M contains an infinite subset consisting of prime divisors of the polynomial h. $\square$

5.2 Basic Galois theory

5.2.1 The Galois correspondence

Theorem 5.2.1. *Any element* $\omega \in K = k(\alpha_1, \ldots, \alpha_n)$ *is a root of an irreducible over k polynomial h all of whose roots belong to K.*

Proof. One can represent ω in the form $\omega = g(\alpha_1, \ldots, \alpha_n)$, where $g \in k[x_1, \ldots, x_n]$. Set $\omega_\sigma = g(\alpha_{\sigma(1)}, \ldots, \alpha_{\sigma(n)})$ and consider the polynomial

$$h(x) = \prod_{\sigma \in S_n} (x - \omega_\sigma).$$

Clearly, all the roots of h belong to K and h is a polynomial over k. Hence ω is a root of an irreducible divisor of h. $\square$

Corollary. *If $p(x)$ is irreducible over k and one of its roots belongs to $K = k(\alpha_1, \ldots, \alpha_n)$, then all the other roots of $p(x)$ belong to K.*

Proof. Let $\omega \in K$ be a root of p and let h be an irreducible over k polynomial all of whose roots belong to K and $h(\omega) = 0$. The polynomials h and p are irreducible over k and have a common root ω. Hence all the roots of p are the roots of h, i.e., belong to K. $\square$

A finite extension K of the field k is called a *normal extension* or a *Galois extension* if any irreducible over k polynomial, one of whose roots belongs to K, factorizes over K into linear factors, i.e., all its roots lie in K.

An example of a non-normal extension is the field $\mathbb{Q}(\sqrt[3]{2})$. This field contains only one of the roots of the polynomial $x^3 - 2$.

By Theorem 5.2.1, the splitting field of any polynomial is a normal extension. The converse statement is also true.

Theorem 5.2.2. *Let $K \supset k$ be a normal extension. Then K is a splitting field of a polynomial over k.*

Proof. Let $K = k(\alpha_1, \ldots, \alpha_n)$ and let f_i be an irreducible polynomial over k with root α_i. Let $f = f_1 \cdots f_n$, and let K' be the splitting field of f over k. On the one hand, the elements $\alpha_1, \ldots, \alpha_n$ are all the roots of f, and so $K \subset K'$. On the other hand, K is a normal field over k and contains a root of an irreducible over k polynomial f_i $(i = 1, \ldots, n)$, and so K contains all the roots of f, and therefore $K \supset K'$. $\square$

Corollary. *Let $K \supset L \supset k$, where K is a normal extension of k, and L an arbitrary intermediate field. Then K is a normal extension of L.*

Proof. The field K is the splitting field of a polynomial f over k. This polynomial can be considered also as a polynomial over L, and so K is the splitting field of a polynomial over L, i.e., K is a normal extension of L. $\square$

If K is a normal extension of k, then the *Galois group* of K over k is the group of automorphisms of K preserving all the elements of k. The Galois group of K over k will be denoted by the symbol $G(K, k)$. If K is the splitting field of a polynomial f, we will also denote the Galois group $G(K, k)$ by the symbol $G_k(f)$.

The elements ω and ω' of the field $K \supset k$ are called *conjugate over k* if ω and ω' are the roots of the same polynomial irreducible over k.

Theorem 5.2.3. *Let K be a normal extension of k. The elements $\omega, \omega' \in K$ are conjugate over k if and only if there exists an automorphism $\sigma \in G(K, k)$ which sends ω into ω'.*

Proof. Let ω be a root of an irreducible over k polynomial p. If $\omega' = \sigma(\omega)$, then $p(\omega') = p(\sigma(\omega)) = \sigma(p(\omega)) = 0$, and therefore ω and ω' are conjugate over k.

Now suppose that ω and ω' are the roots of an irreducible over k polynomial p. To construct the automorphism σ, we start by constructing an isomorphism $\varphi \colon k(\omega) \to k(\omega')$. Any element of the field $k(\omega)$ can be uniquely represented in the form

$$a_0 + a_1\omega + \cdots + a_{n-1}\omega^{n-1}, \text{ where } a_i \in k \text{ and } n = \deg p.$$

The automorphism to be constructed is of the form $\sum_{i=0}^{n-1} a_i\omega^i \mapsto \sum_{i=0}^{n-1} a_i(\omega')^i$. Now, select $\theta \in K \setminus k(\omega)$. Let p_1 be an irreducible over k polynomial with root θ and q_1 an irreducible over $k(\omega)$ divisor of p_1 such that $q_1(\theta) = 0$. Under the isomorphism $\varphi \colon k(\omega) \to k(\omega')$ the irreducible over $k(\omega)$ polynomial q_1 becomes an irreducible over $k(\omega')$ polynomial $\bar{q}_1$. Let θ' be a root of $\bar{q}_1$. The isomorphism $\varphi \colon k(\omega) \to k(\omega')$ can be extended to an isomorphism $k(\omega, \theta) \to k(\omega', \theta')$. This isomorphism is of the form $\sum b_i\theta^i \mapsto \sum \varphi(b_i)(\theta')^i$, where $b_i \in k(\omega)$. Such extensions of isomorphisms enable one to construct an isomorphism of K with a subfield $K' \subset K$ that sends ω to ω'. This isomorphism of fields is, in particular, an isomorphism of linear spaces over k. Therefore, since the dimension of K is finite, it follows that $K' = K$, i.e., we have obtained an automorphism of K. $\square$

Corollary. *If K is a normal extension of k, then the element $\omega \in K$ is invariant with respect to the action of the Galois group $G(K, k)$ if and only if $\omega \in k$.*

Proof. If $\omega \in K$ is invariant with respect to $G(K, k)$, then all its conjugates coincide with it. This means that ω is a root of the polynomial $x - \omega$ with coefficients in k, i.e., $\omega \in k$. $\square$

The property that the extension be normal is very essential. For example, any automorphism of the field $\mathbb{Q}(\sqrt[3]{2})$ is the identity, i.e., the element $\sqrt[3]{2} \notin \mathbb{Q}$ is invariant under all the automorphisms.

A particular feature of normal extensions is that their group of automorphisms is rather large. This makes it possible to establish a one-to-one correspondence between the intermediate fields and subgroups of the Galois group.

Let K be a normal extension of k. Consider an arbitrary intermediate field L, i.e., $k \subset L \subset K$. By Corollary of Theorem 5.2.2 the field K is a normal extension of L, and therefore we can consider the Galois group $G(K, L)$.

Theorem 5.2.4 (The Galois correspondence). a) *There is a one-to-one correspondence between the intermediate fields $k \subset L \subset K$ and the subgroups of the Galois group $G(K, k)$. To the field L, there corresponds the subgroup $G(K, L) \subset G(K, k)$ and to the subgroup $H \subset G(K, k)$ there corresponds the field consisting of the H-invariant elements of K.*

b) *An intermediate field L is a normal extension of k if and only if the subgroup $G(K, L) \subset G(K, k)$ is normal. In this case, we have the exact sequence*[1]

$$0 \to G(K, L) \to G(K, k) \to G(L, k) \to 0.$$

Proof. a) Any automorphism of the field K that preserves the elements of L also preserves the elements of $k \subset L$, i.e., $G(K, L) \subset G(K, k)$.

To the field L, there corresponds the group $G(K, L)$ and to the group $G(K, L)$ there corresponds a field L' consisting of the elements of K invariant with respect to all the automorphisms of K that preserve the elements of L. Clearly, $L' \supset L$. But for the case of normal extensions any element of K invariant with respect to $G(K, L)$ belongs to L (Corollary of Theorem 5.2.3). (An independent proof follows from Theorem 5.1.4 (a) on page 189.)

Therefore $L = L'$, i.e., to every subfield there corresponds a subgroup and this subgroup uniquely determines the subfield.

To the subgroup $H \subset G(K, k)$ there corresponds a field L, and to the field L there corresponds the subgroup $G(K, L) = H'$ consisting of the automorphisms of K that preserve the elements invariant with respect to H. Clearly, $H' \supset H$. But if the subgroup H of $G(K, L)$ is such that all the elements of K invariant with respect to the H-action lie in L, then H coincides with the whole Galois group $G(K, L)$ (Theorem 5.1.4 (b) on page 189). Therefore to each subgroup there corresponds a subfield and this subfield uniquely determines the subgroup.

b) Let L be a normal extension of k. Then any automorphism of K over k sends L into itself (the element of L goes into a conjugate element which again belongs to L). Therefore there is a homomorphism $G(K, k) \to G(L, k)$. Clearly, the group $G(K, L)$ is the kernel of this homomorphism, and hence it is a normal subgroup of $G(K, k)$.

Now, let $G(K, L)$ be a normal subgroup of $G(K, k)$, i.e., if $\varphi \in G(K, L)$ and $\psi \in G(K, k)$, then $\psi^{-1}\varphi\psi \in G(K, L)$. Let $a \in L$. We have to prove that all the elements $a_1, \ldots, a_l$ conjugate to a belong to L. By the hypothesis $a_1, \ldots, a_l \in K$ and $a_i = \psi_i(a)$ for some $\psi_i \in G(K, k)$. If $\varphi \in G(K, L)$, then $\psi_i^{-1}\varphi\psi_i \in G(K, L)$. Therefore $\psi_i^{-1}\varphi\psi_i(a) = a$, i.e., $\varphi(a_i) = a_i$. Hence $a_i \in L$.

The description of the exact sequence

$$0 \to G(K, L) \to G(K, k) \to G(L, k) \to 0$$

---

[1] Recall that a sequence of maps $\cdots \to A \xrightarrow{\alpha} B \xrightarrow{\beta} C \to \ldots$ is *exact in* B if $\mathrm{Im}(\alpha) = \mathrm{Ker}(\beta)$. The sequence is *exact* if it is exact in all its terms except the first and the last.

is as follows. The field L is a normal extension of k, and so any automorphism $\varphi \in G(K, k)$ preserves L, and therefore one can consider restrictions of φ onto L. This gives rise to the homomorphism $G(K, k) \to G(L, k)$. Its kernel consists of the automorphisms of K identical on L. They constitute the group $G(K, L)$. The epimorphic nature of the homomorphism $G(K, k) \to G(L, k)$ follows from the fact that any automorphism of L of k can be extended to an automorphism of K over k. $\square$

If L is the splitting field of g over k and K is the splitting field of f over k, then the exact sequence

$$0 \to G(K, L) \to G(K, k) \to G(L, k) \to 0$$

takes the form

$$0 \to G_L(f) \to G_k(f) \to G_k(g) \to 0.$$

Theorem 5.2.5. *Let L be an arbitrary extension of the field k and let f be a polynomial over k. Then the Galois group $G_L(f)$ is isomorphic to a subgroup of $G_k(f)$.*

Proof. Let $\alpha_1, \ldots, \alpha_n$ be all the roots of f. The automorphism $\sigma \in G_L(f)$ permutes the roots $\alpha_1, \ldots, \alpha_n$ and preserves all the elements of $L \supset k$. Therefore σ preserves the field $k(\alpha_1, \ldots, \alpha_n)$, i.e., to σ we may assign the automorphism $\overline{\sigma} \in G_k(f)$ which is the restriction of σ onto $k(\alpha_1, \ldots, \alpha_n)$.

If $\overline{\sigma} = \mathrm{id}$, then σ preserves all the roots $\alpha_1, \ldots, \alpha_n$. Moreover, by definition, σ preserves all the elements of L, and hence $\sigma = \mathrm{id}$. Therefore the map $\sigma \mapsto \overline{\sigma}$ is a monomorphism, i.e., $G_L(f)$ is isomorphic to a subgroup of $G_k(f)$. $\square$

5.2.2 A polynomial with the Galois group S_5

In order to give an example of equation not solvable by radicals, we need a polynomial whose Galois group is S_n, $n \geq 5$. We are ready to prove that the Galois group of the polynomial $x^5 - 4x + 2$ over $\mathbb{Q}$ is equal to S_5.

The subgroup $G \subset S_n$ is *transitive* if for any two indices $i, j \in \{1, \ldots, n\}$ there exists a permutation $\sigma \in G$ such that $\sigma(i) = j$.

Theorem 5.2.6. *The polynomial f without multiple roots is irreducible if and only if its Galois group is transitive.*

Proof. Let $\alpha_1, \ldots, \alpha_n$ be the roots of f. If f is irreducible over k, then by definition all its roots are conjugate to each other. Hence, by Theorem 5.2.3, there exists an automorphism of the field $k(\alpha_1, \ldots, \alpha_n)$ that sends α_i to α_j.

Now, suppose that the Galois group G of f is transitive. Let f_1 be an arbitrary divisor of f over k and α_1 a root of f_1. Take an automorphism $\sigma_i \in G$ such that $\sigma_i(\alpha_1) = \alpha_i$. Under σ_i the relation $f_1(\alpha_1) = 0$ becomes $f_1(\alpha_i) = 0$, i.e., all the roots of f serve also as roots of f_1. Since f has no multiple roots, we deduce that f_1 is divisible by f, i.e., f is irreducible. $\square$

Theorem 5.2.7. *Let n be a prime and let the transitive subgroup $G \subset S_n$ contain at least one transposition (i,j). Then $G = S_n$.*

Proof. On $\{1, \ldots, n\}$, introduce a relation by setting $i \sim j$ if either $i = j$ or G has transposition (i,j). The identity $(i,j)(j,k)(i,j) = (i,k)$ proves that this relation is an equivalence relation.

Let $E(i)$ be the equivalence class containing i. Let us show that $|E(i)| = |E(j)|$, i.e., all the equivalence classes consist of the same number of elements. Since G is transitive, it has an element σ such that $\sigma(i) = j$. Let $a \in E(i)$, i.e., $(i,a) \in G$. The element $\sigma \cdot (i,a) \cdot \sigma^{-1}$ interchanges $\sigma(a)$ and $\sigma(i)$ leaving the remaining elements fixed, i.e.,

$$\sigma \cdot (i,a) \cdot \sigma^{-1} = \big(\sigma(i), \sigma(a)\big) = \big(j, \sigma(a)\big) \in G.$$

Therefore $\sigma\big(E(i)\big) \subset E(j)$, and hence $|E(i)| \leq |E(j)|$. The inequality $|E(j)| \leq |E(i)|$ is similarly proved.

By the hypothesis n is a prime, and so there is precisely one equivalence class. This means that $G = S_n$. $\square$

Theorem 5.2.8. *Let f be an irreducible polynomial over $\mathbb{Q}$ of prime degree p with precisely two non-real roots. Then the Galois group of f over $\mathbb{Q}$ is S_p.*

Proof. Since f is irreducible, its Galois group $G \subset S_p$ is transitive. The above theorem shows that to prove the statement desired it suffices to prove that G contains a transposition. An example of such a transposition is given by the restriction of the complex conjugation $z \mapsto \bar{z}$ onto the splitting field of f. Indeed, under the complex conjugation, all the real roots of the polynomial remain fixed whereas the two complex roots are interchanged (this implies, in particular, that the splitting field transforms into itself). $\square$

The polynomial $f(x) = x^5 - 4x + 2$ is an example of an irreducible polynomial of degree 5 with precisely two complex roots . The irreducibility of this polynomial follows from Eisenstein's criterion. The number of real roots of f is not less than 3 since

$$f(-2) < 0, \quad f(0) > 0, \quad f(1) < 0, \quad f(2) > 0.$$

On the other hand, it cannot have more than three real roots because otherwise the derivative $f'(x) = 5x^4 - 4$ would have had more than two real roots.

5.2.3 Simple radical extensions

The splitting field K of the polynomial $x^n - c$, where $c \in k$, is called a *simple radical* extension of k.

Theorem 5.2.9. a) *The Galois group $G(K, k)$ of a simple radical extension is solvable.*

b) *If k contains an n-th primitive root of unity, then $G(K, k)$ is a subgroup of the cyclic group $\mathbb{Z}/n\mathbb{Z}$.*

c) *If $c = 1$, then $G(K, k)$ is a subgroup of the multiplicative group $(\mathbb{Z}/n\mathbb{Z})^*$.*

Proof. a) Let α be a root of the polynomial $x^n - c$ and ε an nth primitive root of unity. Then all the roots of $x^n - c$ are of the form $\alpha, \alpha\varepsilon, \ldots, \alpha\varepsilon^{n-1}$, and therefore $K \subset k(\alpha, \varepsilon)$. On the other hand, $\varepsilon = (\alpha\varepsilon)\alpha^{-1} \in K$, and so $k(\alpha, \varepsilon) \subset K$, i.e., $K = k(\alpha, \varepsilon)$.

Let $K = k(\alpha, \varepsilon)$. Then $\sigma(\varepsilon)$ is a root of $x^n - 1$, i.e., $\sigma(\varepsilon) = \varepsilon^a$. Observe that ε^a cannot be a root of $x^m - 1$, where $m < n$, since otherwise the element $\varepsilon = \sigma^{-1}(\varepsilon^a)$ would also have been a root of $x^n - 1$, which contradicts the fact that ε is primitive. Thus, $(a, n) = 1$.

The automorphism σ is completely determined by its values on the generators of K, i.e., $\sigma(\varepsilon) = \varepsilon^a$, $\sigma(\alpha) = \varepsilon^b\alpha$. Therefore such an automorphism σ can be denoted by the symbol

$$\sigma = [a, b], \text{where } a \in (\mathbb{Z}/n\mathbb{Z})^* \text{ and } b \in \mathbb{Z}/n\mathbb{Z}.$$

Observe that if $\sigma_i = [a_i, b_i]$, $i = 1, 2$, then

$$\sigma_1\big(\sigma_2(\varepsilon)\big) = \varepsilon^{a_1 a_2}, \quad \sigma_1\big(\sigma_2(\alpha)\big) = \varepsilon^{a_1 b_2 + b_1}\alpha,$$

i.e.,

$$[a_1, b_1][a_2, b_2] = [a_1 a_2, a_1 b_2 + b_1].$$

Consider a homomorphism $\varphi \colon G(K, k) \to (\mathbb{Z}/n\mathbb{Z})^*$ which assigns $\sigma = [a, b]$ to $a \in (\mathbb{Z}/n\mathbb{Z})^*$. The kernel of this homomorphism consists of the elements of the form $[1, b]$. For such elements the composition law is as follows: $[1, b_1][1, b_2] = [1, b_1 + b_2]$. The kernel and the image of φ are abelian groups, and so $G(K, k)$ is solvable.

b) If $\varepsilon \in k$, then $\sigma(\varepsilon) = \varepsilon$. Hence $\sigma = [1, b] \in \mathbb{Z}/n\mathbb{Z}$.

c) If $c = 1$, we may assume that $\alpha = 1$. In this case $\sigma = [a, 0] \in (\mathbb{Z}/n\mathbb{Z})^*$. $\square$

5.2.4 The cyclic extensions

A normal extension K of the field k is called *cyclic* if the Galois group $G(K, k)$ is cyclic.

Theorem 5.2.10. *If k contains an nth primitive root of unity, then any cyclic extension $K \supset k$ of degree n is of the form $K = k(\beta)$, where β is a root of a polynomial $x^n - c$ irreducible over k.*

Proof. We will need the fact that the characters of the group are linearly independent. Let us recall the precise formulation and proof. Let G be a group, K a field and K^* the multiplicative group of the field, i.e., the set of non-zero elements with respect to multiplication. A *character* of G is an arbitrary homomorphism $G \to K^*$.

Lemma. *Distinct characters of G are linearly independent over K.*

Proof. Let $\{\gamma_1, \ldots, \gamma_n\}$ be a minimal non-empty set of linearly dependent characters, i.e.,

$$\lambda_1 \gamma_1(g) + \cdots + \lambda_n \gamma_n(g) = 0 \tag{1}$$

for all $g \in G$ and certain fixed $\lambda_1, \ldots, \lambda_n \in K^*$. Clearly, $n \geq 2$. The characters γ_1 and γ_n are distinct, and so $\gamma_1(h) \neq \gamma_n(h)$ for some $h \in G$. Let us multiply (1) by $\gamma_n(h)$ and subtract from the result the identity $\lambda_1 \gamma_1(hg) + \cdots + \lambda_n \gamma_n(hg) = 0$. After simplification we obtain

$$\lambda_1 \big(\gamma_n(h) - \gamma_1(h)\big)\gamma_1(g) + \cdots + \lambda_{n-1}\big(\gamma_n(h) - \gamma_{n-1}(h)\big)\gamma_{n-1}(g) = 0.$$

This contradicts the minimality of the set $\{\gamma_1, \ldots, \gamma_n\}$. $\square$

If σ is an automorphism of K, then the restriction of σ onto K^* is a character of K^*. Therefore, if $\sigma_1, \ldots, \sigma_n$ are distinct automorphisms of K and $\alpha_1, \ldots, \alpha_n \in K^*$, then $\alpha_1 \sigma_1(\alpha) + \cdots + \alpha_n \sigma_n(\alpha) \neq 0$ for a certain $\alpha \in K^* \subset K$.

Let us now pass to the proof of the theorem. Let σ be the generator of the cyclic group $G(K, k)$ and ε an nth primitive root of unity which belongs to k. Consider the *Lagrange resolvent*

$$(\varepsilon, \alpha)_\sigma = \alpha + \varepsilon\sigma(\alpha) + \cdots + \varepsilon^{n-1}\sigma^{n-1}(\alpha).$$

The automorphisms $\mathrm{id}, \sigma, \sigma^2, \ldots, \sigma^{n-1}$ are distinct, and so there exists an element $\alpha \in K$ for which $(\varepsilon, \alpha)_\sigma = \beta \neq 0$. It is easy to verify that $\sigma(\beta) = \varepsilon^{-1}\beta$ and $\sigma(\beta^n) = \big(\sigma(\beta)\big)^n = \beta^n$. Therefore $\sigma(\beta) \neq \beta$, i.e., $\beta \notin k$ and $\sigma^i(\beta^n) = \beta^n$ for $i = 1, \ldots, n-1$, i.e., $\beta^n = c \in k$.

Consider the polynomial

$$x^n - c = (x - \beta)(x - \varepsilon\beta) \cdot \ldots \cdot (x - \varepsilon^{n-1}\beta).$$

The field $k(\beta)$ is its splitting field and $k(\beta) \subset K$. Since the automorphisms $\mathrm{id}, \sigma, \sigma^2, \ldots, \sigma^{n-1}$ are distinct automorphisms of $k(\beta)$, it follows that the order of the Galois group of $k(\beta)$ over k is not less than n, i.e., $[k(\beta) : k] \geq n$. On the other hand, $[K : k] = n$, and so $k(\beta) = K$. The Galois group of the polynomial $x^n - c$ is transitive, and so the polynomial is irreducible. $\square$

5.3 How to solve equations by radicals

An extension L of the field k is said to be *radical* if there exists a sequence of intermediate fields

$$k = L_0 \subset L_1 \subset \cdots \subset L_r = L,$$

such that $L_i = L_{i-1}(\beta_i)$, where $\beta_i^{n_i} \in L_{i-1}$. In other words, we consecutively add to the field k the roots of the elements of the fields obtained at the preceding step.

Let f be an irreducible polynomial over k and $\alpha_1, \ldots, \alpha_n$ all its roots. The equation $f(x) = 0$ is said to be *solvable by radicals* if the field $k(\alpha_1, \ldots, \alpha_n)$ is contained in a radical extension of k.

To formulate and prove the criterion of solvability of equations by radicals, we will need the notion of solvable groups. We therefore start with a recapitulation of the basic notions of solvable groups.

5.3.1 Solvable groups

A group G is called *solvable* if there exists a sequence of nested subgroups

$$\{e\} = G_r \subset G_{r-1} \subset \cdots \subset G_0 = G,$$

such that G_i is a normal subgroup in G_{i-1} and the quotient G_{i-1}/G_i is abelian (for $i = 1, \ldots, r$).

In what follows we only deal with finite groups.

For any finite abelian group G, one can construct a sequence of nested subgroups for which all the quotient groups G_{i-1}/G_i are cyclic (moreover, the G_{i-1}/G_i are cyclic groups of prime order). Therefore, in the definition of a solvable group, we may assume that all the quotients G_{i-1}/G_i are cyclic.

In the description of the relation between solvability of a polynomial equation $f(x) = 0$ by radicals and the solvability of the Galois group of f, we use the Galois correspondence which provides an exact sequence $0 \to H \to G \to G/H \to 0$. Usually, for some of these three groups, it is known that they are solvable and we have to decide whether the remaining groups are solvable. For this purpose, we use the following theorem.

Theorem 5.3.1. a) *Any subgroup H of a solvable group G is solvable.*

b) *Let H be a normal subgroup of G such that H and G/H are solvable. Then the group G is solvable.*

c) *Let H be a normal subgroup of a solvable group G, then G/H is solvable.*

Proof. a) Let us show that the sequence of subgroups $H_i = G_i \cap H$ possesses the properties required, i.e., H_i is a normal subgroup in H_{i-1} and the quotient H_{i-1}/H_i is abelian. Since $G_i \subset G_{i-1}$, it follows that $H_i = H_{i-1} \cap G_i$. Therefore H_i is a normal subgroup in H_{i-1} and

$$H_{i-1}/H_i = H_{i-1}/(H_{i-1} \cap G_i) \cong G_i H_{i-1}/G_i \subset G_{i-1}/G_i.$$

b) For solvable groups H and G/H, take the sequences that define their solvability:

$$\{e\} = H_n \subset H_{n-1} \subset \cdots \subset H_0 = H,$$
$$\{H\} = A_m \subset A_{m-1} \subset \cdots \subset A_0 = G/H.$$

We define $G_i = p^{-1}(A_i)$, where $p \colon G \to G/H$ is the natural projection. Clearly $G_m = H$ and $G_0 = G$. Let us show that the sequence of subgroups

$$\{e\} = H_n \subset H_{n-1} \subset \cdots \subset H_0 = H = G_m \subset G_{m-1} \subset \cdots \subset G_0 = G$$

possesses the properties required in the definition of a solvable group, i.e., G_i is a normal subgroup in G_{i-1} and G_{i-1}/G_i is abelian.

The second property follows from the fact that $G_{i-1}/G_i \cong A_{i-1}/A_i$. Let $g_i \in G_i$ and $g_{i-1} \in G_{i-1}$. Then

$$p\big(g_{i-1}^{-1} g_i g_{i-1}\big) = p(g_{i-1})^{-1}\, p(g_i)\, p(g_{i-1}) \in A_i,$$

since A_i is a normal subgroup in A_{i-1}. Therefore $g_{i-1}^{-1} g_i g_{i-1} \in G_i$, i.e., G_i is a normal subgroup in G_{i-1}.

c) For a solvable group G, take the sequence of subgroups that defines solvability:

$$\{e\} = G_r \subset G_{r-1} \subset \cdots \subset G_0 = G,$$

and define $A_i = G_i H/H$. Then the sequence of subgroups

$$\{H\} = A_r \subset A_{r-1} \subset \cdots \subset A_0 = G/H$$

possesses the property required, i.e., A_i is a normal subgroup in A_{i-1} and A_{i-1}/A_i is abelian. Indeed, let $g_i H \in A_i$ and $g_{i-1} H \in A_{i-1}$. Then

$$(g_{i-1}H)g_i H(g_{i-1}H)^{-1} = g_{i-1}H g_i H g_{i-1}^{-1} = g_{i-1} g_i g_{i-1}^{-1} H \in A_i.$$

Moreover,

$$A_{i-1}/A_i \cong G_{i-1}/(G_i H \cap G_{i-1}) \cong (G_{i-1}/G_i)/((G_i H \cap G_{i-1})/G_i),$$

i.e., the group A_{i-1}/A_i is isomorphic to a quotient of an abelian group, and hence A_{i-1}/A_i is an abelian group. $\square$

5.3.2 Equations with solvable Galois group

Using the Galois correspondence and Theorem 5.2.10 on the structure of the cyclic extension, it is not difficult to prove that any equation with solvable Galois group is solvable by radicals.

Theorem 5.3.2. *Let f be a polynomial without multiple roots over k whose Galois group $G_k(f)$ is solvable. Then the equation $f = 0$ is solvable by radicals.*

Proof. If the field k does not contain any dth primitive root of unity, where $d = |G_k(f)|$, then we add to k a dth primitive root ε of unity, i.e., we consider the field $L = k(\varepsilon)$. The Galois group $G_L(f)$ is isomorphic to a subgroup of the solvable group $G_k(f)$, and so $G_L(f)$ is solvable itself and $|G_L(f)|$ divides d. In particular, the field L contains primitive roots of unity of any degree that divides $|G_L(f)|$.

For the solvable group $G_L(f)$, we construct a sequence of subgroups

$$\{e\} = G_r \subset \cdots \subset G_0 = G_L(f),$$

in which the quotients G_{i-1}/G_i are cyclic. The Galois correspondence assigns to this sequence a sequence of fields

$$K = L_r \supset \cdots \supset L_0 = L,$$

where K is the extension field of f over L such that the extension $L_i \supset L_{i-1}$ is normal, and therefore the sequence of fields $K \supset L_i \supset L_{i-1}$ provides an exact sequence

$$0 \to G(K, L_i) \to G(K, L_{i-1}) \to G(L_i, L_{i-1}) \to 0.$$

Therefore $G(L_i, L_{i-1}) \cong G(K, L_{i-1})/G(K, L_i) = G_{i-1}/G_i$ is a cyclic group whose order divides $|G_L(f)|$. Since $L_{i-1} \supset L$ contains a dth primitive root of unity, where $d_i = |G(L_i, L_{i-1})|$, we see that $L_i = L_{i-1}(\beta_i)$, where β_i is a root of the polynomial $x^{d_i} - c_i$, where $c_i \in L_{i-1}$. Therefore L_i is a radical extension of L_{i-1}, and hence K is a radical extension of L. It is also clear that L is a radical extension of k. Therefore the splitting field of the polynomial f is a radical extension of k, i.e., the equation is solvable by radicals. $\square$

5.3.3 Equations solvable by radicals

We have just proved that, if the Galois group of an equation is solvable, then this equation is solvable by radicals. Let us now prove the converse statement.

Theorem 5.3.3. *Let f be a polynomial without multiple roots over a field k and let the equation $f = 0$ be solvable by radicals. Then the Galois group $G_k(f)$ is solvable.*

Proof. By the hypothesis, for some field L containing all the roots of f, there exists a sequence of fields

$$L = L_r \supset \cdots \supset L_0 = k, \tag{1}$$

such that $L_i = L_{i-1}(\beta_i)$, where $\beta_i^{n_i} \in L_{i-1}$. Here the extension $L \supset k$ is not necessarily normal, and therefore we cannot directly apply the Galois correspondence. We therefore begin with the construction of a radical extension $K \supset L$ for which the extension $K \supset k$ is normal.

We use induction on r. For $r = 0$, the statement is obvious. Hence, by the inductive hypothesis, we may assume that we have already constructed a radical extension $K_{r-1} \supset L_{r-1}$ for which the extension $K_{r-1} \supset k$ is normal. Set $K' = K_{r-1}$ and $L' = K'(\beta_r) \supset L$. The induction step consists in the proof of the following statement.

Lemma. *Let $K' \supset k$ be a normal extension and let $L' = K'(\beta)$, where $\beta^n \in K'$. Then there exists a radical extension $K \supset L'$ such that the extension $K \supset k$ is normal.*

Proof. Consider an irreducible over k polynomial $g(x)$ with root $\beta^n \in K'$. The extension $K' \supset k$ is normal, and so all the roots of $g(x)$ lie in K'. Set $h(x) = g(x^n)$ and consider the field K, the splitting field of h over K'. Let us prove that K possesses all the properties required.

1. *The extension $K \supset k$ is normal.* Indeed, the extension $K' \supset k$ is normal, and hence K' is the splitting field over k of a polynomial $\varphi(x)$. In this case K is the splitting field over k of the polynomial $h(x)\varphi(x)$.

2. $K \supset L' = K'(\beta)$. Indeed, by definition $K' \subset K$. We also have $\beta \in K$ since $h(\beta) = g(\beta^n) = 0$.

3. *The extension $K \supset L'$ is radical.* Let $\widehat{\beta}$ be a root of $h(x)$. Then $\widehat{\beta}^n$ is a root of $g(x)$, and hence $\widehat{\beta}^n \in K' \subset L'$. It remains to observe that the field K is obtained by adjoining to L' all the roots $\widehat{\beta}$ of the polynomial $h(x)$. $\square$

Thus, we may assume that the field L in (1) is a normal extension of k. Moreover, we may assume that the numbers n_i are primes and the degree of the extension $L_i \supset L_{i-1}$ is equal to n_i (the adjoining of a root of degree pq can be replaced by the adjoining of a root of degree p with a subsequent adjoining of a root of degree q).

Since the extension $L \supset k$ is normal, the extension $L \supset L_{i-1}$ is also normal. The coefficients of the polynomial $x^{n_i} - \beta_i^{n_i}$ belong to L_{i-1} and its root β_i belongs to L but not to L_{i-1}. Therefore the polynomial $x^{n_i} - \beta_i^{n_i}$ has an irreducible over L_{i-1} divisor with root β_i and this divisor is different from $x - \beta_i$. Hence the field L contains a root of the polynomial $x^{n_i} - \beta_i^{n_i}$ distinct from β_i, and therefore the field L contains an n_ith primitive root of unity (we use the fact that n_i is a prime).

The field L contains primitive roots of unity of all degrees n_i, and so it contains a primitive root ε of unity whose degree is divisible by all the n_i. Set $L_i' = L_i(\varepsilon)$ and consider the sequence of subfields

$$L = L_r' \supset L_{r-1}' \supset \cdots \supset L_0' = L_0(\varepsilon) \supset L_0 = k.$$

From the Galois correspondence we obtain a sequence of subgroups

$$\{e\} = G(L, L_r') \subset G(L, L_{r-1}') \subset \cdots \subset G(L, L_0') = G(L, k(\varepsilon)) \subset G(L, k).$$

The extension $L'_i \supset L'_{i-1}$ is normal because L'_i is the splitting field over L'_{i-1} of the polynomial $x^{n_i} - \beta_i^{n_i}$. Therefore the sequence of fields $L'_{i-1} \subset L'_i \subset L$ provides us with an exact sequence of groups

$$0 \to G(L, L'_i) \to G(L, L'_{i-1}) \to G(L'_i, L'_{i-1}) \to 0.$$

Thus, $G(L, L'_{i-1})/G(L, L'_i) \cong G(L'_i, L'_{i-1})$ is a cyclic group of order n_i. Hence $G(L, k(\varepsilon))$ is a solvable group.

The next step is the proof of solvability of $G(L, k)$. The extension $k(\varepsilon) \supset k$ is normal, so for the sequence of fields $k \subset k(\varepsilon) \subset L$ we obtain an exact sequence of groups

$$0 \to G(L, k(\varepsilon)) \to G(L, k) \to G(k(\varepsilon), k) \to 0.$$

The group $G(k(\varepsilon), k)$ is abelian by Theorem 5.2.9 (c) on page 197, and so $G(L, k)$ is solvable.

The last step is the proof of solvability of the group $G_k(f) = G(N, k)$, where N is the splitting field of the polynomial f over k. The sequence of fields $k \subset N \subset L$ yields an exact sequence of groups

$$0 \to G(L, N) \to G(L, k) \to G(N, k) \to 0.$$

Therefore $G(N, k)$ is a quotient of the solvable group $G(L, k)$, and hence is solvable itself. $\square$

Example. The equation

$$x^5 - 4x + 2 = 0$$

is not solvable by radicals.

Proof. The Galois group of the polynomial $x^5 - 4x + 2 = 0$ over $\mathbb{Q}$ is equal to S_5. (See page 196). It remains to prove that S_5 is non-solvable. Observe, first of all, that if $H \subset G$ is a normal subgroup such that the group G/H is abelian, then for any $x, y \in G$ the element $xyx^{-1}y^{-1}$ belongs to H. In S_5, there is a normal subgroup A_5 consisting of the even permutations. It is easy to verify that any element of A_5 can be represented in the form $xyx^{-1}y^{-1}$, where $x, y \in A_5$. Indeed, any element of A_5 is either a cycle of length 5, or a cycle of length 3, or the product of two transpositions $(ij)(kl)$ with distinct i, j, k, l.

For the cycle (12345), set $x = (12534)$ and $y = (12)(35)$.
For the cycle (123), set $x = (123)$ and $y = (23)(45)$.
For $(12)(34)$, set $x = (14)(23)$ and $y = (123)$.
Therefore, if $H \subset S_5$ is a normal subgroup and the group S_5/H is abelian, then $H = A_5$ (or S_5). But already A_5 has no normal subgroup K such that A_5/K is abelian. $\square$

5.3.4 Abelian equations

In the memoir "On a particular class of algebraically solvable equations" Abel proved three important statements concerning solvability of equations by radicals.

1. If one of the roots of an irreducible polynomial f can be rationally expressed in terms of the other root, then the solution of the equation $f(x) = 0$ reduces to the solution of several equations of lesser degrees.

2. If the roots of an irreducible polynomial f are of the form x_1, $\theta(x_1)$, $\theta^2(x_1) = \theta(\theta(x_1)), \ldots, \theta^{n-1}(x_1)$, where θ is a rational function such that $\theta^n(x_1) = x_1$, then the equation $f(x) = 0$ is solvable by radicals.

3. If the roots of an irreducible polynomial f are of the form x_1, $\theta_2(x_1)$, $\theta_3(x_1), \ldots, \theta_n(x_1)$, where the θ_i are rational functions such that $\theta_i\theta_j(x_1) = \theta_j\theta_i(x_1)$, then the equation $f(x) = 0$ is solvable by radicals. Moreover, if $\deg f = p_1^{n_1} \cdots p_k^{n_k}$, then the solution of $f(x) = 0$ reduces to solution of n_1 equations of degree p_1, n_2 equations of degree p_2, etc.

The polynomial f in statement 3 is called *abelian*, and the equation $f(x) = 0$ is called an *abelian* equation. Clearly, the Galois group of a polynomial g is abelian if and only if the Galois resolvent of g is an abelian polynomial. Therefore Abel's theorem is a particular case of the Galois theorem which reads as follows:

any equation with an abelian Galois group is solvable by radicals.

Nevertheless, the methods of Abel's theorem still retain a certain significance because Abel's solution of Abel equations is rather constructive.

The polynomial f in statement 2 is called a *cyclic abelian* polynomial. The Galois group of such a polynomial is cyclic.

To solve cyclic abelian equations, Abel applied methods developed by Lagrange and Gauss. His contribution consists in the fact that he separated the most general class of equations to which these methods are applicable. Moreover, studying the theory of elliptic functions, Abel found a new interesting example of a cyclic abelian equation, namely, the *lemniscate division equation*. For a modern proof of the fact that the lemniscate division equation is a cyclic abelian equation, see [Pr3].

Let us begin with statement 1. It deals with polynomials of a particular form but with an arbitrary Galois group. Indeed, all the roots of the Galois resolvent of any polynomial can be rationally expressed in terms of one of the roots.

Theorem 5.3.4. a) *Let f be an irreducible polynomial over a field k of characteristic zero, one of whose roots can be rationally expressed in terms of another root. Then all the roots of f can be organized into a table (elucidated in the course of the proof)*

$$x_1^1, \quad x_2^1 = \theta(x_1^1), \quad \ldots, \quad x_p^1 = \theta^{p-1}(x_1^1),$$

$$\ldots\ldots\ldots\ldots\ldots\ldots\ldots\ldots\ldots\ldots\ldots\ldots\ldots$$

$$x_1^m, \quad x_2^m = \theta(x_1^m), \quad \ldots, \quad x_p^m = \theta^{p-1}(x_1^m),$$

where θ is the rational function such that $\theta^p(x_1^i) = x_1^i$ for $i = 1, \ldots, m$.

b) *The problem of solving the equation $f = 0$ reduces to solving an equation $g = 0$ of degree m with coefficients in k and cyclic abelian equations $h_1 = 0, \ldots, h_m = 0$, where h_i is a polynomial of degree p whose coefficients are rationally expressed (over k) in terms of a root y_i of g.*

Proof. a) Let $x_1, \ldots, x_n$ be the roots of f and let $x_2 = \theta(x_1)$, where θ is a rational function. Consider the polynomial $\varphi(x) = \prod_{i=1}^{n} (x - \theta(x_i))$. The coefficients of this polynomial are symmetric functions in $x_1, \ldots, x_n$, and hence belong to k. The polynomials f and φ have a common root $x_2 = \theta(x_1)$. But the polynomial f is irreducible and $\deg f = \deg \varphi$, and so $\theta(x_1), \ldots, \theta(x_n)$ is a permutation of the numbers $x_1, \ldots, x_n$. Thus to each root x_i there corresponds, uniquely, a root x_j, i.e., $x_j = \theta(x_i)$.

Consider all the possible cycles $x_i, \theta(x_i), \theta^2(x_i), \ldots, \theta^{p-1}(x_i)$, where $p > 0$ is the first integer such that $\theta^p(x_i) = x_i$. Clearly, any two cycles, considered as sets, either do not intersect or coincide. It only remains to prove that the length of all the cycles is the same. Let p be the least length of the cycles. If $\theta^p(x) = x$ for all x, then all the cycles are of length p. If $\theta^p(x) \neq x$, then the equations $\theta^p(x) = x$ and $f(x) = 0$ have a common root. From the irreducibility of f it follows that $\theta^p(x_i) = x_i$ for all $i = 1, \ldots, n$, and hence all the cycles are of length p.

b) To avoid cumbersome notations, we assume that $p = 3$ and $m = 4$. In the general case the proof is the same. The table of roots can be expressed as follows:

$$x_1, \quad x_2 = \theta(x_1), \quad x_3 = \theta^2(x_1),$$
$$x_4, \quad x_5 = \theta(x_4), \quad x_6 = \theta^2(x_4),$$
$$x_7, \quad x_8 = \theta(x_7), \quad x_9 = \theta^2(x_7),$$
$$x_{10}, x_{11} = \theta(x_{10}), x_{12} = \theta^2(x_{10}).$$

Let q be an arbitrary symmetric polynomial in three variables over k. Then

$$\varphi(x_1) = q\left(x_1, \theta(x_1), \theta^2(x_1)\right) = q\left(\theta(x_1), \theta^2(x_1), x_1\right) = \varphi(x_2).$$

Similarly, $\varphi(x_1) = \varphi(x_3)$. Thus, if

$$\varphi(x_i) = q\left(x_i, \theta(x_i), \theta^2(x_i)\right),$$

then

$$\varphi(x_1) = \varphi(x_2) = \varphi(x_3) = q_1, \qquad \varphi(x_4) = \varphi(x_5) = \varphi(x_6) = q_2,$$
$$\varphi(x_7) = \varphi(x_8) = \varphi(x_9) = q_3, \qquad \varphi(x_{10}) = \varphi(x_{11}) = \varphi(x_{12}) = q_4.$$

Hence $q_1 + q_2 + q_3 + q_4 = \frac{1}{3} \sum_{i=1}^{12} \varphi(x_i) \in k$. Considering the functions q^2, q^3, q^4 instead of q, we see that $\sum q^2, \sum q^3, \sum q^4 \in k$. This means that the coefficients of the polynomial $\prod(y - q_i)$ lie in k.

If r is one more symmetric polynomial in three variables, we can consider symmetric polynomials rq^l for $l = 0, 1, 2, 3$. Now, determine the r_i in the same way as the q_i. Then the system of equations

$$r_1 q_1^l + r_2 q_2^l + r_3 q_3^l + r_4 q_4^l = R_l, \text{ where } l = 0, 1, 2, 3 \text{ and } R_l \in k,$$

shows that $r_i = \Phi(q_i)$, where Φ is a rational function (the same for all i).

Let $q = t_1 + t_2 + t_3$, $r = t_1 t_2 + t_2 t_3 + t_3 t_1$ and $\tilde{r} = t_1 t_2 t_3$. Then

$$q_1 = x_1 + x_2 + x_3 = y_1$$

(a root of $\prod(y - q_i)$), and

$$r_1 = x_1 x_2 + x_2 x_3 + x_3 x_1 = \Phi(y_1)$$

and $\tilde{r}_1 = x_1 x_2 x_3$; we denote: $\tilde{r}_1 = \tilde{\Phi}(y_1)$. Therefore x_1, x_2, x_3 are the roots of the equation

$$x^3 - y_1 x^2 + \Phi(y_1)x - \tilde{\Phi}(y_1) = 0,$$

where Φ and $\tilde{\Phi}$ are rational functions. This equation is a cyclic abelian one since $x_2 = \theta(x_1)$, $x_3 = \theta^2(x_1)$ and $x_1 = \theta^3(x_1)$. $\square$

Theorem 5.3.5. a) *Any cyclic abelian equation is solvable by radicals.*

b) *The solutions of a cyclic abelian equation of order $n = pm$ reduces to the solution of two cyclic abelian equations of orders p and m.*

Proof. a) Let f be an cyclic abelian polynomial of degree n with roots $x_1, \ldots, x_n$ and let ε be an nth primitive root of unity. Let θ be a rational function such that $x_{k+1} = \theta^k(x_1)$ and $x_1 = \theta^n(x_1)$. Consider the *Lagrange resolvent*

$$(\varepsilon^r, x_1) = x_1 + \varepsilon^r \theta(x_1) + \varepsilon^{2r} \theta^2(x_1) + \cdots + \varepsilon^{(n-1)r} \theta^{(n-1)}(x_1).$$

Since $x_{k+1} = \theta^k(x_1)$, it follows that

$$(\varepsilon^r, x_{k+1}) = \theta^k(x_1) + \varepsilon^r \theta^{k+1}(x_1) + \varepsilon^{2r} \theta^{k+2}(x_1) + \cdots = \sum \varepsilon^{sr} \theta^{k+s}(x_1).$$

Clearly, $\theta^{k+s}(x_1) = x_1$ for $s = n-k$, and therefore $(\varepsilon^r, x_{k+1}) = \varepsilon^{(n-k)r}(\varepsilon^r, x_1)$. In particular, $(\varepsilon^r, x_{k+1})^n = (\varepsilon^r, x_1)^n$. Hence

$$(\varepsilon^r, x_1)^n = (\varepsilon^r, x_2)^n = \cdots = (\varepsilon^r, x_n)^n = \frac{1}{n} \sum (\varepsilon^r, x_i)^n = u_r(\varepsilon),$$

where u_r is a rational function.

Thus

$$x_1 + \varepsilon^r \theta(x_1) + \varepsilon^{2r}\theta^2(x_1) + \cdots + \varepsilon^{(n-1)r}\theta^{(n-1)}(x_1) = \sqrt[n]{u_r(\varepsilon)} \qquad (*)$$

for $r = 1, \ldots, n-1$. Further, for $r = 0$, we see that $x_1 + x_2 + \cdots + x_n = -a_1$, where a_1 is the coefficient of x^{n-1} in the polynomial f. Adding up all the equalities $(*)$ for $r = 0, 1, \ldots, n-1$ we see that the sum of coefficients of x_1 is equal to n, and the sum of coefficients of $\theta^m(x_1)$, where $1 \le m \le n-1$, is equal to

$$1 + \varepsilon^m + \varepsilon^{2m} + \cdots + \varepsilon^{(n-1)m} = 0.$$

Thus,

$$nx_1 = a_1 + \sqrt[n]{u_1(\varepsilon)} + \cdots + \sqrt[n]{u_{n-1}(\varepsilon)}.$$

If we multiply the rth equality $(*)$ by ε^{-kr} we similarly deduce that

$$nx_{k+1} = a_1 + \varepsilon^{-k}\sqrt[n]{u_1(\varepsilon)} + \cdots + \varepsilon^{-k(n-1)}\sqrt[n]{u_{n-1}(\varepsilon)}.$$

It is not difficult to obtain a slightly more precise expression:

$$nx_{k+1} = a_1 + y + A_2 y^2 + \cdots + A_{n-1}y^{n-1},$$

where $y = \varepsilon^{-k}\sqrt[n]{u_1(\varepsilon)}$ and $A_2, \ldots, A_{n-1}$ are constants which rationally depend on ε. Indeed,

$$\frac{(\varepsilon^r, x_{k+1})}{(\varepsilon, x_{k+1})^r} = \frac{\varepsilon^{(n-k)r}(\varepsilon^r, x_1)}{(\varepsilon^{n-k}(\varepsilon^r, x_1))^r} = \frac{(\varepsilon^r, x_1)}{(\varepsilon, x_1)^r} = \frac{\sqrt[n]{u_r(\varepsilon)}}{\left(\sqrt[n]{u_1(\varepsilon)}\right)^r} = A_r.$$

This means that A_r depends rationally on ε and symmetric functions in $x_1, \ldots, x_n$.

b) To avoid cumbersome notations, we assume that $p = 3$ and $m = 4$. In the general case the proof is the same. Let

$$y_1 = x_1 + x_5 + x_9 = x_1 + \theta^4(x_1) + \theta^8(x_1),$$
$$y_2 = x_2 + x_6 + x_{10} = \theta(x_1) + \theta^5(x_1) + \theta^9(x_1),$$
$$y_3 = x_3 + x_7 + x_{11} = \theta^2(x_1) + \theta^6(x_1) + \theta^{10}(x_1),$$
$$y_4 = x_4 + x_8 + x_{12} = \theta^3(x_1) + \theta^7(x_1) + \theta^{11}(x_1).$$

In the situation considered, the conditions of Theorem 5.3.4 (with θ replaced with θ^4) are satisfied, and so x_1, $\theta^4(x_1)$ and $\theta^8(x_1)$ are the roots of a cyclic abelian equation of order $p = 3$ whose coefficients are rational functions in y_1. Further, y_1, y_2, y_3 and y_4 are the roots of an equation of degree $m = 4$ with rational coefficients. It only remains to prove that this equation is cyclic abelian.

Let

$$q_l(x) = \left(\theta(x_1) + \theta^5(x_1) + \theta^9(x_1)\right)\left(x_1 + \theta^4(x_1) + \theta^8(x_1)\right)^l.$$

Then $q_l(x_1) = y_2 y_1^l$. Further,

$$q_l(x_5) = (x_6 + x_{10} + x_2)(x_5 + x_9 + x_1)^l = q_l(x_1).$$

Similarly, $q_l(x_9) = q_l(x_5)$. Similar arguments show that

$$y_3 y_2^l = q_l(x_2) = q_l(x_6) = q_l(x_{10}),$$
$$y_4 y_3^l = q_l(x_3) = q_l(x_7) = q_l(x_{11}),$$
$$y_1 y_4^l = q_l(x_4) = q_l(x_8) = q_l(x_{12}).$$

Hence

$$y_2 y_1^l + y_3 y_2^l + y_4 y_3^l + y_1 y_4^l = \frac{1}{4}\sum_{i=1}^{12} q_l(x_i) \in k.$$

For the system of equations

$$y_2 y_1^l + \cdots + y_1 y_4^l = T_l, \text{ where } l = 0, 1, 2, 3 \text{ and } T_l \in k,$$

we deduce that $y_2 = \varphi(y_1)$, $y_3 = \varphi(y_2)$, $y_4 = \varphi(y_3)$ and $y_1 = \varphi(y_4)$. $\square$

Corollary. *Any cyclic abelian equation of degree 2^n can be solved by quadratic radicals.*

5.3.5 The Abel-Galois criterion for solvability of equations of prime degree

Evariste Galois perished in a duel and had no time to publish his main results in the study of the theory of solvability of polynomial equations by radicals. Still, he managed to publish some of his results. In a short notice in "Bulletin des Sciences mathématiques" (1830) Galois communicated the following result:

In order that an equation of prime degree be solvable by radicals it is necessary and sufficient that given any two of its roots the others would rationally depend on them.

It is interesting to observe that in 1828 Abel wrote to Crelle that he found a criteria for solvability by radicals of the equation of prime degree. Abel formulated his criterion in almost the same words as Galois: *"In any triple of the roots one root should be rationally expressed in terms of the other two"*. No testimony, however, on the existence of Abel's proof has survived.

Theorem 5.3.6. a) *An irreducible over $\mathbb{Q}$ equation $f = 0$ of prime degree p is solvable by radicals if and only if its roots can be numbered so that any permutation σ from the Galois group is of the form $\sigma(i) \equiv ai + b \pmod{p}$, where $a \not\equiv 0 \pmod{p}$.*

b) *An irreducible over $\mathbb{Q}$ equation $f = 0$ of prime degree p is solvable by radicals if and only if all its roots can be rationally expressed in terms of any two of the roots. (In other words, if $\alpha_1, \ldots, \alpha_p$ are all the roots of the equation, then $\mathbb{Q}(\alpha_1, \ldots, \alpha_p) = \mathbb{Q}(\alpha_i, \alpha_j)$ for any distinct i and j).*

Proof. a) For the Galois group indicated, the multiplication law is of the form
$$[a_1, b_1][a_2, b_2] = [a_1 a_2, a_1 b_2 + b_1].$$
The solvability of this group was proved in the theorem on simple radical extensions (Theorem 5.2.9 on page 197). Therefore it remains to prove that, if an irreducible equation of prime degree p is solvable by radicals, then its Galois group consists of transformations of the form indicated.

When proving theorems on equations solvable by radicals, we have shown that, for an equation $f = 0$ solvable by radicals, there exists a sequence of radical extensions of prime degrees
$$L = L'_r \supset L'_{r-1} \supset , \cdots \supset L'_0 = L_0(\varepsilon) \supset L_0 = \mathbb{Q},$$
where ε is an nth primitive root of unity, where $n = [L : \mathbb{Q}]$, the extension $L \supset \mathbb{Q}$ is normal and L contains the splitting field N for f.

In this scenario, $G(N, \mathbb{Q}) = G(L, \mathbb{Q})/G(L, N)$. Therefore it suffices to prove that any automorphism of L over $\mathbb{Q}$ permutes the roots of f in the manner indicated above, i.e., the root number i is replaced by the root number $ai + b$.

We may assume that L'_{r-1} does not contain all the roots of f. The proof of the theorem is based on the fact that f is irreducible over L'_{r-1} and the degree of the extension $L \supset L'_{r-1}$ is equal to p. This means that, until the very last radical extension, the polynomial f remains irreducible whereas at the last step it factorizes into linear factors. The statement required obviously follows from the next lemma.

Lemma. *Let f be irreducible over a field k which contains a qth primitive root of unity, where q is a prime. Further, let $K = k(\beta)$, where $\beta^q \in k$. Then, over K, the polynomial f is either irreducible or factorizes into q irreducible factors of equal degree.*

Proof. Let L be the splitting field for f over k. The field k contains a qth primitive root of unity, and so $L(\beta)$ is the splitting field of f over $k(\beta)$.

The sequences of extensions $L(\beta) \supset k(\beta) \supset k$ and $L(\beta) \supset L \supset k$ yield the exact sequences

$$0 \to G(L(\beta), k(\beta)) \to G(L(\beta), k) \to G(k(\beta), k) \to 0,$$

$$\|$$

$$0 \to \quad G(L(\beta), L) \quad \to G(L(\beta), k) \to \quad G(L, k) \quad \to 0.$$

Therefore

$$\left| G\left(L(\beta), k(\beta)\right) \right| \cdot \left| G\left(k(\beta), k\right) \right| = \left| G(L, k) \right| \cdot \left| G\left(L(\beta), L\right) \right|.$$

By Theorem 5.2.5 on page 195, the groups $G\left(L(\beta), k(\beta)\right)$ and $G(L(\beta), L)$ are subgroups in $G(L, k)$ and $G\left(k(\beta), k\right)$, respectively. Moreover, by Theorem 5.2.9 (b) on page 197, the groups $G\left(L(\beta), L\right)$ and $G\left(k(\beta), k\right)$ are subgroups in $\mathbb{F}_q$. By the hypothesis q is a prime, and so $\mathbb{F}_q$ has no nontrivial subgroups, and therefore

$$\left[G(L, k) : G\left(L(\beta), k(\beta)\right) \right] = 1 \text{ or } q,$$

where the degree of the extension is equal to q only if $G\left(L(\beta), L\right)$ is trivial, i.e., $G\left(L(\beta), k\right) \cong G(L, k)$. In this case $G\left(L(\beta), k(\beta)\right)$ is a normal subgroup in $G(L, k)$ of index q.

Recall that a polynomial (without multiple roots) is irreducible if and only if its Galois group acts transitively on the set of its roots (Theorem 5.2.6 on page 195). Hence we only have to consider the case when $H = G(L(\beta), k(\beta))$ is a normal subgroup of $G = G(L, k)$ of index q. In this case $G/H \cong \mathbb{F}_q$, and so

$$G = \{H, gH, g^2 H, \ldots, g^{q-1} H\} \text{ for some } g \in G.$$

Let $\alpha_1, \ldots, \alpha_n$ be all the roots of f. We define the set $H\alpha_i = \{h(\alpha_i) \mid h \in H\}$. Then $H\alpha_i$ and $H\alpha_j$ either coincide or do not intersect. Therefore the sets $H\alpha_1$ and $gH\alpha_1 = Hg(\alpha_1)$ either coincide or do not intersect.

In the first case, the group H transitively acts on the set of Roots, whereas in the second case the set of roots splits into q subsets of equal cardinality on each of which the group H transitively acts. Let $n = rq$ and let H transitively act on the set $\alpha_{i_1}, \ldots, \alpha_{i_r}$. Then H preserves all the coefficients of the polynomial $(x - \alpha_{i_1}) \cdots (x - \alpha_{i_r})$, and hence this polynomial is irreducible over $k(\beta)$. $\square$

Thus, f is irreducible over L'_{r-1} and it factorizes into linear factors over $L'_r = L'_{r-1}(\beta)$, where $\beta^q \in L'_{r-1}$. This means, in particular, that $q = p$. Therefore $L = L'_r$ is a cyclic extension of L'_{r-1} of degree p and L'_{r-1} contains a pth primitive root of unity. Hence $G(L, L'_{r-1}) = \mathbb{F}_p$. Let σ be a generator of this group. The roots $\alpha_1, \ldots, \alpha_p$ of f can be numbered so that $\sigma(i) = i + 1$, i.e., σ sends α_i into α_{i+1}.

The group $G(L, L'_{r-1})$ is a normal subgroup of $G(L, L'_{r-2})$. Let τ be an arbitrary element of $G(L, L'_{r-2})$. Then $\tau\sigma\tau^{-1} \in G(L, L'_{r-1})$, i.e., $\tau\sigma\tau^{-1} = \sigma^a$ for some a. This means that $\tau\sigma(i) = \sigma^a\tau(i)$, i.e., $\tau(i+1) = \tau(i) + a$. Therefore

$$\tau(2) = \tau(1) + a,$$
$$\tau(3) = \tau(2) + a = \tau(1) + 2a,$$
$$\cdots\cdots\cdots\cdots\cdots\cdots\cdots\cdots\cdots\cdots\cdots\cdots$$
$$\tau(i) = \tau(1) + (i-1)a = ai + (\tau(1) - a) = ai + b,$$

where $b = \tau(1) - a$. The element τ is of the form required.

Let us prove now that if all the elements of $G(L, L'_m)$ are of the form $\tau(i) = ai + b$, then all the elements of $G(L, L'_{m-1})$ are also of the same form. In the proof we use the fact that $G(L, L'_m)$ contains an element such that $\sigma(i) = i + 1$ and $G(L, L'_m)$ is a normal subgroup of $G(L, L'_{m-1})$. Let μ be an arbitrary element of $G(L, L'_{m-1})$. Then $\mu\sigma\mu^{-1} \in G(L, L'_m)$ so $\mu\sigma\mu^{-1}(i) = ai + b$. For $j = \mu^{-1}(i)$, we have $\mu\sigma(j) = a\mu(j) + b$, i.e., $\mu(j + 1) = a\mu(j) + b$. Let us prove, first of all, that $a = 1$. Indeed,

$$\mu(2) = a\mu(1) + b,$$
$$\mu(3) = a\mu(2) + b = a^2\mu(1) + ab + b,$$
$$\mu(4) = a\mu(3) + b = a^3\mu(1) + a^2b + ab + b,$$
$$\cdots\cdots\cdots\cdots\cdots\cdots\cdots\cdots\cdots\cdots\cdots\cdots$$
$$\mu(j) = a^{j-1}\mu(1) + (a^{j-2} + a^{j-3} + \cdots + a + 1)b.$$

Hence, $\mu(1) = \mu(p+1) = a^p\mu(1) + (a^{p-1} + a^{p-2} + \cdots + a + 1)b$, i.e.,

$$(1 - a^p)\mu(1) \equiv (a^{p-1} + a^{p-2} + \cdots + a + 1)b \pmod{p}.$$

Let us multiply both sides of the last congruency by $1 - a$ and use the fact that $a^p \equiv a \pmod{p}$. As a result, we obtain

$$(1 - a)^2\mu(1) \equiv (1 - a)b \pmod{p}.$$

If $a \not\equiv 1 \pmod{p}$, then $\mu(1) \equiv (1 - a)^{-1}b \pmod{p}$. But then the same arguments show that $\mu(2) \equiv (1 - a)^{-1}b \pmod{p}$ which is impossible.

b) If an irreducible equation of prime degree is solvable by radicals, then its Galois group consists of transformations of the form $i \mapsto ai + b$. Any transformation of such a form with two fixed points is the identity. This means that, after adjoining two roots α_i and α_j, the Galois group reduces to the identity transformation, i.e., all the roots belong to $\mathbb{Q}(\alpha_i, \alpha_j)$.

Now suppose that for any two distinct roots α_i and α_j the field $\mathbb{Q}(\alpha_i, \alpha_j)$ contains all the other roots. This means that if the transformation from the Galois group fixes two roots, then it fixes all the other roots, i.e., any nonidentical transformation has no more than one fixed point.

The Galois group G transitively acts on the set $\{\alpha_1, \ldots, \alpha_p\}$ of p elements, and so $|G|$ is divisible by p.

Lemma. *If the number of elements of the group G is divisible by a prime p, then G contains an element of order p.*

Proof. Let $|G| = n = mp$. We use induction on m. For $m = 1$, the statement is obvious. If G has a proper subgroup H whose index $[G : H]$ is not divisible by p, then $|H|$ is divisible by p and we can apply the inductive hypothesis. Therefore we may assume that the index of any proper subgroup is divisible by p.

For any $x \in G$, consider the subgroup $N_x = \{g \in G \mid gxg^{-1} = x\}$ and the class of conjugate elements $G_x = \{gxg^{-1} \mid g \in G\}$. Clearly, $|G_x| = [G : N_x]$. The conjugacy classes either do not intersect or coincide, and so $n = n_1 + \cdots + n_s$, where n_i is the number of elements in the ith conjugacy class. By the hypothesis n_i is either equal to 1 or divisible by p. If $n_i = 1$, then the corresponding element x commutes with all the elements of G, i.e., belongs to the center $Z(G)$. The number of the n_i equal to 1 is divisible by p and distinct from zero (since $n_i = 1$ corresponds to the identity element). Hence $Z(G)$ is an abelian group whose order is divisible by p. Therefore $Z(G)$ has an element of order p. $\square$

Any element σ of order p in $G \subset S_p$ is a cycle of length p. Renumbering the roots, we may assume that $\sigma(i) = i + 1$. Let us show that σ generates a normal subgroup in G. Let $\tau \in G$. Then $\tau\sigma\tau^{-1} \neq \mathrm{id}$ and $(\tau\sigma\tau^{-1})^p = \mathrm{id}$, i.e., $\tau\sigma\tau^{-1}$ is a cycle of length p. Now define $a(i)$ by the formula

$$\tau\sigma\tau^{-1}(i) = i + a(i) = \sigma^{a(i)}(i).$$

A cycle of length p in the group S_p cannot possess fixed points, and so $a(i) \neq 0$ for all i. Therefore the function $a(i)$ on the set of p elements takes not more than $p - 1$ distinct values, and hence it assumes a certain value a at two distinct points, say i and j. This means that the transformation $\sigma^{-a}\tau\sigma\tau^{-1}$ has two fixed points. But we know already that any transformation from G with two fixed points is the identity. Hence $\tau\sigma\tau^{-1}(i) = \sigma^a$.

The equation $\tau\sigma\tau^{-1}(i) = \sigma^a$ implies that $\tau(i) = ai + b$, where $b = \tau(1) - a$. The group consisting of elements of this form is, as we know, solvable. $\square$

Corollary. *If an irreducible over $\mathbb{Q}$ equation of prime degree $p \geq 3$ is solvable by radicals, then the number of its real roots is equal to either 1 or p.*

Proof. The number p is odd, and so any equation of degree p has at least one real root. If an equation of prime degree p which is solvable by radicals has two real roots α_i and α_j, then all its roots belong to $\mathbb{Q}(\alpha_i, \alpha_j) \subset \mathbb{R}$. $\square$

5.4 Calculation of the Galois groups

5.4.1 The discriminant and the Galois group

Theorem 5.4.1. *Let $f \in \mathbb{Z}[x]$ an irreducible monic polynomial of degree n. The Galois group of f over $\mathbb{Q}$ is contained in the alternating subgroup $A_n \subset S_n$ if and only if the discriminant $D(f)$ is a perfect square.*

Proof. Let $\alpha_1, \ldots, \alpha_n$ be all the roots of f. Then $D(f) = \prod_{i<j} (\alpha_i - \alpha_j)^2$.
Set $\delta = \delta(f) = \prod_{i<j} (\alpha_i - \alpha_j)$. If $\sigma \in G_{\mathbb{Q}}(f)$, then

$$\sigma(\delta) = \prod_{i<j} (\sigma(\alpha_i) - \sigma(\alpha_j)) = (-1)^{\sigma}\delta.$$

By the hypothesis f is irreducible. Therefore, in particular, $\delta \neq 0$. Hence all the automorphisms $\sigma \in G_{\mathbb{Q}}(f)$ preserve δ if and only if $G_{\mathbb{Q}}(f) \subset A_n$. On the other hand, all the automorphisms $\sigma \in G_{\mathbb{Q}}(f)$ preserve δ if and only if $\delta \in \mathbb{Q}$. $\square$

Example. Let $f(x) = x^3 + ax^2 + bx + c$ be an irreducible polynomial over $\mathbb{Z}$ and D its discriminant. Then $G_{\mathbb{Q}}(f) = A_3$ if $\sqrt{D} \in \mathbb{Q}$ and $G_{\mathbb{Q}}(f) = S_3$ if $\sqrt{D} \notin \mathbb{Q}$.

Proof. Since f is irreducible, its Galois group $G_{\mathbb{Q}}(f)$ is transitive. In S_3, there is only one transitive group distinct from S_3, namely, A_3. $\square$

5.4.2 Resolvent polynomials

Let $\varphi \in \mathbb{Q}[x_1, \ldots, x_n]$ and

$$G_{\varphi} = \{\sigma \in S_n \mid \sigma\varphi = \varphi \quad \text{(i.e., } \varphi(x_{\sigma(1)}, \ldots, x_{\sigma(n)}) = \varphi(x_1, \ldots, x_n))\}.$$

Under the action of S_n we obtain from φ distinct functions

$$\varphi_1 = \varphi, \quad \varphi_2 = \tau_2\varphi, \ldots, \varphi_m = \tau_m\varphi, \text{ where } m = |S_n|/|G_{\varphi}|.$$

Example 5.4.2. If $\varphi = x_1 x_2 + x_3 x_4$, then G_{φ} consists of the identity element and the permutations

$$(12), \quad (34), \quad (12)(34), \quad (13)(24), \quad (14)(23), \quad (1324), \quad (1423).$$

So in this case $\varphi_2 = x_1 x_3 + x_2 x_4$ and $\varphi_3 = x_1 x_4 + x_2 x_3$.

It is easy to verify that $G_{\varphi_i} = \tau_i G_{\varphi} \tau_i^{-1}$. Indeed, the equality $\sigma\varphi_i = \varphi_i$ is equivalent to $\sigma\tau_i\varphi = \tau_i\varphi$. Therefore $\tau_i^{-1}\sigma\tau_i \in G_{\varphi_i}$, i.e., $\sigma \in \tau_i G_{\varphi_i} \tau_i^{-1}$.

Let $f(x) = x^n + a_{n-1}x^{n-1} + \cdots + a_0$ be a polynomial with integer coefficients and $\varphi \in \mathbb{Z}[x_1, \ldots, x_n]$. Define the polynomials $\varphi_1, \ldots, \varphi_m$ and the respective groups G_{φ_i} as above. The *resolvent* polynomial (of φ and f) is the polynomial

$$\operatorname{Res}(\varphi, f)(x) = \prod_{i=1}^{m} (x - \varphi_i(\alpha_1, \ldots, \alpha_n)),$$

where $\alpha_1, \ldots, \alpha_n$ are all the roots of f.

Since the resolvent polynomial $\operatorname{Res}(\varphi, f)$ has integer coefficients, one can calculate it by approximately computing the roots of f (the coefficients of the resolvent polynomial are calculated with sufficient accuracy and rounded up to the nearest integer).

Theorem 5.4.3. *Let the resolvent polynomial* $\mathrm{Res}(\varphi, f)$ *have no multiple roots. Then the Galois group of* f *over* $\mathbb{Q}$ *is contained in the group conjugate to* G_φ *if and only if* $\mathrm{Res}(\varphi, f)$ *has an integer root.*

Proof. First, suppose that

$$G_{\mathbb{Q}}(f) \subset \tau G_\varphi \tau^{-1} = G_{\varphi_i},$$

where $\varphi_i = \tau\varphi$. Then the number $\varphi_i(\alpha_1, \ldots, \alpha_n)$ which is a root of $\mathrm{Res}(\varphi, f)$ is preserved under the action of all the transformations from the Galois group, and therefore is a rational number. But the numbers $\alpha_1, \ldots, \alpha_n$ are algebraic integers and the coefficients of the polynomial φ_i are integers, so $\varphi_i(\alpha_1, \ldots, \alpha_n)$ is an integer.

Now suppose that $\varphi_i(\alpha_1, \ldots, \alpha_n)$ is an integer. By the hypothesis the resolvent polynomial has no multiple roots. This means that, if

$$\sigma\varphi_i(\alpha_1, \ldots, \alpha_n) = \varphi_j(\alpha_1, \ldots, \alpha_n),$$

then $\tau_j^{-1}\sigma\tau_i \in G_\varphi$. Any element σ of the Galois group preserves the integer $\varphi_i(\alpha_1, \ldots, \alpha_n)$, and hence, for this σ, we have $i = j$, i.e., $\sigma \in \tau_i G_\varphi \tau_i^{-1}$. $\square$

Making use of Theorems 5.4.1 and 5.4.3 we can easily calculate the Galois group of any irreducible polynomial of the form $f(x) = x^4 + a_1 x^3 + a_2 x^2 + a_3 x + a_4$ with integer coefficients. First of all, observe that, up to conjugation, S_4 contains only the following transitive subgroups:

1) The whole group S_4.

2) The alternating group A_4.

3) The dihedral group of order 8 described in Example 5.4.2. (We denote this group by D_4.)

4) Klein's Viergruppe of order 4 consisting, apart from the identity, of the permutations $(12)(34)$, $(13)(24)$ and $(14)(23)$. (We denote this group by V_4.)

5) The cyclic group $\mathbb{F}_4$ generated by the cycle (1234).

There are the following embeddings: $V_4 \subset D_4 \cap A_4$ and $\mathbb{F}_4 \subset D_4$ but $\mathbb{F}_4 \not\subset A_4$.

Let us start our calculation of the Galois group with the calculation of the discriminant D of f and the resolvent polynomial $\mathrm{Res}(\varphi, f)(x)$ for $\varphi = x_1 x_2 + x_3 x_4$. Easy calculations show that

$$\mathrm{Res}(\varphi, f)(x) = x^3 - a_2 x^2 - (a_1 a_3 - 4a_4)x - a_4 a_1^2 - 4a_4 a_2 - a_3^2.$$

It is also easy to verify that $\mathrm{Res}(\varphi, f)(x)$ has no multiple roots. Let, e.g.,

$$\alpha_1\alpha_2 + \alpha_3\alpha_4 = \alpha_1\alpha_3 + \alpha_2\alpha_4.$$

Then $(\alpha_1 - \alpha_4)(\alpha_2 - \alpha_3) = 0$. But f has no multiple roots, and so $\alpha_1 \neq \alpha_4$ and $\alpha_2 \neq \alpha_3$.

If the resolvent polynomial has no integer roots, then the Galois group is equal to either S_4 or A_4. One can distinguish these groups with the help of

D: if D is a perfect square, then the Galois group is equal to A_4. Otherwise, it is equal to S_4.

If the resolvent polynomial has an integer root, then the Galois group is equal to $\mathbb{F}_4$, V_4 or D_4. Of these groups only V_4 is contained in A_4. So if $\sqrt{D}$ is an integer, then the Galois group is equal to V_4; otherwise it is equal to either $\mathbb{F}_4$ or D_4.

One can distinguish which is actually the case — that of $\mathbb{F}_4$ or D_4 — with the help of the resolvent polynomial corresponding to

$$\varphi = x_1 x_2^2 + x_2 x_3^2 + x_3 x_4^2 + x_4 x_1^2,$$

for which $G_\varphi = \mathbb{F}_4$. But in this case the resolvent polynomial (of degree 6) may have multiple roots. For example, the equality

$$\alpha_1 \alpha_2^2 + \alpha_2 \alpha_3^2 + \alpha_3 \alpha_4^2 + \alpha_4 \alpha_1^2 = \alpha_1 \alpha_2^2 + \alpha_2 \alpha_4^2 + \alpha_4 \alpha_3^2 + \alpha_3 \alpha_1^2$$

is equivalent to the equality

$$(\alpha_1 - \alpha_2)(\alpha_3 + \alpha_4) + \alpha_3 \alpha_4 = 0. \tag{1}$$

But if we replace each root α_i with $\alpha_i + a$, then, for some a, identity (1) will be violated.

In the general case one can get rid of the multiple roots of the resolvent polynomial applying a more complicated Tchirnhaus transformation. This is a useful practical device (simple calculations but does not always work), whereas theoretically (difficult calculations but always works) one can use the following statement.

Theorem 5.4.4. *Let $f \in \mathbb{Z}[x]$ be a polynomial of degree n without multiple roots and let $G \subset S_n$ be an arbitrary subgroup. Then there exists a function $\varphi \in \mathbb{Z}[x_1, \ldots, x_n]$ such that $G_\varphi = G$ and the resolvent polynomial $\mathrm{Res}(\varphi, f)$ has no multiple roots.*

Proof. Let $\alpha_1, \ldots, \alpha_n$ be all the roots of f. We have shown in the construction of the Galois resolvent that there exist integers $m_1, \ldots, m_n$ such that the numbers $m_1 \alpha_{\sigma(1)} + \cdots + m_n \alpha_{\sigma(n)}$ are distinct for any $\sigma \in S_n$. We define

$$\psi(t, x_1, \ldots, x_n) = \prod_{\sigma \in G} (t - m_1 \alpha_{\sigma(1)} - \cdots - m_n \alpha_{\sigma(n)}).$$

For any $\tau \in S_n$, consider the polynomial

$$\tau \psi(t, x_1, \ldots, x_n) = \psi(t, x_{\tau(1)}, \ldots, x_{\tau(n)}).$$

The polynomials $\tau_1 \psi$ and $\tau_2 \psi$ are distinct if and only if they are distinct as polynomials in t for the fixed values $x_1 = \alpha_1, \ldots, x_n = \alpha_n$.

Under the action of S_n, we construct from the function $\psi(t, x_1, \ldots, x_n) = \psi$ distinct functions $\psi_1 = \psi$, ψ_2, $\ldots$, ψ_m, and the polynomials in t of the form $\psi_1(t, \alpha_1, \ldots, \alpha_n), \ldots, \psi_m(t, \alpha_1, \ldots, \alpha_n)$ are also distinct. Therefore there exists $t_0 \in \mathbb{Z}$ for which the numbers $\psi_1(t_0, \alpha_1, \ldots, \alpha_n), \ldots, \psi_m(t_0, \alpha_1, \ldots, \alpha_n)$ are distinct. But then the polynomial

$$\varphi(x_1, \ldots, x_n) = \psi(t_0, x_1, \ldots, x_n) =$$
$$= \prod_{\sigma \in G} (t_0 - m_1 \alpha_{\sigma(1)} - \cdots - m_n \alpha_{\sigma(n)})$$

is the one to be constructed. Indeed, if $\tau \notin G$, the polynomials $\varphi(x_1, \ldots, x_n)$ and $\tau\varphi(x_1, \ldots, x_n)$ are distinct since they have distinct values at the point $(x_1, \ldots, x_n) = (\alpha_1, \ldots, \alpha_n)$. $\square$

5.4.3 The Galois group modulo p

Let $f(x) = x^n + a_1 x^{n-1} + \cdots + a_n$ be an irreducible polynomial with integer coefficients. For any prime p, consider the polynomial $f \pmod{p}$ over $\mathbb{F}_p$ by replacing the coefficients of f by the corresponding residue classes modulo p. The polynomial $f \pmod{p}$ can turn out to be reducible and the structure of its factorization into irreducible factors is closely connected with the structure of the Galois group of f over $\mathbb{Q}$. We now study this connection and apply it to the calculation of certain Galois groups.

We only consider the primes p for which the polynomial $f \pmod{p}$ has no multiple roots, i.e., for which the polynomials $f \pmod{p}$ and $f' \pmod{p}$ are relatively prime. The latter condition is equivalent to the fact that p does not divide $R(f, f') = \pm D(f)$.

In what follows we assume that $D(f)$ is not divisible by p.

The roots of the polynomial $f \pmod{p}$ do not necessarily belong to $\mathbb{F}_p$, but there exists a finite extension of $\mathbb{F}_p$ which contains all the roots of $f \pmod{p}$. In order to construct such an extension, it suffices to show how to adjoin to an arbitrary finite field $\mathbb{F}_q$ a root of an irreducible over $\mathbb{F}_q$ polynomial h. It is easy to verify that the quotient ring

$$K = \mathbb{F}_q[x]/h(x)\mathbb{F}_q[x]$$

is a field. Indeed, if the polynomial $g(x) \in \mathbb{F}_q[x]$ is not divisible by $h(x)$, then it is relatively prime to $h(x)$, and so $u(x)g(x) + v(x)h(x) = 1$ for some $u(x)$ and $v(x) \in \mathbb{F}_q[x]$. Hence $u \equiv g^{-1} \pmod{h(x)}$. Clearly $K = \mathbb{F}_q(\alpha)$, where α is the image of x under the canonical projection. Clearly $h(\alpha) = 0$, i.e., α is a root of h.

Having adjoined to $\mathbb{F}_p$ all the roots $\alpha_1, \ldots, \alpha_n$ of the polynomial $f \pmod{p}$ we obtain a field $\mathbb{F}_p(\alpha_1, \ldots, \alpha_n) \cong \mathbb{F}_{p^r}$. Let Gal be the Galois group of f over $\mathbb{Q}$, and Gal_p the group of automorphisms of $\mathbb{F}_p(\alpha_1, \ldots, \alpha_n)$ over $\mathbb{F}_p$. Any such automorphism sends a root α_i into a root α_j, and any permutation of roots uniquely determines an automorphism. Therefore, having numbered the roots, we may assume that Gal_p is a subgroup of S_n. Observe that the roots of the polynomials f and $f \pmod{p}$ belong to distinct sets, and between the roots of these polynomials there is no natural one-to-one correspondence.

Theorem 5.4.5. a) *Under an appropriate numbering of the roots of f and f (mod p) the group Gal_p becomes a subgroup of* Gal.

b) Gal_p *is a cyclic group of order* $r = [\mathbb{F}_p(\alpha_1, \ldots, \alpha_n) : \mathbb{F}_p]$.

c) *If f (mod p) is the product of several irreducible factors whose degrees are $n_1, \ldots, n_k$, then Gal_p contains the product of (non-intersecting) cycles whose lengths are $n_1, \ldots, n_k$.*

Proof. a) One can construct a polynomial similar to the Galois resolvent by replacing integers $m_1, \ldots, m_n$ by the indeterminates $u_1, \ldots, u_n$. Namely, let $\beta_1, \ldots, \beta_n$ be all the roots of f. Consider the polynomial

$$F(x, u_1, \ldots, u_n) = \prod_{\sigma \in G} (x - u_1 \beta_{\sigma(1)} - \cdots - u_n \beta_{\sigma(n)})$$

and let $G(x, u_1, \ldots, u_n)$ be its irreducible over $\mathbb{Z}$ divisor divisible by

$$x - u_1 \beta_1 - \cdots - u_n \beta_n.$$

In the same way as for the Galois resolvent, we prove that the Galois group Gal consists of the elements which correspond to the linear divisors of G.

Over $\mathbb{F}_p$, the polynomial G (mod p) factorizes into irreducible factors G_1 (mod p), ..., G_l (mod p). Any permutation of the roots $\alpha_1, \ldots, \alpha_n$ of the polynomial f (mod p) which belongs to Gal_p sends G_i (mod p) into the same polynomial G_i (mod p). Therefore the same permutation of the roots $\beta_1, \ldots, \beta_n$ cannot send $G(x, u_1, \ldots, u_n)$ into another irreducible factor of $F(x, u_1, \ldots, u_n)$ since F (mod p) has no multiple divisors.

b) In the field of characteristic p, we have $(x + y)^p = x^p + y^p$. Therefore, if $x \neq y$, then $x^p - y^p = (x - y)^p \neq 0$. Hence the map $x \mapsto x^p$ is an automorphism of the field $\mathbb{F}_{p^r}$ which preserves the elements of $\mathbb{F}_p$. The powers of this automorphism send x to $x^p, x^{p^2}, \ldots, x^{p^r} = x$. The field $\mathbb{F}_{p^r}$ is obtained from the field $\mathbb{F}_p$ by adjoining a root ζ of the polynomial $x^{q-1} - 1$, where $q = p^r$. Here $\zeta^a \neq \zeta^b$ if $0 < a < b \leq q - 1$. Therefore the automorphisms

$$x \mapsto x^p, \quad x \mapsto x^{p^2}, \quad \ldots, \quad x \mapsto x^{p^r} = x$$

are distinct. On the other hand, the degree of the extension $\mathbb{F}_{p^r} \supset \mathbb{F}_p$ is equal to r, and hence ζ satisfies an algebraic equation of degree r over $\mathbb{F}_p$. Any automorphism of $\mathbb{F}_{p^r}$ over $\mathbb{F}_p$ is uniquely determined by the image of ζ, and therefore the number of distinct automorphisms does not exceed r.

c) The group Gal_p is cyclic, and so it is generated by an element σ. Let us represent this element as a product of non-intersecting cycles:

$$\sigma = (1\,2\,\ldots\,j)(j{+}1\,\ldots)\cdots(\ldots\,n).$$

The group Gal_p transitively acts on the elements of each cycle, and so the cycles correspond to the irreducible factors of f (mod p). $\square$

Example. The Galois group of the polynomial $x^5 - x - 1$ over $\mathbb{Q}$ is equal to S_5.

Proof. Modulo 2, the polynomial considered factorizes into irreducible factors $x^2 + x + 1$ and $x^3 + x^2 + 1$, and therefore its Galois group contains an element of the form $(ij)(klm)$, where $\{i, j, k, l, m\}$ is a permutaion of $\{1, 2, 3, 4, 5\}$.

Modulo 3, the polynomial $x^5 - x - 1$ is irreducible. Indeed, if this polynomial were reducible, it would have had a factor of degree 1 or 2. The product of all irreducible polynomials of degree 1 or 2 over $\mathbb{F}_3$ is equal to $x^9 - x$ (see Theorem 3.3.7 on page 99). Therefore $x^5 - x - 1$ should have a common divisor either with the polynomial $x^5 - x$ or with a polynomial $x^5 + x$. Neither is possible. Therefore the Galois group contains the cycle (12345).

The Galois group also contains the element $\left((ij)(klm) \right)^3 = (ij)$. Considering conjugations of (ij) by $(12345)^a$, we obtain the transpositions $(i+a, j+a)$. For $a = j - i$, we consecutively obtain transpositions (ij), (jk), (kl), (lm), (mi) which generate the whole group S_5. $\square$

Frobenius proved that, if the Galois group of an irreducible polynomial f of degree n contains an element representable as the product of cycles whose lengths are equal to $n_1, \ldots, n_k$, then there exist infinitely many primes p for which the polynomial f (mod p) factorizes into irreducible factors of degrees $n_1, \ldots, n_k$. He even calculated the density of such primes p. For Frobenius's density theorem and its generalization — Chebotaryov's density theorem — see [J], [Ch], [Al] and [Se1].

6

Ideals in Polynomial Rings

6.1 Hilbert's basis theorem and Hilbert's theorem on zeros

6.1.1 Hilbert's basis theorem

Hilbert's basis theorem appeared in his famous paper [Hi2]. In this work he suggested totally new methods, using which he managed to prove the existence of a finite basis for the invariants of forms. Previously, in 1868, Gordan proved the existence of a finite basis only for binary forms, and this was performed by a very labour-consuming case-checking. Hilbert, on the contrary, managed to solve a number of central problems of invariant theory. His methods, however, were not constructive and this prompted Gordan to complain: "This is not mathematics. This is theology!"

Let K be a field (for example, $\mathbb{Q}$, $\mathbb{R}$ or $\mathbb{C}$) or the ring $\mathbb{Z}$. Let $K[x_1, \ldots, x_n]$ be a polynomial ring in n indeterminates with coefficients in K.

Theorem 6.1.1 (Hilbert). *Let $M \subset K[x_1, \ldots, x_n]$ be an arbitrary subset. Then there exists a finite set of polynomials $m_1, \ldots, m_r \in M$, such that any polynomial $m \in M$ can be represented in the form $m = \lambda_1 m_1 + \cdots + \lambda_r m_r$, where $\lambda_i \in K[x_1, \ldots, x_n]$.*

It is convenient to formulate Hilbert's theorem in terms of ideals. Then it will be easier to prove.

A subset $I \subset K[x_1, \ldots, x_n]$ is called an *ideal* if the following two conditions hold:

1) $a, b \in I \Longrightarrow a + b \in I$;
2) $a \in I$, $f \in K[x_1, \ldots, x_n] \Longrightarrow fa \in I$.

For any set $M \subset K[x_1, \ldots, x_n]$, we can consider the ideal $I(M)$ generated by M which consists of all sums of the form $\lambda_1 m_1 + \cdots + \lambda_r m_r$, where $\lambda_i \in K[x_1, \ldots, x_n]$ and $m_i \in M$.

A collection $\{a_\alpha \mid a_\alpha \in I\}$ is called a *basis* of the ideal I if any element $a \in I$ can be represented in the form $a = \lambda_1 a_{\alpha_1} + \cdots + \lambda_t a_{\alpha_t}$, where $\lambda_i \in$

$K[x_1, \ldots, x_n]$. The ideal I is said to be *finitely generated* if it possesses a finite basis.

To prove Theorem 6.1.1, it suffices to prove that the ideal $I(M)$ is finitely generated. Indeed, in this case any element of $M \subset I(M)$ can be expressed in terms of a finite collection of elements $a_1, \ldots, a_s \in I$, and each of these elements by definition can be expressed in terms of a finite collection of elements of M.

Theorem 6.1.2 (Hilbert's basis theorem). *Any ideal of $K[x_1, \ldots, x_n]$ is finitely generated.*

Proof. First, we observe that any ideal in the considered rings K is finitely generated. Indeed, if K is a field, then any nonzero ideal coincides with K and is generated by 1. If $K = \mathbb{Z}$, then any ideal is of the form $m\mathbb{Z}$ and is generated by m (to prove this, consider the smallest positive element of the ideal).

Let $L_n = K[x_1, \ldots, x_n]$ for $n \geq 1$ and $L_0 = K$. Then $K[x_1, \ldots, x_{n+1}] = L_n[x]$, where $x = x_{n+1}$. As we have observed, for $n = 0$ any ideal of L_n is finitely generated. Therefore it suffices to prove that, if any ideal of $L = L_n$ is finitely generated, then any ideal I of the ring $L[x]$ is also finitely generated.

Step 1. *The leading coefficients of the polynomials which belong to the ideal $I \subset L[x]$, together with zero, constitute an ideal J in L.*

Indeed, let $f(x) = ax^n + \cdots$ and $g(x) = bx^m + \cdots$ be polynomials in I. We may assume that $m \leq n$. Then the polynomial $f(x) + x^{n-m}g(x)$ belongs to I and its leading coefficient is equal to $a + b$. It is also clear that if $\lambda \in L$ and $\lambda \neq 0$, then the leading coefficient of λf is equal to λa.

Let us begin the construction of a finite basis of the ideal I in $L[x]$ by selecting a finite basis $a_1, \ldots, a_r$ of the ideal J in L. The elements $a_1, \ldots, a_r$ are the leading coefficients of some polynomials $f_1, \ldots, f_r \in I$.

Step 2. *There exists a positive integer n such that any polynomial in I is the sum of a polynomial whose degree is less than n and a polynomial of the form $\lambda_1 f_1 + \cdots + \lambda_r f_r$, where $\lambda_i \in L[x]$.*

Let us prove that we can take n to be the greatest of the degrees of the polynomials $f_1, \ldots, f_r$. In I, take an arbitrary polynomial $f(x) = ax^N + \cdots$ of degree $N \geq n$. By definition, $a \in J$, and hence $a = \sum \lambda_i a_i$ for some $\lambda_i \in L[x]$. Consider the polynomial

$$g(x) = f(x) - \sum \lambda_i x^{N - \deg f_i} f_i.$$

The coefficient of x^N in g is $a - \sum \lambda_i a_i = 0$, and hence $\deg g \leq N - 1$. If $N - 1 \geq n$ we can repeat this construction, and so on.

Step 3. *There exists a finite set of polynomials $g_1, \ldots, g_s \in I$ such that any polynomial in I of degree less than n can be represented in the form $\lambda_1 g_1 + \cdots + \lambda_s g_s$, where $\lambda_i \in L[x]$.*

The coefficients of x^{n-1} in polynomials in I whose degrees are less than $n-1$ constitute an ideal of the ring L. Let $b_1, \ldots, b_k$ be a basis of this ideal and $g_1, \ldots, g_k$ the polynomials of degree $n-1$ in I whose leading coefficients are $b_1, \ldots, b_k$, respectively. In I, take an arbitrary polynomial h of degree $n-1$. Let b be the leading coefficient of this polynomial. Then $b = \lambda_1 b_1 + \cdots + \lambda_k b_k$, where $\lambda_i \in L[x]$. Therefore the degree of $h - \lambda_1 g_1 - \cdots - \lambda_k g_k$ does not exceed $n-2$. Thus, up to elements of the ideal generated by $g_1, \ldots, g_k$, we have replaced the polynomial of degree $n-1$ by a polynomial of degree no higher than $n-2$. We can similarly select polynomials $g_{k+1}, \ldots, g_l$ so that, up to elements of the ideal generated by them, any polynomial of degree $n-2$ is equal to a polynomial of degree no higher than $n-3$, and so on. $\square$

6.1.2 Hilbert's theorem on zeros

Hilbert's theorem on zeros appeared in another famous paper by Hilbert on invariant theory [Hi4]. This theorem is sometimes called *Hilbert's theorem on roots*. Its German name, *Nullstellensatz*, is also widely used in the English mathematical literature.

Theorem 6.1.3 (Hilbert's Nullstellensatz). *Let*

$$f, f_1, \ldots, f_r \in \mathbb{C}[x_1, \ldots, x_n],$$

where f vanishes at all the common zeros of the polynomials $f_1, \ldots, f_r$. Then, for a certain positive integer q, the polynomial f^q belongs to the ideal generated by $f_1, \ldots, f_r$, i.e., $f^q = g_1 f_1 + \cdots + g_r f_r$ for some $g_1, \ldots, g_r \in \mathbb{C}[x_1, \ldots, x_n]$.

Proof. We first prove one particular case of Hilbert's Nullstellensatz from which we can deduce the general case. Namely, consider the case when $f = 1$.

Theorem 6.1.4. *Let the polynomials $f_1, \ldots, f_r \in \mathbb{C}[x_1, \ldots, x_n]$ have no common zeros. Then there exist polynomials $g_1, \ldots, g_r \in \mathbb{C}[x_1, \ldots, x_n]$ such that*

$$g_1 f_1 + \cdots + g_r f_r = 1.$$

Proof. Let $I(f_1, \ldots, f_r)$ be the ideal in $K = \mathbb{C}[x_1, \ldots, x_n]$ generated by $f_1, \ldots, f_r$. Suppose that there are no polynomials $g_1, \ldots, g_r$ such that $g_1 f_1 + \cdots + g_r f_r = 1$. Then $I(f_1, \ldots, f_r) \neq K$.

Step 1. *Let I be a nontrivial maximal ideal of K, let $I \supset I(f_1, \ldots, f_r)$. Then the ring $A = K/I$ is a field.*

It suffices to verify that any nonzero element in K/I has an inverse. If $f \notin I$, then $I + fK$ is an ideal strictly containing I, and so $I + fK = K$. This means, in particular, that there exist polynomials $a \in I$ and $b \in K$ such that $a + bf = 1$. Then the class $\bar{b} \in K/I$ is the inverse of $\bar{f} \in K/I$.

Let α_i be the image of x_i under the canonical projection

$$p : \mathbb{C}[x_1, \ldots, x_n] \to \mathbb{C}[x_1, \ldots, x_n]/I = A.$$

Then $A = \mathbb{C}[\alpha_1, \ldots, \alpha_n]$. Therefore A is a finitely generated algebra over $\mathbb{C}$ which at the same time is a field.

Step 2. *If a finitely generated algebra $A = \mathbb{C}[\alpha_1, \ldots, \alpha_n]$ over $\mathbb{C}$ is a field, then A coincides with $\mathbb{C}$.*

We need the following auxiliary statement.

Lemma. *In $A = \mathbb{C}[\alpha_1, \ldots, \alpha_n]$, there exist elements $y_1, \ldots, y_k$, algebraically independent over $\mathbb{C}$, such that any element $a \in A$ satisfies a normed algebraic equation over $\mathbb{C}[y_1, \ldots, y_k]$, i.e.,*

$$a^l + b_1 a^{l-1} + \cdots + b_l = 0, \quad where \quad b_1, \ldots, b_l \in \mathbb{C}[y_1, \ldots, y_k].$$

Proof. We use induction on n. If the elements $\alpha_1, \ldots, \alpha_n$ are algebraically independent, the statement is obvious. Let $f(\alpha_1, \ldots, \alpha_n) = 0$ be an algebraic relation between them. If f is a polynomial of degree m whose coefficient of x_n^m does not vanish, then

$$\alpha_n^m + b_1 \alpha_n^{m-1} + \cdots + b_m = 0, \quad b_1, \ldots, b_m \in \mathbb{C}[\alpha_1, \ldots, \alpha_{n-1}].$$

It remains to use the induction hypothesis.

If the coefficient of x_n^m is zero, we make the change of variables

$$x_n = \xi_n \text{ and } x_i = \xi_i + a_i \xi_n \text{ for } i = 1, \ldots, n-1.$$

Let us try to select the numbers $a_i \in \mathbb{C}$ sp that the polynomial

$$g(\xi_1, \ldots, \xi_{n-1}, \xi_n) = f(x_1, \ldots, x_n) = f(\xi_1 + a_1 \xi_n, \ldots, \xi_{n-1} + a_{n-1}\xi_n, \xi_n)$$

has a nonzero coefficient of ξ_n^m. This coefficient is equal to

$$g_m(0, \ldots, 0, 1) = f_m(a_1, \ldots, a_{n-1}, 1),$$

where f_m and g_m are homogeneous components of the highest degree of f and g, respectively. Clearly, the nonzero homogeneous polynomial $f_m(x_1, \ldots, x_n)$ cannot be identically zero for $x_n = 1$. $\square$

Now we can concentrate on the proof that A coincides with $\mathbb{C}$. Select the elements $y_1, \ldots, y_k \in A$ as in Noether's normalization lemma. Let us show that any nonzero element x is invertible in $B = \mathbb{C}[y_1, \ldots, y_k]$, i.e., B is a field. By assumption A is a field, and so x is invertible in A. Moreover, by Noether's lemma the element x^{-1} satisfies the equation

$$(x^{-1})^l + b_1(x^{-1})^{l-1} + \cdots + b_l = 0, \quad b_1, \ldots, b_l \in B.$$

Multiplying both sides of this equation by x^{l-1} we obtain

$$x^{-1} = -b_1 - b_2 x - \cdots - b_l x^{l-1} \in B.$$

The field $B = \mathbb{C}[y_1, \ldots, y_k]$ is a ring of polynomials in k indeterminates over $\mathbb{C}$, but for $k \neq 0$ the polynomial ring cannot be a field. Hence $B = \mathbb{C}$. Any element of the field A is a root of a polynomial

$$(x^{-1})^l + b_1 (x^{-1})^{l-1} + \cdots + b_l, \quad b_1, \ldots, b_l \in B = \mathbb{C}.$$

Therefore $A = \mathbb{C}$.

Step 3. *The polynomials $f_1, \ldots, f_r$ vanish at the point*

$$(\alpha_1, \ldots, \alpha_n) \in \mathbb{C}^n.$$

Indeed, under the canonical projection

$$p \colon \mathbb{C}[x_1, \ldots, x_n] \to \mathbb{C}[x_1, \ldots, x_n]/I = A = \mathbb{C}$$

the element x_i transforms into $\alpha_i \in \mathbb{C}$, and so the polynomial $\varphi(x_1, \ldots, x_n)$ transforms into $\varphi(\alpha_1, \ldots, \alpha_n)$. Since the polynomials $f_1, \ldots, f_r$ belong to the ideal I, the canonical projection annihilates them.

Thus, having assumed that $I(f_1, \ldots, f_r) \neq \mathbb{C}[x_1, \ldots, x_n]$, we deduce that $f_1, \ldots, f_r$ have a common zero. This contradicts the assumption of the theorem. □

Following [Ra1] let us show now how to deduce the general Hilbert's Nullstellensatz from Theorem 6.1.4. For $f = 0$, the statement is obvious. We therefore assume that $f \neq 0$. Let us add to the indeterminates $x_1, \ldots, x_n$ a new indeterminate $x_{n+1} = z$ and consider the polynomials $f_1, \ldots, f_r, 1 - zf$. They have no common zeros, and so

$$1 = h_1 f_1 + \cdots + h_r f_r + h(1 - zf),$$

where $h_1, \ldots, h_r, h$ are some polynomials in $x_1, \ldots, x_n, z$. Set $z = \dfrac{1}{f}$. After reducing to a common denominator we obtain

$$f^q = g_1 f_1 + \cdots + g_r f_r,$$

where $g_1, \ldots, g_r$ are some polynomials in $x_1, \ldots, x_n$. This is a relation of the form required. □

Remark. If the coefficients of the polynomials $f, f_1, \ldots, f_r$ are real and f vanishes at all the common complex roots of $f_1, \ldots, f_r$, then there exist polynomials $g_1, \ldots, g_r$ with real coefficients such that $f^q = g_1 f_1 + \cdots + g_r f_r$.

Indeed, by Hilbert's Nullstellensatz the equality $f^q = h_1 f_1 + \cdots + h_r f_r$ holds for some polynomials $h_1, \ldots, h_r$ with complex coefficients. Let $h_j = g_j + i p_j$, where g_j and p_j are polynomials with real coefficients. But then $f^q = g_1 f_1 + \cdots + g_r f_r$.

Any set of homogeneous polynomials has a trivial common zero — the origin. Therefore, for homogeneous polynomials, an analogue of the set of polynomials without common zeros is the set of polynomials without common *nontrivial* zeros.

For homogeneous polynomials the following analogue of Theorem 6.1.4 holds.

Theorem 6.1.5. *Let $F_1, \ldots, F_r \in \mathbb{C}[x_1, \ldots, x_n]$ be homogeneous polynomials without common nontrivial zeros. Then the ideal $I(F_1, \ldots, F_r)$ generated by them contains all the homogeneous polynomials of degree $d \geq d_0$, where d_0 is a fixed number.*

Proof. By assumption the only common zero of the polynomials is the origin. Therefore the linear polynomials $x_1, \ldots, x_n$ vanish at all the common zeros of the polynomials $F_1, \ldots, F_r$. By Hilbert's Nullstellensatz $x_i^{p_i} \in I(F_1, \ldots, F_r)$ for some p_i. Set $d_0 = (p_1 - 1) + \cdots + (p_n - 1) + 1$. Then any monomial $X_d = x_1^{a_1} \cdot \ldots \cdot x_n^{a_n}$ of degree $d = a_1 + \cdots + a_n \geq d_0$ is divisible by $x_i^{p_i}$ for some i. Therefore $X_d \in I(F_1, \ldots, F_r)$. $\square$

A simple direct proof of Theorem 6.1.5 is given in the paper by Cartier and Tate [Ca3].

6.1.3 Hilbert's polynomial

Let us first recall certain definitions from commutative algebra. A *module* over a ring A is an Abelian group M on which A acts, i.e., for any $a \in A$ and $m \in M$ there is determined an element $am \in M$ such that, for any $a, b \in A$ and $m, n \in M$, we have

$$a(m + n) = am + an, \quad (ab)m = a(bm),$$
$$(a + b)m = am + bm, \quad 1m = m.$$

For example, if A is a field, then a module over A is just a vector space over A.

A module M over a ring A is said to be *finitely generated* if any element $m \in M$ can be represented in the form $m = \sum a_i m_i$, where $m_1, \ldots, m_n$ is a fixed finite set of elements of M.

A ring A is said to be *graded* if it is of the form $A = \bigoplus_{i=0}^{\infty} A_i$, where the A_i are the additive subgroups of A and $A_i A_j \subset A_{i+j}$, and where $A_i A_j$ consists of sums of elements $a_i a_j$ such that $a_i \in A_i$, $a_j \in A_j$. A *graded module* over a graded ring A is a module $M = \bigoplus_{i=0}^{\infty} M_i$, where the M_i are additive subgroups of M such that $A_i M_j \subset M_{i+j}$, and where $A_i M_j$ consists of sums of elements $a_i m_j$ such that $a_i \in A_i$, $m_j \in M_j$.

In this section the main example of a graded ring is $A = K[x_0, \ldots, x_n]$, where K is a field; the homogeneous component A_i consists of all homogeneous polynomials of degree i.

An ideal $I \subset K[x_0, \ldots, x_n]$ is said to be *homogeneous* if all the homogeneous components of any element of I also belong to I (the homogeneous component of degree i of a polynomial $f \in K[x_0, \ldots, x_n]$ is the sum of all its terms of degree i). It is easy to verify that I is homogeneous if and only if it is generated by homogeneous polynomials $f_1, \ldots, f_k$. Indeed, if I is homogeneous, then the homogeneous components of the polynomials that generate I lie in I and generate it. If the ideal I is generated by homogeneous polynomials $f_1, \ldots, f_k$, then any element $g \in I$ can be first expressed in the form $g = \sum h_\alpha f_\alpha$, and then one can represent every polynomial h_α as the sum of its homogeneous components. As a result, every homogeneous component g_i of g will be represented in the form $g_i = \sum x_\beta f_\beta$, and so $g_i \in I$.

An ideal I is homogeneous if and only if it can be represented in the form $I = \bigoplus_{i=0}^{\infty} I_i$, where $I_i = I \cap A_i$. Therefore the quotient ring $M = A/I$ is a graded module over A whose grading is *induced*, i.e., is given by $M_i = A_i/(I \cap A_i)$.

Let us elucidate this. Let $g, h \in A = K[x_0, \ldots, x_n]$ be some polynomials, and g_i and h_i their homogeneous components of degree i. The classes $g + I$ and $h + I$ coincide if and only if the classes $g_i + I \cap A_i$ and $h_i + I \cap A_i$ coincide for all i. Therefore

$$M = A/I = \bigoplus_{i=0}^{\infty} A_i/(I \cap A_i) = \bigoplus_{i=0}^{\infty} M_i.$$

The action of A on M is as follows: for $g \in A$ and $f + I \in M$, the element $g(f + I) \in M$ is defined as $gf + I \in M$. Clearly, we have

$$A_i\big(A_j/(I \cap A_j)\big) \subset A_{i+j}/(I \cap A_{i+j}).$$

Theorem 6.1.6 (Hilbert). *Let K be a field $A = K[x_0, \ldots, x_n]$ and let $M = \bigoplus_{i=0}^{\infty} M_i$ be a finitely generated graded A-module. Then there exists a polynomial $p_M(t)$ of degree $\leq n$ such that for all sufficiently large i the dimension of M_i, as a vector space over K, is equal to $p_M(i)$.*

Proof. We use induction on n. The starting point of the induction is $n = -1$, i.e., $A = K$. In this case, M is a finite dimensional vector space over K, and so $M_i = 0$ for sufficiently large i. Therefore $p_M = 0$.

Now let $n \geq 0$ and suppose that the statement holds for the modules over the ring $A' = K[x_0, \ldots, x_{n-1}]$, where $A' = K$ for $n = 0$. Set $x = x_n$ and consider the A-modules $M' = \{m \in M \mid xm = 0\}$ and $M'' = M/xM$. These modules are finitely generated over A and annihilated by multiplication by x, i.e., $xM' = 0$ and $xM'' = 0$. Hence M' and M'' are finitely generated A'-modules. Therefore, by the induction hypothesis, for sufficiently large i, we have $\dim M_i' = p_1(i)$ and $\dim M_i'' = p_2(i)$, where p_1 and p_2 are polynomials of degree not higher than $n - 1$.

For every positive integer i, we have the exact sequence

$$0 \to M_i' \to M_i \overset{\times x}{\to} M_{i+1} \to M_i'' \to 0,$$

where the map $M_i \overset{\times x}{\to} M_{i+1}$ is the multiplication by x. Therefore

$$\dim M_i' - \dim M_i + \dim M_{i+1} - \dim M_{i+1}'' = 0,$$

i.e.,

$$\dim M_{i+1} - \dim M_i = \dim M_{i+1}'' - \dim M_i'.$$

For sufficiently large i, we have

$$\dim M_{i+1}'' - \dim M_i' = p_2(i+1) - p_1(i) = q(i),$$

where $q(i)$ is a polynomial of degree no higher than $n - 1$.

Let $f(i) = \dim M_i$. For sufficiently large i, we have $f(i+1) - f(i) = q(i)$, where q is a polynomial of degree no higher than $n - 1$. Therefore f is a polynomial of degree no higher than n for sufficiently large i. Set

$$x^{(m)} = x(x-1) \cdot \ldots \cdot (x - m + 1).$$

It is easy to verify that

$$(x+1)^{(m)} - x^{(m)} = m x^{(m-1)}.$$

The polynomials $x^{(0)} = 1, x^{(1)}, \ldots, x^{(n-1)}$ constitute a basis of the space of polynomials of degree no higher than $n - 1$, and so we can represent q in the form $q(x) = \sum_{s=0}^{n-1} a_s x^{(s)}$. Thus, the polynomial $f_0(x) = \sum_{s=0}^{n-1} \frac{a_s}{s+1} x^{(s+1)}$ satisfies the relation $f_0(i+1) - f_0(i) = q(i)$. It is also clear that the function $c(i) = f(i) - f_0(i)$ satisfies the relation $c(i+1) - c(i) = 0$ for sufficiently large i, and so $c(i) = c$ is a constant. Therefore $f(x) = \sum_{s=0}^{n-1} \frac{a_s}{s+1} x^{(s+1)} + c$ is a polynomial of degree no higher than n. $\square$

The polynomial $p_M(i)$ is called *Hilbert's polynomial* of the module M. Clearly, this polynomial is integer-valued (see page 85), and so it can be represented in the form

$$p_M(i) = c_0 \binom{i}{m} + c_1 \binom{i}{m-1} + \cdots + c_m,$$

where $c_0, \ldots, c_m$ are integers and $m \leq n$. We assume that $c_0 \neq 0$. If $M = \mathbb{C}[x_0, \ldots, x_n]/I$, where I is a homogeneous ideal, then, under certain natural restrictions, the numbers c_0 and m have the following geometric interpretation.

To a homogeneous ideal I there corresponds an algebraic set $V(I)$ in the projective space $\mathbb{C}P^n$, namely

$$V(I) = \{(a_0, \ldots, a_n) \in \mathbb{C}P^n \mid f(a_0, \ldots, a_n) = 0 \text{ for any } f \in I\}.$$

The ideal I is said to be *prime* if $fg \in I$ implies either $f \in I$ or $g \in I$. For a prime ideal I, the algebraic set $V(I)$ is irreducible, i.e., it cannot be represented as a union of $V(I_1)$ and $V(I_2)$, where I_1 and I_2 are nontrivial homogeneous ideals. The restriction mentioned above is that I should be a prime ideal. Then m coincides with the dimension of the projective algebraic variety $V(I)$ and c_0 coincides with the degree of this variety (the *degree of a variety* of dimension m in $\mathbb{C}P^n$ is defined as the number of intersection points of this variety with the generic subspace of dimension $n - m$, in other words, with almost all such subspaces). For the proof of this statement, see [Mu2].

Example 1. If $M = \mathbb{C}[x_0, \ldots, x_n]$, then $m = n$ and $c_0 = 1$.

Indeed, M_i consists of homogeneous polynomials of degree i in $n + 1$ indeterminates. We assign to a monomial $x_0^{i_0} \cdots x_n^{i_n}$ the sequence consisting of i_0 zeros and one unit, followed by i_1 zeros and one unit, and so on, and the sequence ends with i_n zeros. This sequence consists, therefore, of $i + n$ numbers among which there are i zeros and n units. The total number of such sequences is

$$\binom{i+n}{n} = \frac{(i+n) \cdot \ldots \cdot (i+1)}{n!} = \frac{i^n}{n!} + \cdots = \binom{i}{n} + \cdots$$

Therefore $p_M(i) = \dim M_i = \binom{i}{n} + \cdots$.

Example 2. If $M = A/f_d A$, where $A = \mathbb{C}[x_0, \ldots, x_n]$, and f_d is a homogeneous polynomial of degree d, then $m = n - 1$ and $c_0 = d$.

Indeed, multiplication by f_d yields the exact sequence

$$0 \to H_{i-d} \xrightarrow{\times f_d} H_i \to M_i \to 0,$$

where H_i is the space of homogeneous polynomials of degree i in $n + 1$ indeterminates. In Example 1 it is shown that $\dim H_i = \binom{i+n}{n}$, and so

$$\dim M_i = \dim H_i - \dim H_{i-d} = \binom{i+n}{n} - \binom{i-d+n}{n} =$$

$$= d\frac{i^{n-1}}{(n-1)!} + \cdots = c_0\binom{i}{m} + \cdots,$$

where $c_0 = d$ and $m = n - 1$.

Let M be a graded finitely generated A-module and $p_M(i) = c_0\binom{i}{m} + \cdots$ its Hilbert's polynomial. Then the number $m = \dim M$ will be called the *dimension* of M and $c_0 = \deg M$ its *degree*. As we have already mentioned (and examples 1 and 2 support), if $I \subset \mathbb{C}[x_0, \ldots, x_n]$ is a homogeneous prime ideal and $M = \mathbb{C}[x_0, \ldots, x_n]/I$, the number $\dim M$ is the dimension of the variety $V(I) \subset \mathbb{C}P^n$ and $\deg M$ is the degree of this variety.

Let us discuss now certain properties of the degree and the dimension of graded A-modules M which we will need in section 6.1.4. We are especially interested in the case when $M = A/I$, where I is a homogeneous prime ideal. In this case M is a ring without zero divisors, i.e., an integer domain.

On the other hand, the homogeneous ideal I is prime if and only if the A-module $M = A/I$ possesses the following property: if $f \notin I$, then $fm \neq 0$ for $m \neq 0$ (here $f \in A$ and $m \in M$). In the general case the A-module M is said to be *integral* if for any $f \in A$ either $fM = 0$ or $fm \neq 0$ for $m \neq 0$.

In the rest of this section we will assume that $A = K[x_0, \ldots, x_n]$, where K is a field and A is endowed with the natural grading $A = \overset{\infty}{\underset{i=0}{\oplus}} A_i$, where A_i is the set of homogeneous polynomials of degree i. Let M be the graded finitely generated A-module such that $\dim M \geq 0$, i.e., $p_M \neq 0$.

Theorem 6.1.7. *Let M be an integral module and let $f \in A_d$ be a homogeneous polynomial of degree d such that $fM \neq 0$. Then*

$$\dim(M/fM) = \dim M - 1 \quad and \quad \deg(M/fM) = d \deg M.$$

In geometric terms, this statement looks as follows: a hypersurface of degree d singles out a subvariety of dimension $m-1$ and degree dr in a projective algebraic variety of dimension m and degree r.

Proof. Let $M' = \{m \in M \mid fm = 0\}$. The multiplication by f gives us the exact sequence

$$0 \to M'_{i-d} \to M_{i-d} \overset{\times f}{\to} M_i \to (M/fM)_i \to 0.$$

From the hypothesis of the theorem it follows that $M' = 0$. Therefore, for sufficiently large i, we have

$$p_{M/fM}(i) = p_M(i) - p_M(i - d).$$

Let $\dim M = m$ and $\deg M = r$. Then

$$p_M(i) - p_M(i - d) = r\binom{i}{m} - r\binom{i-d}{m} + \cdots =$$
$$\frac{r}{m!}(i^m - (i-d)^m) + \cdots =$$
$$= \frac{rmd}{m!}i^{m-1} + \cdots = \frac{rd}{(m-1)!}i^{m-1} + \cdots,$$

i.e., $p_{M/fM}(i) = dr\binom{i}{m-1} + \cdots$ as was required. $\square$

A submodule $S \subset M$ is said to be *homogeneous* if it is generated by homogeneous elements (i.e., the elements from homogeneous summands M_i). An equivalent condition is that $S = \overset{\infty}{\underset{i=0}{\oplus}} S_i$, where $S_i = S \cap M_i$. The quotient module M/S inherits the natural grading: $M/S = \overset{\infty}{\underset{i=0}{\oplus}} (M_i/S_i)$.

Theorem 6.1.8. *Let p be a prime which does not divide $\deg M$. Then $\deg M$ has a homogeneous submodule S such that the quotient $N = M/S$ satisfies the following conditions:*
  (a) $\dim N = \dim M$;
  (b) $\deg N$ *is not divisible by p;*
  (c) *the module N is integral.*

In geometric terms this corresponds to separation of an irreducible component of the maximal dimension from an arbitrary algebraic set consisting of several components.

Proof. Properties (a) and (b) hold for $S = 0$. In addition, M is finitely generated over $A = K[x_0, \ldots, x_n]$, and so any increasing sequence of submodules of M stabilizes. Therefore there exists a maximal homogeneous submodule S with properties (a) and (b). Let us show that this maximal submodule S also possesses property (c), i.e., the module $N = M/S$ is integral.

We have to prove that if $f \in A$, then either $fn = 0$ for all $n \in N$ or $fn \neq 0$ for all $n \neq 0$. It suffices to prove this statement for any homogeneous f. Indeed, for an arbitrary polynomial we could then deduce the statement desired as follows. Let us decompose f and n into homogeneous constituents: $f = f_s + f_{s+1} + \cdots$ and $n = n_t + n_{t+1} + \cdots$, where $f_s \neq 0$ and $n_t \neq 0$. Suppose that $fn = 0$, i.e.,

$$f_s n_t = 0, \quad f_{s+1} n_t + f_s n_{t+1} = 0, \quad f_{s+2} n_t + f_{s+1} n_{t+1} + f_s n_{t+2} = 0, \ldots .$$

Since $n_t \neq 0$, we consecutively get $f_s N = 0$, $f_{s+1} N = 0$, $f_{s+2} N = 0, \ldots$ Hence, $fN = 0$.

Now, let f be a homogeneous polynomial of degree d and $fN \neq 0$. Set $N' = \{n \in N \mid fn = 0\}$. Then multiplication by f yields the exact sequence

$$0 \to N'_{i-d} \to N_{i-d} \overset{\times f}{\to} N_i \to (N/fN)_i \to 0,$$

and hence

$$0 \to N_{i-d}/N'_{i-d} \to N_i \to (N/fN)_i \to 0.$$

Therefore, for large i, we have

$$p_N(i) = p_{N/fN}(i) + p_{N/N'}(i - d). \tag{1}$$

The condition $fN \neq 0$ means that $fM + S \neq S$. Since S is maximal, the module

$$N/fN = (M/S)/f(M/S) \cong M/(S + fM)$$

cannot simultaneously possess properties (a) and (b). Therefore either (i) $\dim(N/fN) < \dim N = \dim M$ or (ii) $\dim(N/fN) = \dim M$ and $\deg(N/fN)$ is divisible by p.

In case (i) by formula (1)

$$p_{N/N'}(i - d) = \frac{ri^n}{n!} + \cdots,$$

where $n = \dim N$, $r = \deg N$.

In case (ii)

$$p_{N/N'}(i - d) = \frac{(r - r_1)i^n}{n!} + \cdots,$$

where $r_1 = \deg(N/fN)$. Since r is not divisible by p and r_1 is divisible by p, it follows that in both cases $\dim(N/N') = n = \dim M$ and the number $\deg(N/N')$, which is either r or $r - r_1$, is not divisible by p. Therefore the module N/N' possesses properties (a) and (b). The module N/N' is of the form M/S', where S' is a homogeneous submodule of M containing S. From the maximality of S it follows that $S' = S$, i.e., $N' = 0$, as was required. $\square$

Theorem 6.1.9. *Let $I \subset A = K[x_0, \ldots, x_n]$ be a homogeneous prime ideal and suppose that the dimension of the integral A-module $M = A/I$ is equal to zero and its degree is equal to $r \neq 0$, i.e., $p_M(i) = r \neq 0$. Then*
   a) *$xM \neq 0$ for some $x = x_j$;*
   b) *$L = M/(x - 1)M$ is a field which is an extension of degree r of K.*

Proof. a) If $x_j M = 0$ for $j = 0, \ldots, n$, then $x_j \in I$ for all j, and therefore $M = K$. Thus, $p_M(i) = 0$ which contradicts the assumption that $p_M(i) = r \neq 0$.

b) Fix $x = x_j$ such that $xM \neq 0$. Then $xm \neq 0$ for $m \neq 0$ since M is an integral module. This means that the map $M \xrightarrow{\times x} M$ is monomorphic. For sufficiently large i, we have $\dim M_i = p_M(i) = r$, and so $\dim M_{i+1} = \dim M_i$. Hence, for sufficiently large i, the map $M_i \xrightarrow{\times x} M_{i+1}$ is one-to-one.

In M, consider a non-homogeneous submodule $(x - 1)M$. Let

$$\pi \colon M \to M/(x - 1)M = L$$

be the natural projection. Then $\pi(xm) = \pi(m)$. Clearly, $x\pi(m) = \pi(xm)$. Hence $L \xrightarrow{\times x} L$ is the identity map.

Let, for definiteness sake, $M_i \xrightarrow{\times x} M_{i+1}$ be one-to-one for $i \geq a$. Let us show then that $L = \pi(M_a) = \pi(M_{a+1}) = \cdots$. Let us express $m \in M$ in the form $m = m_0 + \cdots + m_a + m_{a+1} + \cdots + m_k$, where $m_s \in M_s$. Clearly,

$$\pi(m_0 + \cdots + m_a) = \pi(x^a m_0 + x^{a-1} m_1 + \cdots + m_a) = \pi(m'),$$

where $m' = x^a m_0 + x^{a-1} m_1 + \cdots + m_a \in M_a$. Moreover,

$$m_{a+1} = xm_{a,1}, \quad m_{a+2} = x^2 m_{a,2}, \quad, \ldots, \quad m_k = x^{k-a} m_{a,k-a},$$

where $m_{a,s} \in M_a$. Therefore

$$\pi(m_{a+1} + \cdots + m_k) = \pi(m_{a,1} + \cdots + m_{a,k-a}) = \pi(m''),$$

where $m'' = m_{a,1} + \cdots + m_{a,k-a} \in M_a$. Thus the natural projection $M_a \to L$ is onto.

On the other hand, for obvious reasons, this projection is monomorphic: if $\pi(m_a) = 0$, then $m_a = (x-1)m$ but no homogeneous element $m_a \neq 0$ can be represented in the form $(x-1)m$. Therefore the projection $M_a \to L$ is one-to-one and $\dim L = \dim M_a = r$.

In the situation considered, L is an r-dimensional algebra over K (i.e., a commutative ring and simultaneously a linear space over K). Let us show that L has no zero divisors. Let $l', l'' \in L$ and $l'l'' = 0$. We have shown above that $l' = \pi(m')$ and $l'' = \pi(m'')$ for some $m', m'' \in M_a$. Hence $0 = l'l'' = \pi(m'm'')$, where $m'm'' \in M_{2a}$. For $b \geq a$, the projection $M_b \to L$ is an isomorphism, and therefore $m'm'' = 0$. By assumption the ideal I is prime, and so in the ring $M = A/I$ there are no zero divisors. Hence either $m' = 0$ or $m'' = 0$, i.e., either $l' = 0$ or $l'' = 0$.

Now it is easy to show that L is a field, i.e., any nonzero element $l \in L$ is invertible. Indeed, the map $x \mapsto lx$, where $x \in L$, is a linear map $L \to L$ with the zero kernel. In the finite-dimensional case, any such map is an isomorphism, and so, in particular, $lx = 1$ for some $x \in L$. $\square$

6.1.4 The homogeneous Hilbert's Nullstellensatz for p-fields

The following statement was first proved in Hilbert's paper [Hi4] though many mathematicians of 19th century already applied it, albeit without proper justification.

Theorem 6.1.10. *Let K be an algebraically closed field, and, for $n \geq 1$, let $A = K[x_0, \ldots, x_n]$. Then any homogeneous polynomials $f_1, \ldots, f_n \in A$ have a common zero distinct from the origin.*

A similar statement holds also for so-called p-fields among which we encounter in particular, the field of real numbers $\mathbb{R}$.

Let p be a prime. The field K is called a p-*field* if the degree of any finite extension of K is of the form p^s. In particular, any algebraically closed field is a p-field for all primes p, and $\mathbb{R}$ is a 2-field.

Theorem 6.1.11. *Let K be a p-field, and let $A = K[x_0, \ldots, x_n]$, where $n \geq 1$. Then any homogeneous polynomials $f_1, \ldots, f_n \in A$ whose degrees are not divisible by p have a common non-trivial zero.*

Proof. [H. Fendrich; see [Pf2], ch. 4] We use the results of the preceding section, namely Theorems 6.1.7-6.1.9.

Let us start with the construction of a sequence of integral finitely generated graded A-modules $M_0 = A$, $M_1 = A/I_1, \ldots, M_n = A/I_n$ such that $\dim M_i = n - i$, $\deg M_i$ is not divisible by p and the homogeneous prime ideal I_i contains the polynomials $f_0, \ldots, f_i$. The module $M_0 = A$ satisfies these conditions since $\dim M_0 = n$ and $\deg M_0 = 1$ (see Example 1 on page 227).

Suppose that the modules $M_0, \ldots, M_i$ $(i \geq 0)$ are already constructed. Let us show how to construct M_{i+1} given $M_i = A/I_i$ and f_{i+1}. There are two cases to consider.

Case 1: $f_{i+1} \notin I_i$, i.e., $f_{i+1} M_i \neq 0$. Set

$$N_{i+1} = M_i/f_{i+1} M_i \cong A/(I_i + f_{i+1} A).$$

By Theorem 6.1.7,

$$\dim N_{i+1} = \dim M_i - 1 = n - (i+1)$$

and

$$\deg N_{i+1} = \deg f_{i+1} \deg M_i.$$

The number $\deg N_{i+1}$ is not divisible by p since neither $\deg f_{i+1}$ nor $\deg M_i$ is divisible by p.

Case 2: $f_{i+1} \in I_i$. From the condition $\dim M_i \geq 0$ it follows that $x \notin I_i$, i.e., $x M_i \neq 0$ for some $x = x_j$. Indeed, if $x_0, \ldots, x_n \in I_i$, then either $M_i = K$ or $M_i = 0$, and so $p_{M_i} = 0$.

Set $N_{i+1} = M_i/x M_i$. By Theorem 6.1.7, $\dim N_{i+1} = \dim M_i - 1 = n - (i+1)$ and $\deg N_{i+1} = \deg M_i$ since the degree of the polynomial x is equal to 1.

In both cases we have obtained a module N_{i+1} but it is not necessarily an integral one. To obtain an integral module, we use Theorem 6.1.8. By this theorem N_{i+1} has an homogeneous submodule S_{i+1} such that the quotient

$$M_{i+1} = N_{i+1}/S_{i+1} = A/I_{i+1}$$

possesses all the properties desired: $\dim M_{i+1} = \dim N_{i+1} = n - (i + 1)$, $\deg M_{i+1}$ is not divisible by p and M_{i+1} is integral; here I_{i+1} is a homogeneous ideal containing $f_1, \ldots, f_{i+1}$.

The dimension of the last of the constructed modules M_n is equal to zero, And so we can apply Theorem 6.1.9 to it. As a result, we obtain a field $L = M_n/(x_j - 1)M_n \cong A/I$. Here I is an inhomogeneous prime ideal which possesses two properties important for us: (1) $x_j \equiv 1 \pmod{I}$ and (2) $I \supset I_n \ni f_1, \ldots, f_n$.

The field L is an extension of K of degree $\deg M_n$. But on the one hand, $\deg M_n$ is not divisible by p while, on the other hand, the degree of any extension of K is of the form p^s. Hence, $L = K$.

Let $a_i \in K$ be the image of x_i under the natural projection

$$A = k[x_0, \ldots, x_n] \to A/I = L = K.$$

Property (1) implies that $a_j = 1$, and so $a = (a_0, \ldots, a_n) \neq 0$. Property (2) implies that $f_1(a) = \cdots = f_n(a) = 0$. $\square$

Theorem 6.1.11 enables to prove purely algebraically, for the case of polynomial functions, the well-known *Borsuk–Ulam* theorem on the common zero of odd functions on the sphere.

Theorem 6.1.12. *Let* $q_1, \ldots, q_n \in \mathbb{R}[x_1, \ldots, x_{n+1}]$ *be odd polynomials, i.e.,* $q_i(-x) = -q_i(x)$. *Then these polynomials have a common zero on the unit sphere* $x_1^2 + \cdots + x_{n+1}^2 = 1$.

Proof. Let us pass from q_i to the homogeneous polynomial $\widetilde{q}_i$ with the help of an extra indeterminate x_0. Under this passage the monomial $x_1^{a_1} \cdot \ldots \cdot x_n^{a_n}$ in q_i is replaced by $x_0^{m_0} x_1^{m_1} \cdot \ldots \cdot x_n^{m_n}$, where $m_0 = \deg q_i - m_1 - \cdots - m_n$. The degrees of all terms of any odd polynomial are odd, and so the numbers $\deg q_i$ and $m_1 + \cdots + m_n$ are odd, and therefore m_0 is even. Having replaced x_0^2 by $x_1^2 + \cdots + x_{n+1}^2$ we obtain from $\widetilde{q}_i$ a homogeneous polynomial f_i of odd degree. By Theorem 6.1.11 the polynomials $f_1, \ldots, f_n$ have a common zero $a = (a_1, \ldots, a_{n+1}) \neq 0$. For all $t \in \mathbb{R}$ the point ta is also a common zero of the homogeneous polynomials $f_1, \ldots, f_n$, and so we may assume that $a_1^2 + \cdots + a_{n+1}^2 = 1$. Then $q_i(a) = \widetilde{q}_i(1, a) = f_i(a) = 0$, as was required. $\square$

From Theorem 6.1.12 we can derive the usual Borsuk–Ulam theorem for odd continuous functions $g_1, \ldots, g_n$ by approximating these functions by polynomials.

6.2 Gröbner bases

In solutions of various computational problems related to ideals in polynomial rings, Gröbner bases are very convenient. Bruno Buchberger introduced this notion in his thesis [Bu1] written under the scientific guidance of Wolfgang Gröbner, see also [Bu2]. Buchberger also suggested a convenient algorithm for calculating Gröbner bases; this made Gröbner bases an effective computational tool.

Our exposition of the theory of Gröbner bases is largely based on the first chapter of the book [Ad]. For further details with various aspects of the theory of Gröbner bases, see the book [Bu3].

6.2.1 Polynomials in one variable

In case of polynomials in one variable over a field K, the algorithm for finding a basis of the ideal is based on the division with residue of one polynomial by another one. The first step in division with residue is performed as follows. Let

$$f(x) = a_n x^n + \cdots + a_0 \text{ and } g(x) = b_m x^m + \cdots + b_0, \text{ where } n \geq m.$$

Set

$$f_1(x) = f(x) - \frac{a_n x^n}{b_m x^m} g(x).$$

If $\deg f_1 \geq \deg g$, we apply the same procedure to f_1, and so on. Finally we obtain $f = qg + r$, where $\deg r < \deg g$ (or $r = 0$). Here the polynomials q and r are uniquely defined.

Theorem 6.2.1. *Any ideal I in the ring $K[x]$ of polynomials in one variable is a principal one, i.e., is generated by one element.*

Proof. In I, select a polynomial g of the least degree. Let $f \in I$. Then $f = qg + r$, where $\deg r < \deg g$. But $r = f - qg \in I$, and hence $r = 0$. This means that I is generated by g. $\square$

Let $I(f_1, \ldots, f_n)$ be the ideal generated by $f_1(x), \ldots, f_n(x)$. The polynomial $g(x)$ which generates this ideal is denoted by $(f_1, \ldots, f_n)$ and is called the *greatest common divisor* (GCD) of the polynomials $f_1(x), \ldots, f_n(x)$. The greatest common divisor possesses the following properties:

(1) $f_1, \ldots, f_n$ are all divisible by g;

(2) if $f_1, \ldots, f_n$ are all divisible by a polynomial h, then h is divisible by g.

Property (1) follows from the fact that $f_1, \ldots, f_n \in I(f_1, \ldots, f_n) = I(g)$. Property (2) follows from the fact that $g \in I(f_1, \ldots, f_n)$, i.e.,

$$g = u_1 f_1 + \cdots + u_n f_n, \text{ where } u_1, \ldots, u_n \in K[x].$$

Properties (1) and (2) determine the polynomial g uniquely up to a constant factor. Indeed, if the polynomials g_1 and g_2 are divisible by each other, they are proportional.

The greatest common divisor (f_1, f_2) of two polynomials f_1 and f_2 can be found with the help of Euclid's algorithm.

From properties (1) and (2) it follows easily that

$$(f_1, f_2, \ldots, f_n) = \big(f_1, (f_2, \ldots, f_n)\big).$$

This remark reduces the calculation of the greatest common divisor of n polynomials to the calculation of the greatest common divisor of two polynomials.

6.2.2 Division of polynomials in several variables

To determine the division with residue for polynomials in several variables, we have to fix an order in the set of monomials. In what follows we assume that the monomials are ordered *lexicographically*, i.e., the monomial $x^\alpha = x_1^{\alpha_1} \cdot \ldots \cdot x_n^{\alpha_n}$ is greater than $x^\beta = x_1^{\beta_1} \cdot \ldots \cdot x_n^{\beta_n}$ if $\alpha_1 = \beta_1, \ldots, \alpha_k = \beta_k$ and $\alpha_{k+1} > \beta_{k+1}$ (perhaps $k = 0$).

The expression $f = a_\alpha x^\alpha + \cdots$ will mean that $a_\alpha x^\alpha$ is the highest term of f, i.e., x^α is the highest monomial entering f.

Let $f = a_\alpha x^\alpha + \cdots$ and $g = b_\beta x^\beta + \cdots$ be two polynomials in n variables. If a term $c_\gamma x^\gamma$ of f is divisible by x^β, we define

$$f_1 = f - \frac{c_\gamma x^\gamma}{b_\beta x^\beta} g.$$

If a term of f_1 is divisible by x^β we apply to f_1 a similar transformation etc.

For this process to converge after finitely many steps, we should proceed, for example, as follows. For the term $c_\gamma x^\gamma$ we take the highest of all the monomials of f divisible by x^β. In this way the order of the highest term of f divisible by x^β will be strictly decreasing. Clearly, any strictly decreasing sequence of monomials in n variables is finite. Indeed, after finitely many steps, first x_1 vanishes, then after finitely many steps x_2 vanishes, etc.

Similarly, one can define division with residue of a polynomial by several polynomials. As a result, we obtain a representation

$$f = u_1 f_1 + \cdots + u_s f_s + r,$$

where the polynomial r has no terms divisible by the highest monomial of the polynomials $f_1, \ldots, f_s$. In this case we say that r is the *residue after division* of f by polynomials $f_1, \ldots, f_s$. Observe that r is not uniquely defined. One of the possible definitions of the Gröbner basis consists precisely in the fact that $f_1, \ldots, f_s$ is a Gröbner basis if the residue after division of any polynomial f by $f_1, \ldots, f_s$ is uniquely defined.

6.2.3 Definition of Gröbner bases

We say that (nonzero) polynomials $g_1, \ldots, g_t \in I$ constitute a *Gröbner basis* of the ideal I if the highest term of any (nonzero) polynomial $f \in I$ is divisible by the highest term of one of the polynomials $g_1, \ldots, g_t$.

Theorem 6.2.2. *The polynomials $g_1, \ldots, g_t$ form a Gröbner basis of an ideal I if and only if one of the following equivalent conditions holds:*

(a) $f \in I \iff$ *the residue after division of f by $g_1, \ldots, g_t$ is equal to 0;*

(b) $f \in I \iff f = \sum h_i g_i$ *and the highest monomial of f is equal to the highest of the products of the highest monomials of h_i and g_i;*

(c) *The ideal $L(I)$ generated by the highest terms of the elements of I is also generated by the highest terms of the polynomials $g_1, \ldots, g_t$.*

Proof. First, we prove that if $g_1, \ldots, g_t$ is a Gröbner basis of I, then condition (a) holds. It suffices to prove that if r is the residue after division of $f \in I$ by $g_1, \ldots, g_t$, then $r = 0$. Clearly, $r = f - \sum h_i g_i \in I$. Therefore, if $r \neq 0$, the highest term of r is divisible by the highest term of one of the polynomials $g_1, \ldots, g_t$ which contradicts the definition of r.

(a) $\Longrightarrow$ (b) By definition of division with residue, $f = \sum h_i g_i + r$, where the highest monomial of f is equal to the highest of the products of the highest monomials of h_i and g_i. Condition (a) implies that if $f \in I$, then $r = 0$.

(b) $\Longrightarrow$ (c) If $f = ax^\alpha + \cdots \in I$, then $f = \sum h_i g_i$, where $h_i = b_i x^{\beta_i} + \cdots$ and $g_i = c_i x^{\gamma_i} + \cdots$ and all the monomials $x^{\beta_i} x^{\gamma_i}$ are not greater than x^α. Therefore $ax^\alpha = \sum_i b_i x^{\beta_i} c_i x^{\gamma_i}$, where the sum runs over the i for which $x^{\beta_i} x^{\gamma_i} = x^\alpha$. Since $c_i x^{\gamma_i} \in L(I)$, it follows that $ax^\alpha \in L(I)$.

It remains to prove that if (c) holds, then $g_1, \ldots, g_t$ is a Gröbner basis. Let $f = ax^\alpha + \cdots \in I$. Then

$$ax^\alpha = \sum_i b_i x^{\beta_i} c_i x^{\gamma_i},$$

where the $c_i x^{\gamma_i}$ are the highest terms of some of the polynomials $g_1, \ldots, g_t$. Clearly, x^α is divisible by x^{γ_i} for some i. $\square$

Corollary. *If $g_1, \ldots, g_t$ is a Gröbner basis of an ideal I, then the polynomials $g_1, \ldots, g_t$ generate I.*

This follows from (a).

Theorem 6.2.3. *Every nonzero ideal $I \subset K[x_1, \ldots, x_n]$ possesses a Gröbner basis.*

Proof. Consider an ideal $L(I)$ generated by the highest monomials X_α of all the polynomials $g_\alpha = a_\alpha X_\alpha + \cdots \in I$. Clearly, $f \in L(I)$ if and only if any monomial of f is divisible by a monomial X_α. By Hilbert's basis theorem, the ideal $L(I)$ is generated by finitely many monomials $f_1, \ldots, f_k$. Every monomial of any of these polynomials is divisible by a monomial X_α. As a result, we obtain a finite set of monomials $X_1, \ldots, X_t$ which generate the ideal $L(I)$. These monomials are highest monomials of the polynomials $g_1, \ldots, g_t$. By Theorem 6.2.2 (c) the polynomials $g_1, \ldots, g_t$ generate a Gröbner basis of I. $\square$

We will say that polynomials $g_1, \ldots, g_t$ constitute a *Gröbner basis* if they constitute a Gröbner basis of the ideal which they generate.

Theorem 6.2.4. *Nonzero polynomials $g_1, \ldots, g_t$ constitute a Gröbner basis if and only if the residue after the division of any polynomial f by $g_1, \ldots, g_t$ is uniquely determined.*

Proof. First, suppose that the polynomials $g_1, \ldots, g_t$ constitute a Gröbner basis. Let r_1 and r_2 be residues after division of f by $g_1, \ldots, g_t$. Then the polynomials $f - r_1$ and $f - r_2$ belong to the ideal I generated by $g_1, \ldots, g_t$.

Therefore $r_1 - r_2 = (f - r_2) - (f - r_1) \in I$. By definition of the Gröbner basis, the highest monomial of the polynomial $r_1 - r_2$ is divisible by the highest monomial of one of the polynomials $g_1, \ldots, g_t$. On the other hand, neither r_1 nor r_2 has terms divisible by the highest monomials of $g_1, \ldots, g_t$. Hence $r_1 - r_2 = 0$.

Now suppose that the residue after division of any polynomial f by $g_1, \ldots, g_t$ is uniquely determined. We have to prove that if $f \in I$, then the residue r after division of f by $g_1, \ldots, g_t$ is zero.

First let us show that, if a is a number, then the polynomials f and $f - ax^\alpha g_i$, where x^α is a monomial, give identical residues after division by $g_1, \ldots, g_t$. Recall that, during the division with residue, an elementary transformation consists in annihilating a monomial $c_\gamma x^\gamma$ of f by replacing f with $f - dx^\delta g_i$.

Let $g_i = bx^\beta + \cdots$. If, for one of the polynomials f and $f - ax^\alpha g_i$, the coefficient of $x^{\alpha + \beta}$ vanishes, then the polynomial with a nonzero coefficient of this monomial can be reduced by an elementary transformation to a polynomial with the zero coefficient.

If the coefficients of $x^{\alpha + \beta}$ for both polynomials f and $f - ax^\alpha g_i$ are nonzero, then both polynomials can be reduced by an elementary transformation to the polynomial $f - cx^\alpha g_i$ with a nonzero coefficient of $x^{\alpha + \beta}$.

In all the cases the polynomials f and $f - ax^\alpha g_i$ can be reduced by an elementary transformation to the same polynomial. Hence, after division by $g_1, \ldots, g_t$, their residues are identical (we make use of the assumption on the uniqueness of the residue).

Now it is easy to prove the desired result. If $f \in I$, then $f = \sum h_i g_i$. Having expressed each polynomial h_i as the sum of monomials, we obtain $f = \sum a_\alpha x^\alpha g_{i_\alpha}$. The polynomials f and $f - \sum a_\alpha x^\alpha g_{i_\alpha} = 0$ give the same residues after division by $g_1, \ldots, g_t$. But, for the zero polynomial, the residue after division is equal to zero. Hence the same holds for f. $\square$

6.2.4 Buchberger's algorithm

None of the preceding definitions of the Gröbner basis allowed us to determine in finitely many steps whether the set $g_1, \ldots, g_t$ is a Gröbner basis or not. Let us give at last a definition which enables us to deal with this.

Let $f = ax^\alpha + \cdots$, $g = bx^\beta + \cdots$ and let x^γ be the least common multiple of x^α and x^β. Set

$$S(f, g) = \frac{x^\gamma}{ax^\alpha} f - \frac{x^\gamma}{bx^\beta} g.$$

We construct $S(f, g)$ so that the highest terms of two of its constituents cancel.

Theorem 6.2.5 (Buchberger). *The polynomials $g_1, \ldots, g_t$ constitute a Gröbner basis if and only if, for all $i \neq j$, the residue after division of $S(g_i, g_j)$ by $g_1, \ldots, g_t$ is zero.*

Proof. If $g_1, \ldots, g_t$ constitute a Gröbner basis, then by Theorem 6.2.2 (a) the polynomial $S(g_i, g_j)$ which belongs to the ideal I generated by the polynomials gives residue 0 after division by them. Therefore we only have to prove that if the residue after division of $S(g_i, g_j)$ by $g_1, \ldots, g_t$ is 0 for all $i \neq j$, then the polynomials $g_1, \ldots, g_t$ form a Gröbner basis, i.e., any polynomial $f \in I$ can be represented in the form $f = \sum h_i g_i$, where the highest monomial of f is equal to the highest of the products of the highest monomials of h_i and g_i (see Theorem 6.2.2 (b)).

Let us first prove one auxiliary statement.

Lemma. *Let $f_1, \ldots, f_s$ be polynomials with the same highest monomial x^α. If the highest monomial of $f = \sum \lambda_i f_i$, where λ_i are numbers, is strictly smaller than x^α, then $f = \sum_{i<j} \mu_{ij} S(f_i, f_j)$.*

Proof. By the hypothesis $f_i = a_i x^\alpha + \cdots$ and $f_j = a_j x^\alpha + \cdots$, and so $S(f_i, f_j) = \frac{f_i}{a_i} - \frac{f_j}{a_j}$. It is also clear that

$$f = \sum \lambda_i f_i = \lambda_1 a_1 \left(\frac{f_1}{a_1} - \frac{f_2}{a_2} \right) + (\lambda_1 a_1 + \lambda_2 a_2) \left(\frac{f_2}{a_2} - \frac{f_3}{a_3} \right) + \cdots$$

$$\cdots + (\lambda_1 a_1 + \cdots + \lambda_{s-1} a_{s-1}) \left(\frac{f_{s-1}}{a_{s-1}} - \frac{f_s}{a_s} \right) + (\lambda_1 a_1 + \cdots + \lambda_s a_s) \frac{f_s}{a_s}.$$

It remains to observe that $\lambda_1 a_1 + \cdots + \lambda_s a_s = 0$. Indeed, the coefficient of the monomial x^α in the polynomial $f = \sum \lambda_i f_i$ equals exactly $\lambda_1 a_1 + \cdots + \lambda_s a_s$ and, by assumption, the highest monomial of f is strictly smaller than x^α. □

There are several ways to represent $f = ax^\alpha + \cdots \in I$ in the form

$$f = \sum h_i g_i. \tag{*}$$

Let $h_i = b_i x^{\beta_i} + \cdots$ and $g_i = c_i x^{\gamma_i} + \cdots$. Denote the highest of the monomials $x^{\beta_i} x^{\gamma_i}$, where $i = 1, \ldots, t$, by x^δ. Select a representation (*) so that the monomial x^δ is the minimal one. We have to prove that in this case $x^\delta = x^\alpha$. Clearly, x^α cannot be higher than x^δ. Suppose that x^δ is higher than x^α. We may assume that $x^{\beta_i} x^{\gamma_i} = x^\delta$ for $i = 1, \ldots, M$ and, for $i = M+1, \ldots, t$, the monomial x^δ is higher than $x^{\beta_i} x^{\gamma_i}$.

Consider the polynomial $g = \sum_{i=1}^{M} b_i x^{\beta_i} g_i$. The coefficients of x^δ in this polynomial and in f coincide. Hence g is a linear combination of polynomials with the highest monomial x^δ and all the highest monomials cancel each other. In this case, thanks to the Lemma,

$$g = \sum \mu_{ij} S \left(x^{\beta_i} g_i, x^{\beta_j} g_j \right), \tag{1}$$

where summation runs over pairs i, j such that $1 \le i \le j \le M$. The highest monomials of $x^{\beta_i} g_i$ and $x^{\beta_j} g_j$ coincide, and so

$$S\left(x^{\beta_i} g_i, x^{\beta_j} g_j\right) = \frac{x^\delta}{c_i x^{\gamma_i}} g_i - \frac{x^\delta}{c_j x^{\gamma_j}} g_j = \frac{x^\delta}{c_j x^{\gamma_{ij}}} S(g_i, g_j),$$

where $x^{\gamma_{ij}}$ is the least common multiple of x^{γ_i} and x^{γ_j}.

By assumption, the residue after division of $S(g_i, g_j)$ by $g_1, \ldots, g_t$ is equal to 0. The polynomial $S\left(x^{\beta_i} g_i, x^{\beta_j} g_j\right)$ is divisible by $S(g_i, g_j)$, and hence the residue after its division by $g_1, \ldots, g_t$ is also zero. The algorithm of division with residue gives a representation

$$S\left(x^{\beta_i} g_i, x^{\beta_j} g_j\right) = \sum h_{ij\nu} g_\nu,$$

where the highest of the products of the highest monomials of the polynomials $h_{ij\nu}$ and g_ν coincides with the highest monomial of $S\left(x^{\beta_i} g_i, x^{\beta_j} g_j\right)$. The latter monomial is strictly less than x^δ. Let us substitute the obtained representation of $S\left(x^{\beta_i} g_i, x^{\beta_j} g_j\right)$ into (1) and then substitute the obtained representation of g into $f = g + \cdots$ As a result, we obtain a representation of f that contradicts the assumption on the minimality of x^δ. This contradiction shows that $x^\delta = x^\alpha$. $\square$

With the help of Theorem 6.2.5 it is easy to show that the following algorithm enables us to find a Gröbner basis of the ideal generated by polynomials $f_1, \ldots, f_s$. Let us calculate the residues after division of the polynomials $S(f_i, f_j)$ by $f_1, \ldots, f_s$ and add all the nonzero residues to the collection $f_1, \ldots, f_s$. Let us repeat this procedure for the obtained set of polynomials, etc. Clearly, this sequence of operations will terminate after finitely many steps, and Theorem 6.2.5 ensures that as a result we obtain a Gröbner basis of the ideal generated by $f_1, \ldots, f_s$. This algorithm for calculating a Gröbner basis is called *Buchberger's algorithm*.

6.2.5 A reduced Gröbner basis

For the same ideal, Buchberger's algorithm leads to distinct finite results depending on the choice of generators of the ideal and the sequence of operations. One can however modify the algorithm so that the final result only depends on the ideal I itself; this modification also belongs to Buchberger.

First of all, let us ensure that the number of elements of the Gröbner basis is uniquely determined. We call a Gröbner basis $g_1, \ldots, g_t$ *minimal* if $g_i = x^{\alpha_i} + \cdots$ and the monomials x^{α_i} and x^{α_j} are not divisible by each other for $i \ne j$.

Any ideal I has a minimal Gröbner basis.

Indeed, let $g_1, \ldots, g_t$ be a Gröbner basis of I. We may assume that $g_i = x^{\alpha_i} + \cdots$. If x^{α_1} is divisible by x^{α_2}, then already $g_2, \ldots, g_t$ is a Gröbner basis of I. Indeed, if $f = x^\alpha + \cdots \in I$, then, by definition of the Gröbner basis, x^α is divisible by x^{α_i} for some i. But x^{α_1} is divisible by x^{α_2}, and therefore x^α is divisible by x^{α_i} for $i \geq 2$. This means that $g_2, \ldots, g_t$ is a Gröbner basis of I. Consecutively, deleting the polynomials whose highest monomials are divisible by the highest monomials of other polynomials of the basis, we can pass from an arbitrary Gröbner basis $g_1, \ldots, g_t$ to a minimal Gröbner basis.

Theorem 6.2.6. *If $g_1, \ldots, g_t$ and $f_1, \ldots, f_s$ are two minimal Gröbner bases of the same ideal I, then $s = t$ and the highest monomials of the polynomials g_i and $f_{\sigma(i)}$, where σ is a permutation of indices, coincide.*

Proof. Let $g_i = x^{\alpha_i} + \cdots$ and $f_j = x^{\beta_j} + \cdots$ On the one hand, $f_1 \in I$ and $g_1, \ldots, g_t$ is a Gröbner basis of I. Therefore x^{β_1} is divisible by x^{α_i} for some i. After renumbering we may assume that $i = 1$. On the other hand, $g_1 \in I$ and $f_1, \ldots, f_s$ is a Gröbner basis of I. Therefore x^{α_1} is divisible by x^{β_j} for some j, and hence x^{β_1} is divisible by x^{β_j}. From the minimality of the Gröbner basis $f_1, \ldots, f_s$, it follows that $j = 1$. The monomials x^{α_1} and x^{β_1} are divisible by each other, and so $x^{\beta_1} = x^{\alpha_1}$.

Similar arguments show that x^{β_2} is divisible by x^{α_i}. From the minimality of the Gröbner basis $f_1, \ldots, f_s$ it follows that $x^{\alpha_i} \neq x^{\beta_1}$, i.e., $i \neq 1$. Therefore, after renumbering, we obtain $x^{\alpha_2} = x^{\beta_2}$, etc. Clearly, the sets of polynomials $g_1, \ldots, g_t$ and $f_1, \ldots, f_s$ should be exhausted simultaneously, i.e., $s = t$. $\square$

Now we can ensure that, not only the number of the elements in the Gröbner basis is uniquely defined, but also the elements themselves. Call a Gröbner basis $g_1, \ldots, g_t$ *reduced* if $g_i = x^{\alpha_i} + \cdots$ and the residue after division of g_i by $g_1, \ldots g_{i-1}, g_{i+1}, \ldots, g_t$ coincides with g_i, i.e., none of the monomials that enter g_i is divisible by x^{α_j} for $j \neq i$.

Obviously, any reduced basis is also a minimal one. With the help of a minimal Gröbner basis $g_1, \ldots, g_t$, we can construct a reduced basis of I generated by the polynomials $g_1, \ldots, g_t$ as follows. Let h_1 be the residue after division of g_1 by $g_2, \ldots, g_t$; let h_2 be the residue after division of g_2 by $h_1, g_3, \ldots, g_t$; let h_3 be the residue after division of g_3 by $h_1, h_2, g_4, \ldots, g_t$; etc., let h_t be the residue after division of g_t by $h_1, h_2, \ldots, h_{t-1}$.

Then $h_1, \ldots, h_t$ is the reduced Gröbner basis of I. Indeed, the minimality of the Gröbner basis $g_1, \ldots, g_t$ implies that the highest monomials of the polynomials h_i and g_i coincide for all i. Hence $h_1, \ldots, h_t$ is a minimal Gröbner basis of I. Besides, after division of g_i by $h_1, \ldots h_{i-1}, g_{i+1}, \ldots, g_t$ we get the residue h_i which does not contain the terms divisible by the highest monomials of the polynomials $h_1, \ldots h_{i-1}$ and $g_{i+1}, \ldots, g_t$. The latter monomials coincide with the highest monomials of the polynomials $h_{i+1}, \ldots, h_t$. Therefore $h_1, \ldots h_t$ is a reduced Gröbner basis.

Theorem 6.2.7 (Buchberger). *For any ideal I, there exists precisely one reduced Gröbner basis.*

Proof. We have just proved the existence of a reduced Gröbner basis. It remains to prove its uniqueness. Let $f_1, \ldots, f_t$ and $g_1, \ldots, g_s$ be two reduced Gröbner bases of I. The reduced bases are minimal, and so by Theorem 6.2.6 we have $s = t$ and we may assume that the highest monomials of f_i and g_i coincide. Suppose that $f_i - g_i \neq 0$. Then the highest monomial of $f_i - g_i \in I$ is divisible by the highest monomial of some polynomial g_j. Here $j \neq i$ since the highest monomial of $f_i - g_i$ is strictly less than the highest monomial of g_i. On the other hand, if the highest monomial of g_j divides the highest monomial of $f_i - g_i$, then it should divide some monomial of one of the polynomials f_i and g_i. But this contradicts the fact that the bases $f_1, \ldots, f_t$ and $g_1, \ldots, g_t$ are reduced (recall that the highest monomials of g_j and f_j coincide). $\square$

7

Hilbert's Seventeenth Problem

7.1 The sums of squares: introduction

7.1.1 Several examples

It is not difficult to prove that any polynomial $p(x)$ with real coefficients which takes non-negative values for all $x \in \mathbb{R}$ can be represented as the sum of squares of two polynomials with real coefficients. Indeed, the roots of a polynomial with real coefficients can be divided into the real roots and pairs of complex conjugate ones. Therefore

$$p(x) = a \prod_{j=1}^{s} (x - z_j)(x - \overline{z_j}) \prod_{k=1}^{t} (x - \alpha_k)^{m_k},$$

where $\alpha_k \in \mathbb{R}$. If $p(x) \geq 0$ for all $x \in \mathbb{R}$, then $a \geq 0$ and all the numbers m_k are even, and so the real roots also split into pairs. Hence

$$p(x) = \left(\sqrt{a} \prod_{j=1}^{l} (x - z_j) \right) \left(\sqrt{a} \prod_{j=1}^{l} (x - \overline{z_j}) \right),$$

where some of the z_j can be real. Let

$$\sqrt{a} \prod_{j=1}^{l} (x - z_j) = q(x) + ir(x),$$

where q and r are polynomials with real coefficients. Then

$$\sqrt{a} \prod_{j=1}^{l} (x - \overline{z_j}) = q(x) - ir(x).$$

As a result, we obtain $p(x) = (q(x))^2 + (r(x))^2$.

V.V. Prasolov, *Polynomials*, Algorithms and Computation in Mathematics 11,
DOI 10.1007/978-3-642-03980-5_7, © Springer-Verlag Berlin Heidelberg 2004

For polynomials in several indeterminates, a similar statement is not always true, i.e., there exist *non-negative* polynomials — by which we mean polynomials with real coefficients whose values are always non- negative for real values of the variables — that cannot be represented as the sum of squares of polynomials with real coefficients. Hilbert was the first to prove this in 1888 [Hi1] but he did not give an explicit example of such a polynomial. The first simple example was given by T. Motzkin in 1967.

Example 7.1.1. [Mo] The polynomial

$$F(x, y) = x^2 y^2 (x^2 + y^2 - 3) + 1$$

is non-negative but it cannot be represented as the sum of squares of polynomials with real coefficients.

Proof. First let us verify that $F(x, y) \geq 0$. If $x = 0$ or $y = 0$, then $F(x, y) = 1$. We therefore consider $xy \neq 0$. In this case, x^2, y^2 and $x^{-2} y^{-2}$ are positive and their product is equal to 1. Hence

$$x^2 + y^2 + x^{-2} y^{-2} \geq 3,$$

and therefore $x^2 y^2 (x^2 + y^2 - 3) + 1 \geq 0$ as required.

Now, suppose that $F(x, y) = \sum f_j(x, y)^2$, where f_j are polynomials with real coefficients. Then $\sum f_j(x, 0)^2 = F(x, 0) = 1$. Hence $f_j(x, 0) = c_j$ is a constant, and therefore $f_j(x, y) = c_j + y g_j(x, y)$. Similar arguments show that $f_j(x, y) = c'_j + x g'_j(x, y)$. Clearly, $c_j = c'_j$ and $f_j(x, y) = c_j + xy h_j(x, y)$. Thus,

$$x^2 y^2 (x^2 + y^2 - 3) + 1 = x^2 y^2 \sum h_j^2 + 2xy \sum c_j h_j + \sum c_j^2,$$

i.e.,

$$x^2 y^2 (x^2 + y^2 - 3) - x^2 y^2 \sum h_j^2 = 2xy \sum c_j h_j + \sum c_j^2 - 1.$$

All the monomials on the right-hand side of this equality are of degree no higher than 3, and all the monomials on the left-hand side of this equality are of degree no less than 4. Indeed,

$$\deg h_j = \deg f_j - 2 \leq \frac{1}{2} \deg F - 2 = 1.$$

Hence, $x^2 y^2 (x^2 y^2 - 3) - x^2 y^2 \sum h_j^2 = 0$, and therefore $x^2 + y^2 - 3 = \sum h_j^2$. A contradiction since $x^2 + y^2 - 3 < 0$ for $x = y = 0$. $\square$

Example 7.1.2. (R. M. Robinson, 1973) The polynomial

$$S(x, y) = x^2 (x^2 - 1)^2 + y^2 (y^2 - 1)^2 - (x^2 - 1)(y^2 - 1)(x^2 + y^2 - 1)$$

is non-negative but it cannot be represented as the sum of squares of polynomials with real coefficients.

Proof. First let us verify that $S(x,y) \geq 0$. This is obvious for the points lying in the non-shaded part of the plane on Fig.7.1 since for any such point either $x^2 + y^2 - 1 \leq 0$ and $(x^2 - 1)(y^2 - 1) \geq 0$, or $x^2 + y^2 - 1 \geq 0$ and $(x^2 - 1)(y^2 - 1) \leq 0$. But $S(x,y)$ can be xpressed differently, namely,

$$S(x,y) = (x^2 + y^2 - 1)(x^2 - y^2)^2 + (x^2 - 1)(y^2 - 1).$$

This expression makes it clear that $S(x,y) \geq 0$ for the points lying in the shaded domain since for any such point $x^2 + y^2 - 1 \geq 0$ and $(x^2 - 1)(y^2 - 1) \geq 0$.

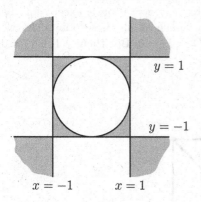

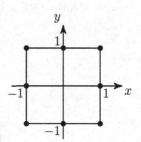

FIGURE 7.1   FIGURE 7.2

Now, suppose that $S(x,y) = \sum f_j(x,y)^2$. The function S vanishes at 8 points depicted on Fig.7.2. Therefore, at these points, each of the functions f_j vanishes. But $\deg f_j \leq \frac{1}{2} \deg S = 3$ and if a curve of degree no greater than 3 passes through the 8 points indicated, it necessarily passes through the origin as well, as we will prove in a moment. Thus, $f_j(0,0) = 0$ for all j, and hence $S(0,0) = 0$. But, obviously, $S(0,0) = 1$. The contradiction obtained shows that $S(x,y)$ cannot be represented as the sum of squares of polynomials.

The proof of the fact that any cubic curve passing through the 8 intersection points of the lines p_i and q_j $(i,j = 1,2,3)$ must pass through the 9th point can be found in the book [Pr3]. For the configuration of the points considered, we can give a simpler proof. Let us ascribe weight 1 to the points $(\pm 1, \pm 1)$, weight -2 to the points $(\pm 1, 0)$ and $(0, \pm 1)$, and weight 4 to the origin $(0,0)$. Consider the sum over all these points of the values of the function $x^p y^q$ multiplied by the corresponding weights. If $pq = 0$, the sum is zero. If $p > 0$ and $q > 0$, only the points $(\pm 1, \pm 1)$ give a non-zero contribution to the sum. Moreover, the sum is non-zero only if both p and q are even. But the polynomial f_j of degree not greater than 3 has no such monomials. Therefore the weighted sum of the values of f_j over the 9 points considered is zero. In particular, if f_j vanishes at 8 of the points, then it vanishes at the 9th point as well. □

Example 7.1.3. (R. M. Robinson, 1973) The polynomial

$$Q(x, y, z) = x^2(x - 1)^2 + y^2(y - 1)^2 + z^2(z - 1)^2 + 2xyz(x + y + z - 2)$$

is non-negative but it cannot be represented as the sum of squares of polynomials.

Proof. Suppose that $Q(x, y, z) = \sum f_j(x, y, z)^2$. Then the degree of each polynomial f_j does not exceed 2. Since the function $Q(x, y, z)$ vanishes at all the points (x, y, z) with the coordinates $x, y, z = 0$ or 1, except for the point $(1, 1, 1)$, then the functions f_j also vanish at all these points. As we will establish shortly, this implies that $f_j(1, 1, 1) = 0$. But then $Q(1, 1, 1) = 0$ whereas, evidently, $Q(1, 1, 1) = 2$.

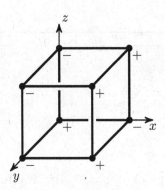

FIGURE 7.3

Let us ascribe to the 8 points considered weights ± 1 as indicated in Fig.7.3 and consider the weighted sum over these points of the values $f_j(x, y, z)$. It is easy to verify that the sum considered vanishes for the following functions: $1, x, xy, x^2$. Hence it is equal to zero for the function $f_j(x, y, z)$ as well, since $\deg f_j \leq 2$. Hence, if the function f_j vanishes at any seven of the 8 points considered, it vanishes in the 8th point as well.

Let us prove now that $Q(x, y, z) \geq 0$. To this end, let us express Q variously as

$$\begin{aligned} Q &= x^2(x - 1)^2 + (y(y - 1) - z(z - 1))^2 + Q_x \\ &= y^2(y - 1)^2 + (z(z - 1) - x(x - 1))^2 + Q_y \\ &= z^2(z - 1)^2 + (x(x - 1) - y(y - 1))^2 + Q_z, \end{aligned}$$

where

$$Q_x = 2yz(x + y - 1)(x + z - 1),$$
$$Q_y = 2xz(x + y - 1)(y + z - 1),$$
$$Q_z = 2xy(x + z - 1)(y + z - 1).$$

It suffices to prove that at any point (x, y, z) one of the functions Q_x, Q_y, Q_z is non-negative. But these functions cannot be simultaneously negative since their product is the square of the polynomial

$$2\sqrt{2}xyz(x + y - 1)(x + z - 1)(y + z - 1). \ \Box$$

Example 7.1.4. (Anneli Lax, Peter D. Lax, 1978) The form

$$A(x) = A_1(x) + A_2(x) + A_3(x) + A_4(x) + A_5(x),$$

where $x = (x_1, x_2, x_3, x_4, x_5)$ and $A_i(x) = \prod_{j \neq i} (x_i - x_j)$, is non-negative but it cannot be represented as a sum of squares of forms.

Remark. The form $A(x)$ only depends on differences of the variables, and so it can be represented as a form in four indeterminates. On dividing this new form by the 4th power of one of the indeterminates, we obtain a polynomial of degree 4 in 3 variables.

First we verify that $A(x) \geq 0$. The value of $A(x)$ does not vary under any permutation of the variables, and so we may assume that

$$x_1 \geq x_2 \geq x_3 \geq x_4 \geq x_5.$$

In this case

$$A_1(x) + A_2(x) =$$
$$= (x_1 - x_2)\left((x_1 - x_3)(x_1 - x_4)(x_1 - x_5) - (x_2 - x_3)(x_2 - x_4)(x_2 - x_5)\right) \geq 0.$$

We similarly prove that $A_4(x) + A_5(x) \geq 0$. It is also clear that $A_3(x)$ is the product of two non-positive and two non-negative factors, and so $A_3(x) \geq 0$.

Now suppose that $A(x) = \sum Q_j(x)^2$, where the Q_j are quadratic forms. If any x_i is equal to any other x_k, then $A(x) = 0$, and hence $Q_j(x) = 0$. Therefore the quadric $Q_j(x) = 0$ in $\mathbb{R}P^4$ contains a projective line $x_1 = x_2, x_3 = x_4 = x_5$. Under permutation of coordinates we obtain 10 lines of this form (to determine a line one should select 2 coordinates of 5). These lines intersect a generic hyperplane at 10 points and the quadric $Q_j(x) = 0$ should pass through these points. But in three-dimensional space the quadric does not pass through 10 generic points. Therefore one should expect that the form Q_j vanishes identically. We will establish shortly that this is indeed the case, and therefore reach a contradiction.

Let $Q(x) = \sum c_{ij}x_i x_j$ with $c_{ij} = c_{ji}$. By the hypothesis

$$0 = Q(s,s,t,t) = (c_{11} + 2c_{12} + c_{22})s^2 +$$
$$+2(c_{13} + c_{14} + c_{15} + c_{23} + c_{24} + c_{25})st +$$
$$+(c_{33} + c_{44} + c_{55} + 2c_{34} + 2c_{35} + 2c_{45})t^2.$$

Hence

$$c_{11} + 2c_{12} + c_{22} = 0, \tag{1}$$

$$c_{13} + c_{14} + c_{15} + c_{23} + c_{24} + c_{25} = 0, \tag{2}$$

$$c_{33} + c_{44} + c_{55} + 2c_{34} + 2c_{35} + 2c_{45} = 0. \tag{3}$$

Moreover, similar equalities obtained under any permutation of indices also hold. In particular, (1) implies that

$$c_{33} + 2c_{34} + c_{44} = 0. \tag{4}$$

Subtracting (4) from (3) we obtain

$$c_{55} + 2c_{35} + 2c_{45} = 0.$$

Let $c_{55} = \lambda$. Then, for distinct i and j different from 5, we have $c_{i5} + c_{j5} = -\dfrac{\lambda}{2}$. Hence $c_{15} = c_{25} = c_{35} = c_{45} = -\dfrac{\lambda}{4}$. Similarly, $c_{21} = c_{31} = c_{41} = c_{51} = a_{15} = -\dfrac{\lambda}{4}$, and so on. As a result, we get $c_{ii} = \lambda$ and $c_{ij} = -\dfrac{\lambda}{4}$ for $i \neq j$. But then (2) implies that $\lambda = 0$, i.e., $Q(x) = 0$ for all x.

7.1.2 Artin-Cassels-Pfister theorem

In subsection 7.1.1 we gave several examples of non-negative polynomials that cannot be represented as the sum of squares of polynomials. In what follows we will show that any non-negative polynomial can be represented as the sum of squares of rational functions. But for polynomials in one variable, the difference between representations as the sum of squares of polynomials and as the sum of squares of rational functions is inessential, as the following statement shows.

Theorem 7.1.5. *Let K be a field of characteristic distinct from 2 and let $f(x)$ be a polynomial over K. Suppose that*

$$f(x) = \alpha_1 r_1(x)^2 + \cdots + \alpha_n r_n(x)^2,$$

where $\alpha_i \in K$ and $r_i(x)$ are rational functions over K. Then

$$f(x) = \alpha_1 p_1(x)^2 + \cdots + \alpha_n p_n(x)^2,$$

where $p_i(x)$ are polynomials over K.

This theorem has a long history. In 1927, Artin [Ar1] proved that

$$f(x) = \beta_1 p_1(x)^2 + \cdots + \beta_m p_m(x)^2,$$

where m is a number not necessarily equal to n. Next, in 1964, Cassels [Ca5] showed that one may assume that $m = n$. And, in 1965, Pfister [Pf1] proved that one may assume that $\beta_i = \alpha_i$.

Theorem 7.1.5 can be applied as well to any polynomial $f(x_1, \ldots, x_n)$ in n indeterminates over a field L. For this, we take, for example, $x = x_1$ and allow $K = L(x_2, \ldots, x_n)$ to be the field of rational functions in indeterminates $x_2, \ldots, x_n$ over L. As a result, we see that in the representation of f as the sum of squares of rational functions we can remove any of the indeterminates in the denominators of these rational functions. But we cannot remove all the indeterminates simultaneously.

Proof. For $n = 1$, the statement is obvious, and so in what follows we assume that $n > 1$ and $\alpha_i \neq 0$ for all i. It is convenient to carry out the proof in terms of quadratic forms over the field $K(x)$. Let $v = (v_1, \ldots, v_n)$ be a vector with coordinates from $K(x)$. Define

$$\varphi(u, v) = \sum \alpha_i u_i v_i.$$

We have to prove that if $f \in K[x]$ and $f = \varphi(u, u)$, where $u_i \in K(x)$, then $f = \varphi(w, w)$, where $w_i \in K[x]$. The quadratic form $\varphi(u, u)$ can be either isotropic, (i.e., $\varphi(u, u) = 0$ for some $u \neq 0$) or anisotropic (i.e., $\varphi(u, u) \neq 0$ for all $u \neq 0$).

Case 1: the form $\varphi(u, u)$ is isotropic. In this case we will not even need the condition $f = \varphi(u, u)$ for $u_i \in K(x)$. In other words, for any polynomial f, there exists a vector u with coordinates from $K[x]$ such that $f = \varphi(u, u)$.

In the equality $\varphi(u, u) = 0$, we may assume that u is a polynomial: indeed, it suffices to reduce the rational functions u_i to the common denominator. We may also assume that the polynomials $u_1, \ldots, u_n$ are relatively prime in totality. Then there exist polynomials $v_1, \ldots, v_n$ such that $u_1 v_1 + \cdots + u_n v_n = 1$.

Indeed, we first represent $\mathrm{GCD}(f_1, f_2)$ in the form $u_1 f_1 + u_2 f_2$; then we represent $\mathrm{GCD}(f_1, f_2, f_3)$ in the form $(u_1 f_1 + u_2 f_2) g_1 + u_3 f_3$, and so on. Having divided each polynomial v_i by the number $2\alpha_i$, we obtain a vector v such that $\varphi(u, v) = \frac{1}{2}$. But since

$$\varphi(u, v + \lambda u) = \varphi(u, v) \quad \text{and} \quad \varphi(v + \lambda u, v + \lambda u) = \varphi(v, v) + \lambda,$$

and since we can replace v by $v - \varphi(v, v)u$, we may assume that $\varphi(v, v) = 0$. The identity

$$\varphi(fu + v, fu + v) = f^2 \varphi(u, u) + 2f\varphi(u, v) + \varphi(v, v) = f$$

shows that any polynomial f can be expressed as $f = \varphi(w, w)$, where $w = fu + v$.

Case 2: the form $\varphi(u, u)$ *is anisotropic.* In this case we need the condition $f = \varphi(u, u)$ for $u_i \in K(x)$. This is clear from the following example: $K = \mathbb{R}$, $f(x) = -1$ and $\varphi(u, u) = u_1^2 + \cdots + u_n^2$.

Let us multiply both sides of the equality $f = \varphi(u, u)$ by the common denominator of the rational functions $u_1, \ldots, u_n$. As a result, we obtain an equality of the form $\alpha_1 u_1^2 + \cdots + \alpha_n u_n^2 = f u_0^2$, where $u_0, \ldots, u_n$ are polynomials. Among all the equalities of this form, we may select the equality with the least degree r of the polynomial u_0. We have to prove that $r = 0$. Suppose that $r = \deg u_0 > 0$. Let us divide u_i by u_0 with a residue, and therefore find a polynomial v_i such that $\deg(u_i - u_0 v_i) \leq r - 1$.

In addition to vectors $u = (u_1, \ldots, u_n)$ and $v = (v_1, \ldots, v_n)$, we consider vectors $\widetilde{u} = (u_0, \ldots, u_n)$ and $\widetilde{v} = (v_0, \ldots, v_n)$, where $v_0 = 1$. Let us also consider the form

$$\widetilde{\varphi}(\widetilde{x}, \widetilde{y}) = \varphi(x, y) - f x_0 y_0.$$

By the hypothesis $\widetilde{\varphi}(\widetilde{u}, \widetilde{u}) = \varphi(u, u) - f u_0^2 = 0$. Moreover, $\widetilde{\varphi}(\widetilde{v}, \widetilde{v}) = \varphi(v, v) - f$, since $v_0 = 1$. Therefore the equality $\widetilde{\varphi}(\widetilde{v}, \widetilde{v}) = 0$ contradicts the condition $r > 0$. In what follows we assume that $\widetilde{\varphi}(\widetilde{v}, \widetilde{v}) \neq 0$.

This means, in particular, that the vectors $\widetilde{u}$ and $\widetilde{v}$ are not proportional. Therefore the vector

$$\widetilde{w} = \widetilde{\varphi}(\widetilde{v}, \widetilde{v})\widetilde{u} - 2\widetilde{\varphi}(\widetilde{u}, \widetilde{v})\widetilde{v}$$

is nonzero and $\widetilde{\varphi}(\widetilde{w}, \widetilde{w}) = 0$ since

$$\widetilde{\varphi}(\lambda\widetilde{u} - \mu\widetilde{v}, \lambda\widetilde{u} - \mu\widetilde{v}) = \mu\left(-2\lambda\widetilde{\varphi}(\widetilde{u}, \widetilde{v}) + \mu\widetilde{\varphi}(\widetilde{v}, \widetilde{v})\right) = 0$$

for $\lambda = \widetilde{\varphi}(\widetilde{v}, \widetilde{v})$ and $\mu = \widetilde{\varphi}(\widetilde{u}, \widetilde{v})$.

Thus we have constructed a nonzero vector $\widetilde{w} = (w_0, w)$ with polynomial coordinates such that $\varphi(w, w) - f w_0^2 = 0$. To reach a contradiction, it suffices to verify that $\deg w_0 < r$. Since $v_0 = 1$, we obtain

$$\widetilde{w}_0 = \widetilde{\varphi}(\widetilde{v}, \widetilde{v})u_0 - 2\widetilde{\varphi}(\widetilde{u}, \widetilde{v})v_0 = \left(\sum_{i=1}^n \alpha_i v_i^2 - f\right)u_0 - 2\left(\sum_{i=1}^n \alpha_i u_i v_i - f u_0\right)$$

$$= \sum_{i=1}^n \alpha_i\left(v_i^2 u_0 - 2u_i v_i + \frac{u_i^2}{u_0}\right) - \sum_{i=1}^n \frac{\alpha_i u_i^2}{u_0} + f u_0 = \frac{1}{u_0}\sum_{i=1}^n \alpha_i(u_i - u_0 v_i)^2$$

because $\sum_{i=1}^n \alpha_i u_i^2 = f u_0^2$.

Recall that $\deg(u_i - u_0 v_i) \leq r - 1$. Hence

$$\deg w_0 = \deg\left(\sum_{i=1}^n \alpha_i(u_i - u_0 v_i)^2\right) - \deg u_0 \leq 2(r - 1) - r = r - 2. \quad \square$$

With the help of Theorem 7.1.5 we can indicate a non-negative polynomial in n indeterminates that cannot be represented as the sum of n squares of rational functions.

Theorem 7.1.6. *The polynomial $x_1^2 + \cdots + x_n^2 + 1$ cannot be represented as the sum of n squares of rational functions in indeterminates $x_1, \ldots, x_n$ over $\mathbb{R}$.*

Proof. Set $K = \mathbb{R}(x_1, \ldots x_{n-1})$ and $x = x_n$ in the conditions of Theorem 7.1.5. Suppose that we have represented the polynomial $x_1^2 + \cdots + x_n^2 + 1$ as the sum of n squares of rational functions in $x_1, \ldots, x_n$. This means that the polynomial $x^2 + d$, where $d = x_1^2 + \cdots + x_{n-1}^2 + 1 \in K$, can be represented as the sum of n squares of elements of the field $K(x)$. In this case, by Theorem 7.1.5, the polynomial $x^2 + d$ can be represented as the sum of n squares of the elements from $K[x]$, i.e.,

$$x^2 + d = \sum_{i=1}^{n} \left(a_{i0} + a_{i1}x + a_{i2}x^2 + \cdots \right)^2.$$

Obviously, in such a representation, we have $a_{i2} = a_{i3} = \cdots = 0$. Therefore

$$x^2 + d = \sum_{i=1}^{n} (a_i x + b_i)^2, \quad a_i, b_i \in K.$$

Let us substitute in this identity the value $x = c$ such that $c^2 = (a_n c + b_n)^2$, i.e., $(1 \pm a_n)c \pm b_n = 0$ (for $a_n \neq \pm 1$, we may take any sign, and, for $a_n = \pm 1$, only one of the signs will do). As a result, we obtain $d = \sum_{i=1}^{n-1} (a_i c + b_i)^2$, where $c, a_i, b_i \in K$. In other words, the polynomial $x_1^2 + \cdots + x_{n-1}^2 + 1$ admits the representation as the sum of $n-1$ squares of rational functions in $x_1, \ldots, x_{n-1}$ over $\mathbb{R}$. Repeating similar arguments we will obtain finally that the polynomial $x_1^2 + 1$ is the square of a rational function in x_1 over $\mathbb{R}$. A contradiction. $\square$

7.1.3 The inequality between the arithmetic and geometric means

The *inequality between the arithmetic and geometric means* consists in the following. If $x_1, \ldots, x_n$ are non-negative numbers, then

$$\frac{x_1 + x_2 + \cdots + x_n}{n} \geq \sqrt[n]{x_1 x_2 \cdots x_n}.$$

Let us replace x_i by t_i^{2n}. Then this inequality becomes

$$P(t_1, \ldots, t_n) = \frac{t_1^{2n} + t_2^{2n} + \cdots + t_n^{2n}}{n} - t_1^2 t_2^2 \cdot \ldots \cdot t_n^2 \geq 0.$$

Theorem 7.1.7 (Hurwitz [Hu]). *The polynomial $P(t_1, \ldots, t_n)$ can be represented as the sum of squares of polynomials.*

Proof. Let $y_i = t_i^2$. For any function $f(y_1, \ldots, y_n)$, we define

$$Sf(y_1, \ldots, y_n) = \sum_{\sigma \in S_n} f(y_{\sigma(1)}, \ldots, y_{\sigma(n)}).$$

For example,

$$\left.\begin{aligned}
Sy_1^n &= (n-1)!(y_1^n + \cdots + y_n^n), \\
&\qquad\cdots\cdots\cdots\cdots\cdots\cdots\cdots \\
Sy_1 \cdot \ldots \cdot y_n &= n!\, y_1 \cdot \ldots \cdot y_n.
\end{aligned}\right\} \tag{1}$$

Consider the functions

$$\varphi_1 = S\left((y_1^{n-1} - y_2^{n-1})(y_1 - y_2)\right),$$
$$\varphi_2 = S\left((y_1^{n-2} - y_2^{n-2})(y_1 - y_2)y_3\right),$$
$$\varphi_3 = S\left((y_1^{n-3} - y_2^{n-3})(y_1 - y_2)y_3 y_4\right),$$
$$\cdots\cdots\cdots\cdots\cdots\cdots\cdots\cdots\cdots\cdots\cdots\cdots$$
$$\varphi_{n-1} = S\left((y_1 - y_2)(y_1 - y_2)y_3 y_4 \cdot \ldots \cdot y_n\right).$$

It is easy to verify that

$$\varphi_1 = Sy_1^n + Sy_2^n - Sy_1^{n-1} - Sy_2^{n-1}y_1 = 2Sy_1^n - 2Sy_1^{n-1}y_2.$$

Similarly,

$$\varphi_2 = 2Sy_1^{n-1}y_2 - 2Sy_1^{n-2}y_2 y_3,$$
$$\varphi_3 = 2Sy_1^{n-2}y_2 y_3 - 2Sy_1^{n-3}y_2 y_3 y_4,$$
$$\cdots\cdots\cdots\cdots\cdots\cdots\cdots\cdots\cdots\cdots\cdots\cdots$$
$$\varphi_{n-1} = 2Sy_1^2 y_2 y_3 \cdot \ldots \cdot y_{n-1} - 2Sy_1 y_2 \cdot \ldots \cdot y_n.$$

Hence

$$\varphi_1 + \varphi_2 + \cdots + \varphi_{n-1} = 2Sy_1^n - 2Sy_1 y_2 \ldots y_n.$$

Taking into account relations (1) we obtain

$$\frac{y_1^n + y_2^n + \cdots + y_n^n}{n} - y_1 y_2 \cdot \ldots \cdot y_n = \frac{1}{2n!}(\varphi_1 + \varphi_2 + \cdots + \varphi_{n-1}),$$

i.e.,

$$\frac{t_1^{2n} + t_2^{2n} + \cdots + t_n^{2n}}{n} - t_1^2 t_2^2 \cdot \ldots \cdot t_n^2 = \frac{1}{2n!}(\varphi_1 + \varphi_2 + \cdots + \varphi_{n-1}),$$

where

$$\begin{aligned}
\varphi_k &= S\left((y_1^{n-k} - y_2^{n-k})(y_1 - y_2)y_3 y_4 \cdot \ldots \cdot y_{k+1}\right) \\
&= S\left((y_1 - y_2)^2(y_1^{n-k-1} + y_1^{n-k-2}y_2 + \cdots + y_2^{n-k-1})y_3 y_4 \cdot \ldots \cdot y_{k+1}\right) \\
&= S\left((t_1^2 - t_2^2)^2(t_1^{2(n-k-1)} + t_1^{2(n-k-2)}t_2^2 + \cdots + t_2^{2(n-k-1)})t_3^2 t_4^2 \cdot \ldots \cdot t_{k+1}^2\right).
\end{aligned}$$

Thus φ_k is the sum of squares of polynomials in $t_1, \ldots, t_n$. □

7.1.4 Hilbert's theorem on non-negative polynomials $p_4(x, y)$

Let p_k be a polynomial of degree k. In section 7.1.1 we gave examples of non-negative polynomials $p_6(x, y)$ and $p_4(x, y, z)$ that cannot be represented as the sums of squares of polynomials. For polynomials $p_2(x_1, \ldots, x_n)$, there are no such examples. Indeed, to any polynomial $p_2(x_1, \ldots, x_n)$, there corresponds the quadratic form

$$F_2(y_1, \ldots, y_{n+1}) = y_{n+1}^2 p_2 \left(\frac{y_1}{y_{n+1}}, \ldots, \frac{y_n}{y_{n+1}} \right)$$

and any quadratic form can be represented in the form

$$f_1^2 + \cdots + f_k^2 - f_{k+1}^2 - \cdots - f_{n+1}^2,$$

where $f_1, \ldots, f_{n+1}$ are linear forms. Clearly, the polynomial p_2 is non-negative only if $f_{k+1} = \cdots = f_{n+1} = 0$.

It is much more difficult to prove that any non-negative polynomial $p_4(x, y)$ can be represented as the sum of squares of polynomials.

Theorem 7.1.8 (Hilbert). *Any non-negative polynomial $p_4(x, y)$ can be represented as the sum of three squares of polynomials.*

We will give two proofs of this theorem. The first proof is simpler but it only enables us to prove a weaker statement, namely, we will show that $p_4(x, y)$ can be represented as the sum of several (not necessarily three) squares of polynomials. The second one is the original Hilbert's proof.

It is more convenient to give both proofs not for polynomials but for homogeneous forms $F_4(x, y, z)$.

First proof. (Choi–Lam) We will prove that any non-negative homogeneous form $F_4(x, y, z)$ can be represented as the sum of squares of homogeneous forms. The first part of the proof concerns forms of any degree in any number of indeterminates.

To a pair of forms P and Q of degree n in m indeterminates, we can assign the form $\lambda P + \mu Q$, i.e., the set of all such forms is naturally endowed with the structure of a linear space. The origin of this linear space is obviously the zero form.

The non-negative forms constitute a closed convex cone C with the vertex at the origin O. Clearly, if Q is a non-zero form and $Q \in C$, then $-Q \notin C$. Therefore any plane passing through O and Q intersects C in a (closed) angle $Q_1 O Q_2$ whose value is strictly less than π. The form Q is a convex linear combination of the forms Q_1 and Q_2, i.e. $Q = \lambda_1 Q_1 + \lambda_2 Q_2$, where $\lambda_1, \lambda_2 \geq 0$ and $\lambda_1 + \lambda_2 = 1$.

Let us draw hyperplanes of support to the cone C passing through the rays $O Q_1$ and $O Q_2$. They intersect C in certain convex cones of strictly lesser dimension. Consider the section of each of these cones by a plane passing through O and Q. After several such operations we will necessarily arrive at the cones of dimension 1 (rays).

A point A of a closed convex cone C is called *extremal* if there exists a hyperplane of support intersecting the cone C only along the ray OA. In other words,

the point A is extremal if it is not the inner point of the segment whose end-points belong to C but do not lie on the ray OA.

The above described construction shows that

any non-negative homogeneous form Q is a convex linear combination of extremal non-negative forms.

So far we have considered forms of any degree in any number of indeterminates. The following statement only holds for forms of degree 4 in three indeterminates.

Lemma 7.1.9. *Any non-negative homogeneous form $T(x, y, z) \neq 0$ of degree 4 can be represented in the form*

$$T = q^2 + T_1,$$

where $q \neq 0$ is a quadratic form and T_1 a non-negative form.

Corollary. *Any extremal non-negative form $T(x, y, z)$ of degree 4 is a total square.*

The corollary obviously follows from Lemma: for an extremal non-negative form T, the decomposition $T = q^2 + T_1$ must be trivial, i.e., q^2 and T_1 should be proportional.

In turn, the corollary obviously implies the theorem. Indeed, the convex linear combination of extremal non-negative forms of degree 4 in 3 indeterminates is a sum of squares of quadratic forms.

Proof. Let $Z(T)$ be the set of zeros of the form T considered up to proportionality, i.e., the set of zeros of this form in $\mathbb{R}P^2$.

Case 1: $Z(T) = \emptyset$. On the unit sphere $x^2 + y^2 + z^2 = 1$ the function attains a minimal value $\mu > 0$, and so $T(x, y, z) \geq \mu(x^2 + y^2 + z^2)^2$ for all (x, y, z).

Case 2: $Z(T)$ consists of one point; without loss of generality we may assume that $T(1, 0, 0) = 0$. In this case the coefficient of x^4 is equal to 0, and so

$$T(x, y, z) = x^3(\alpha_1 y + \alpha_2 z) + x^2 f(y, z) + 2xg(y, z) + h(y, z).$$

If $\alpha_1 \neq 0$ and $\alpha_2 \neq 0$, then as $x \to \pm\infty$ we can obtain negative values of T. Hence,

$$T(x, y, z) = x^2 f + 2xg + h.$$

It is also clear that $f \geq 0$ and $h \geq 0$.

In the decomposition

$$fT = (xf + g)^2 + (fh - g^2)$$

the form $fh - g^2$ is non-negative. Indeed, if $fh - g^2 < 0$ at (a, b), then $f(a, b) \neq 0$. Setting $x = -\dfrac{g(a, b)}{f(a, b)}$ we see that $fT < 0$ at (x, a, b).

Any non-negative quadratic form $f(x, y)$ can be represented either as the square of a linear form or as the sum of two squares of linear forms. Accordingly, consider two possibilities.

(a) $f = f_1^2$, where $f_1 = \alpha y + \beta z$. At the point $(-\beta, \alpha)$ we have $fh - g^2 = -g^2 \leq 0$. Hence $g(-\beta, \alpha) = 0$ since the form $fh - g^2$ is non-negative. Therefore $g = f_1 g_1$, and hence

$$fT \geq (xf + g)^2 = (xf_1^2 + f_1 g_1)^2 = f_1^2(xf_1 + g_1)^2 = f(xf_1 + g_1)^2$$

and thus $T \geq (xf_1 + g_1)^2$.

(b) $f = f_1^2 + f_2^2$, where f_1 and f_1 are linear forms without non-trivial (i.e., distinct from the origin) common zeros. Then $f(y, z) > 0$ for $(y, z) \neq (0, 0)$. Suppose that $fh - g^2 = 0$ for $(y, z) = (a, b) \neq (0, 0)$. Then $T = 0$ for $(x, y, z) = \left(-\dfrac{g(a, b)}{f(a, b)}, a, b \right)$. This is a contradiction since, by the hypothesis, T has only one zero in $\mathbb{R}P^2$, namely $(1, 0, 0)$.

Thus $fh - g^2 > 0$ for $(y, z) \neq (0, 0)$, and therefore $\dfrac{fh - g^2}{f^3} \geq \mu > 0$ on the unit circle. Hence $fh - g^2 \geq \mu f^3$ for all (y, z). Therefore

$$T \geq \frac{fh - g^2}{f} \geq \mu f^2 = (\sqrt{\mu} f)^2.$$

Case 3: $Z(T)$ *contains no less than two points*; without loss of generality we may assume that $T(1, 0, 0) = T(0, 1, 0) = 0$. As in Case 2, the form T cannot contain the terms with x^4 and x^3 nor can it contain terms with y^4 and y^3. Hence

$$T(x, y, z) = x^2 f(y, z) + 2xz g(y, z) + z^2 h(y, z).$$

In the decomposition

$$fT = (xf + zg)^2 + z^2(fh - g^2).$$

the form $fh - g^2$ is non-negative.

While considering subcase (a) of Case 2 we did not use the fact that the form T has precisely one zero. Therefore, if $f = f_1^2$ (or $h = h_1^2$), we may apply the same arguments. It remains to consider the case when $f > 0$ and $g > 0$. Again, consider two possibilities.

(a) $fh - g^2$ has a non-trivial zero (a, b). Let $\alpha = -\dfrac{g(a, b)}{f(a, b)}$ and define

$$T_1(x,y,z) = T(x + \alpha z, y, z) = x^2 f + 2xz(g + \alpha f) + z^2(h + 2\alpha g + \alpha^2 f).$$

At (a, b), we have

$$h + 2\alpha g + \alpha^2 f = h + 2\frac{-g}{f}g + \frac{g^2}{f^2}f = h - g^2 f = \frac{hf - g^2}{f^2} = 0.$$

Therefore $h + 2\alpha g + \alpha^2 f = h_1^2$. Thus $T_1(x, y, z) \geq (zh_1)^2$, and therefore

$$T(x,y,z) = T_1(x - \alpha z, y, z) \geq (zh_1(x - \alpha z, y, z))^2.$$

b) $fh - g^2 > 0$. Then

$$\frac{fh - g^2}{(y^2 + z^2)f} \geq \mu > 0,$$

and therefore

$$fT = (xf + zg)^2 + z^2(fh - g^2) \geq z^2(fh - g^2) \geq \mu z^2(y^2 + z^2)f.$$

As a result, we obtain $T \geq (\sqrt{\mu}zy)^2 + (\sqrt{\mu}z^2)^2 \geq (\sqrt{\mu}z^2)^2$. $\square$

$\square$

Second proof. (Hilbert) The main idea of this proof is to consider the set A consisting of the real forms in three indeterminates that can be represented in the form $f^2 + g^2 + h^2$, where f, g, h are *real* quadratic forms without non-trivial common zeros over the field of *complex* numbers.

Lemma 7.1.10. *The set A is open.*

Proof. The coefficients a_{ijk} of the form $\sum a_{ijk} x^i y^j z^k$ can be considered as coordinates in $\mathbb{R}^n$. Therefore the map

$$\Phi: (f, g, h) \mapsto F = f^2 + g^2 + h^2$$

is an algebraic map $\mathbb{R}^{18} \to \mathbb{R}^{15}$ (the quadratic form in 3 indeterminates is determined by 6 coefficients and the form of degree 4 is determined by 15 coefficients).

It suffices to prove that if $(f, g, h) \in A$, then the rank of the differential $d\Phi$ of the map Φ at point (f, g, h) is equal to 15, i.e., $\dim \ker \Phi = 3$. Clearly,

$$d\Phi(u, v, w) = 2(uf + vg + wh),$$

where $(u, v, w) \in \mathbb{R}^{18}$ is a triple of quadratic forms and $uf + vg + wh$ is a form of degree 4.

The quadratic forms f, g, h have no non-trivial common zeros over $\mathbb{C}$. Let us prove that the equation

$$uf + vg + wh = 0 \tag{1}$$

$(u, v, w$ are quadratic forms) then implies that

$$u = \nu g - \mu h, \quad v = \lambda h - \nu f, \quad w = \mu f - \lambda g$$

for some $\lambda, \mu, \nu \in \mathbb{C}$.

It suffices to prove that

$$u = \nu_1 g - \mu_1 h, \quad v = \lambda_1 h - \nu_2 f, \quad w = \mu_2 f - \lambda_2 g$$

because (1) then gives

$$(\lambda_1 - \lambda_2)hg + (\mu_2 - \mu_1)fh + (\nu_1 - \nu_2)fg = 0.$$

The curves $f = 0, g = 0$ and $h = 0$ are distinct. So, on the curve $f = 0$, there is a point which does not belong to either $g = 0$ or $h = 0$. Having considered the values of f, g and h at this point, we obtain $\lambda_1 = \lambda_2$. We similarly prove that $\mu_1 = \mu_2$ and $\nu_1 = \nu_2$.

Let us now prove, for example, that $w = \mu_2 f - \lambda_2 g$. By Hilbert's Nullstellensatz the ideal generated by f, g and h contains a power of any polynomial since these forms have no common zeros. In particular, x^n can be represented for some n in the form

$$x^n = rf + sg + th, \tag{2}$$

where r, s, t are forms of degree $n - 2$. Consider equation (2) with the minimal n. From (1) and (2) we deduce that

$$uft + vgt + wht = 0, \quad x^n w = rfw + sgw + thw$$

and therefore

$$x^n w = (rw - ut)f + (sw - vt)g = af + bg, \tag{3}$$

where a and b are forms of degree n.

If $n = 0$, we get the equality desired.

If $n > 0$, we obtain a contradiction to the minimality of n. Indeed, for $x = 0$, the equality (3) becomes

$$a_0 f_0 + b_0 g_0 = 0,$$

where $a_0 = a(0, y, z)$, etc. Since f_0 and g_0 have no common zeros, $a_0 = d_0 g_0$ and $b_0 = -d_0 f_0$ for a polynomial $d_0(y, z)$. Set $d(x, y, z) = d_0(y, z)$ and consider the polynomials $a_1 = a - dg$ and $b_1 = b + dg$. Clearly,

$$a_1 f + b_1 g = af + bg = x^n w$$

and $a_1(0, y, z) = b_1(0, y, z) = 0$, i.e., a_1 and b_1 are divisible by x. Dividing by x we obtain an equality of the form $a_2 f + b_2 g = x^{n-1} w$ which contradicts the fact that n is minimal.

Thus, the kernel of the map

$$d\Phi \colon (u, v, w) \mapsto 2(uf + vg + wh)$$

consists of the vectors of the form

$$(\nu g - \mu h, \lambda h - \nu f, \mu f - \lambda g) = \lambda(0, h, -g) + \mu(-h, 0, f) + \nu(g, -f, 0),$$

and therefore the dimension of the kernel is 3. $\square$

Lemma 7.1.11. *Let $F \in \overline{A} \setminus A$, where $\overline{A}$ is the closure of A. Then either F has a nontrivial real zero or, over $\mathbb{C}$, the curve $F = 0$ has at least two double points.*

Proof. Clearly, $F = f^2 + g^2 + h^2$, where f, g and h have a common non-trivial zero (a, b, c). If this zero is not real, then the points (a, b, c) and $(\overline{a}, \overline{b}, \overline{c})$ are two distinct double points on the curve $F = 0$. Indeed, these points are zeros of the functions f, g, h, and so they are zeros of multiplicity 2 of the functions f^2, g^2, h^2. Hence, they are zeros of multiplicity 2 of the function $F = f^2 + g^2 + h^2$. $\square$

Let us now move on to the proof of the theorem proper. We have to prove that any non-negative form lies in A. The open set A is bounded by the surface $\partial A = \overline{A} \setminus A$. Let F_1 be an arbitrary non-negative form of degree 4 in 3 indeterminates. If $F_1 \in A$ we have nothing more to prove. So let us assume that $F_1 \notin A$. Then the segment $F_0 F_1$, where $F_0 \in A$ is an arbitrary point, should intersect ∂A at a point F_t. It suffices to prove that we can select F_0 so that F_t coincides with F_1 (then $F_1 = F_t \in \partial A \subset \overline{A}$)). Assume that F_t is an interior point of the segment $F_0 F_1$. We can select F_0 so that F_t has a non-trivial real zero. Indeed, by Lemma 7.1.11 the forms that belong to ∂A and do not have non-trivial real zeros correspond to curves with two double points, and such forms constitute a set of codimension no less than 2. Indeed, the curve $F = 0$ has a double point if the system of equations

$$F = 0, \quad F_x = 0, \quad F_y = 0, \quad F_z = 0$$

has a solution. The first equation can be disregarded since

$$xF_x + yF_y + zF_z = nF,$$

where n is the degree of F (in our case $n = 4$). The curves $F_x = 0$ and $F_y = 0$ intersect at $(n-1)^2$ points. The curve $F = 0$ has k double points if the curve $F_z = 0$ passes through the k intersection points of the curves $F_x = 0$ and $F_y = 0$. This imposes k algebraic relations on the coefficients of F.

Thus, the form $F_t = (1 - t)F_0 + tF_1$ has a non-trivial real zero. But this contradicts the fact that $F_t \neq F_1$. Indeed, $F_0 > 0$ and $F_1 \geq 0$, and so, for $t \neq 1$, the form $(1 - t)F_0 + tF_1$ takes strictly positive values at all points distinct from the origin. $\square$

7.2 Artin's theory

This section is mainly devoted to Artin's solution of Hilbert's seventeenth problem on the representability of a non-negative polynomial as the sum of squares of rational functions. Artin's proof gives no estimates of the sufficient number of these rational functions in the representation of a polynomial in n indeterminates. This estimate is due to Pfister: a non-negative polynomial in n indeterminates can be represented as the sum of 2^n squares of rational functions. Pfister's theory will be discussed in Section 7.3.

In Sections 7.2.1 and 7.2.2 we give necessary preliminaries from the theory of real fields; the results of these sections are due to Artin and Shreier [Ar2]. The solution proper of Hilbert's seventeenth problem is contained in 7.2.3. This proof is based on a theorem of Sylvester which enables us to calculate the number of real roots of a polynomial in terms of the index of a quadratic form (see page 33). Our exposition follows [Sc1].

7.2.1 Real fields

A field K is said to be *ordered* if it is split into three non-intersecting subsets

$$K = N \cup \{0\} \cup P$$

such that $N = -P$ (the subset N of *negative* numbers and the subset P of *positive* numbers), where the sum and the product of two positive numbers are positive.

For any ordered field, we may set $x - y > 0$ if $x - y \in P$ (we also write $x \geq y$ if either $x - y \in P$ or $x = y$).

Set

$$|a| = \begin{cases} a & \text{for } a > 0, \\ 0 & \text{for } a = 0, \\ -a & \text{for } a < 0. \end{cases}$$

It is easy to verify that $|ab| = |a| \cdot |b|$ and $|a + b| \leq |a| + |b|$.

In any ordered field we have $1 > 0$ since the opposite inequality $-1 > 0$ leads to a contradiction because $1 = (-1)(-1) > 0$ but if $a > 0$, then $-a < 0$ for any $a \neq 0$. In particular, the characteristic of any ordered field is equal to zero since $1 + \cdots + 1 > 0$ for any nonzero number of summands. In any ordered field $x^2 > 0$. Indeed, both inequalities $x > 0$ and $-x > 0$ imply the same inequality $x^2 > 0$.

There is only one ordering of the field $\mathbb{Q}$, namely, $\dfrac{p}{q} > 0$ if and only if $pq > 0$. Indeed, the numbers $\dfrac{p}{q}$ and pq are obtained from each other by multiplication by $q^{\pm 2} > 0$.

Any field L in which -1 is the sum of squares (and the characteristic of L is distinct from 2) is an example of a field that cannot be ordered. Indeed, any element a of such a field L is the sum of squares

$$a = \left(\frac{1+a}{2}\right)^2 + (-1)\left(\frac{1-a}{2}\right)^2.$$

The field K is called *formally real* if -1 cannot be represented as the sum of squares of of elements from K. An equivalent condition: if $b_1^2 + \cdots + b_n^2 = 0$, where $b_1, \ldots, b_k \in K$, then $b_1 = \cdots = b_n = 0$. For brevity, we will call formally real fields just *real* ones.

The characteristic of any real field is equal to 0. Indeed, if the characteristic is equal to p, then $-1 = \underbrace{1^2 + \cdots + 1^2}_{p-1}$.

Theorem 7.2.1. *Let K be a real field, and let $a \in K$.*

a) If a is the sum of squares of elements from K, then $K(\sqrt{a})$ is real.

b) If $K(\sqrt{a})$ is not real, then $-a$ is the sum of squares of some elements from K.

Proof. To prove a), we suppose on the contrary that $K(\sqrt{a})$ is not real. Then, in particular, $K(\sqrt{a}) \neq K$, i.e., $\sqrt{a} \notin K$. Moreover, -1 is the sum of squares of some elements from $K(\sqrt{a})$, i.e., there exist elements $b_i, c_i \in K$ such that

$$-1 = \sum(b_i + c_i\sqrt{a})^2 = \sum b_i^2 + 2\sqrt{a}\sum b_i c_i + a\sum c_i^2. \tag{1}$$

Formula (1) shows that, if $\sum b_i c_i \neq 0$, then $\sqrt{a} \in K$. Therefore

$$-1 = \sum b_i^2 + a\sum c_i^2. \tag{2}$$

Let a be the sum of squares of some elements from K. Formula (2) shows that in this case our supposition that $K(\sqrt{a})$ is not real leads to a contradiction. This proves statement a).

To prove b), we write (2) in the form

$$-a = \frac{1 + \sum b_i^2}{\sum c_i^2}.$$

It remains to observe that if p and q are sums of squares, then so is $\frac{p}{q} = \frac{pq}{q^2}$. $\square$

Corollary. *For a real field K, one of the fields $K(\sqrt{a})$ and $K(\sqrt{-a})$ is necessarily real.*

Proof. If $K(\sqrt{a})$ is not real, then $-a$ is the sum of squares, and so $K(\sqrt{-a})$ is a real field. $\square$

Theorem 7.2.2. *Let K be a real field and let $f \in K[x]$ be an irreducible polynomial of odd degree. Then the field $K(\alpha)$, where α is a root of f, is real.*

Proof. Let $n = \deg f$. Suppose that the field $K(\alpha)$ is not real. Then

$$-1 = \sum g_i(\alpha)^2,$$

where g_i are polynomials of degree no higher than $n - 1$. Since α is a root of $1 + \sum g_i(x)^2$, the latter is divisible by f, i.e.,

$$-1 = \sum g_i(x)^2 + h(x)f(x),$$

where h is a polynomial. Suppose if possible that $h = 0$. Then $-1 = \sum g_i(x)^2$. If $\max_i(\deg g_i) = m > 0$, then the sum of squares of the coefficients of x^m in the polynomials g_i is equal to zero. If $g_i = c_i \in K$, then $-1 = \sum c_i^2$. Both versions contradict the fact that K is real, and so $h \neq 0$.

The degree of the polynomial $\sum g_i(x)^2$ is even and does not exceed $2n - 2$. Therefore the degree of h is odd and does not exceed $n - 2$. The polynomial h has an irreducible factor h_1 whose degree is odd and does not exceed $n - 2$. Let β be a root of h_1. Then $-1 = \sum g_i(\beta)^2$, i.e., -1 is the sum of squares of elements from $K(\beta)$. Repeating for h_1 the same arguments as for f, we see that -1 is the sum of squares of elements from the field $K(\gamma)$, where γ is a root of an irreducible polynomial of odd degree that does not exceed $n - 4$, etc. Thus we have a contradiction. $\square$

The field K is called *real closed* if it is real and any real algebraic extension of it coincides with K.

Theorem 7.2.2 implies that in any real closed field any polynomial of odd degree has a root.

A *real closure* of a real field K is any real closed field R algebraic over K.

Any real field K has a real closure. Indeed, consider the partially ordered set of all real fields algebraic over K. By Zorn's lemma this set has at least one maximal element, R. Clearly, R is a real closed field.

Theorem 7.2.3. *The real closed field R has precisely one ordering, namely, the nonzero element $a \in R$ is positive if and only if it is a square.*

Proof. Clearly, the equalities $a = t_1^2$ and $-a = t_2^2$, where $t_1, t_2 \in R$, cannot be satisfied simultaneously since otherwise $-1 = \left(\dfrac{t_1}{t_2}\right)^2$. Therefore it suffices to prove that $\pm a = t^2$, where $t \in R$.

If $a \neq t^2$, then the field $R(\sqrt{a})$ is a proper extension of R, so it is not real. In this case, by Theorem 7.2.1 (b), the element $-a$ is the sum of squares of elements from R. Then by Theorem 7.2.1 (a) the field $R(\sqrt{-a})$ is real, and so it coincides with R. This means that $\sqrt{-a} \in R$, i.e., $-a = t^2$, where $t \in R$. $\square$

Theorem 7.2.4. *Let K be a real field and suppose that the element $a \in K$ cannot be represented as the sum of squares. Then there exists an ordering of K for which $a < 0$.*

Proof. By Theorem 7.2.1 (b) the field $K(\sqrt{-a})$ is real. Let R be a real closure of $K(\sqrt{-a})$. By Theorem 7.2.3 the field R has an ordering under which the element $-a = (\sqrt{-a})^2$ is positive, i.e., a is negative. The restriction of this ordering onto $K \subset R$ is the ordering desired. $\square$

Denote the algebraic closure of R by $\overline{R}$.

Theorem 7.2.5. *The field R is real closed if and only if $\overline{R} \neq R$ and $\overline{R} = R(\sqrt{-1})$.*

Proof. First, suppose that R is real closed. Then the equation $x^2 + 1 = 0$ has no solutions in R. So $\overline{R} \neq R$. Let us prove that $\overline{R} = R(\sqrt{-1})$.

For brevity, set $i = \sqrt{-1}$. First we show that any second order equation has a root in $R(i)$. The roots of the quadratic $x^2 + 2px + q$ are given by the formula $x_{1,2} = -p \pm \sqrt{p^2 - q}$, and so it suffices to prove that, if $a, b \in K$, then $\sqrt{a + ib} \in K(i)$. In other words, we have to select $c, d \in K$, so that $(c + di)^2 = a + bi$, i.e., $c^2 - d^2 = a$ and $2cd = b$. Clearly, $a^2 + b^2 = (c^2 + d^2)^2$. Therefore

$$c^2 = \frac{\sqrt{a^2 + b^2} + a}{2}, \quad d^2 = \frac{\sqrt{a^2 + b^2} - a}{2}$$

looks like an appropriate choice. Let us verify first of all that numbers c and d thus defined actually belong to R. Since $a^2 \geq 0$ and $b^2 \geq 0$, then $a^2 + b^2 \geq 0$, so $\sqrt{a^2 + b^2} \in R$ by Theorem 7.2.3. Further, $\sqrt{a^2 + b^2} \geq \sqrt{a^2} \geq \pm a$ (here $\sqrt{a^2 + b^2}$ and $\sqrt{a^2}$ are non-negative numbers). Therefore $\sqrt{a^2 + b^2} \pm a \geq 0$, i.e., c and d belong to R. The equality $c^2 - d^2 = a$ is automatically satisfied and the equality $2cd = b$ will be satisfied if we correctly select the signs of c and d.

Let us now consider an arbitrary polynomial f irreducible over R. Let us express its degree n in the form $n = 2^m q$, where q is odd. We prove by induction on m that f has a root in $R(i)$. For $m = 0$ this follows from Theorem 7.2.2. Now suppose that $m > 0$ and the statement required is already proved for $1, \ldots, m - 1$.

Let $\alpha_1, \ldots, \alpha_n$ be all the roots of f; they belong to some extension of R. Select $c \in R$ so that all the numbers $\alpha_k \alpha_l + c(\alpha_k + \alpha_l)$, where $k \neq l$, are distinct. These numbers are the roots of a polynomial g of degree $\dfrac{n(n-1)}{2}$ with coefficients from R (the coefficients belong to R since they are symmetric functions in $\alpha_1, \ldots, \alpha_n$). The number $\deg g = \dfrac{n(n-1)}{2}$ is of the form $2^{m-1}q(n-1)$, where $q(n-1)$ is odd. We can therefore apply to g the induction hypothesis. Without loss of generality we may assume that the root $\alpha_1 \alpha_2 + c(\alpha_1 + \alpha_2)$ of g belongs to $R(i)$.

Let us prove that $R(\alpha_1 \alpha_2, \alpha_1 + \alpha_2) = R(\alpha_1 \alpha_2 + c(\alpha_1 + \alpha_2))$. To this end, consider the polynomial F with the roots $\alpha_k \alpha_l$ and the polynomial G with the roots $\alpha_k + \alpha_l$. The coefficients of F and G belong to R. Let

$$\theta = \alpha_1 \alpha_2 + c(\alpha_1 + \alpha_2).$$

Clearly, $G(\alpha_1 + \alpha_2) = 0$ and

$$F(\theta - c(\alpha_1 + \alpha_2)) = F(\alpha_1\alpha_2) = 0,$$

i.e., $\alpha_1 + \alpha_2$ is a common root of the polynomials $G(x)$ and $F(\theta - cx)$ with coefficients in $R(\theta)$. The condition $\theta - c(\alpha_k + \alpha_l) \neq \alpha_k\alpha_l$ for $k, l \neq 1, 2$ implies that $\alpha_1 + \alpha_2$ is the only common root of these polynomials. Moreover, this common root is a simple one since it is a simple root of G. Hence, the greatest common divisor of $G(x)$ and $F(\theta - cx)$ is of the form $x - (\alpha_1 + \alpha_2)$. The coefficients of these polynomials lie in $R(\theta)$, so $\alpha_1 + \alpha_2 \in R(\theta)$ and

$$\alpha_1\alpha_2 = \theta - c(\alpha_1 + \alpha_2) \in R(\theta) = R(\alpha_1\alpha_2 + c(\alpha_1 + \alpha_2)).$$

So we have proved that $R(\alpha_1\alpha_2, \alpha_1 + \alpha_2) \subset R(\alpha_1\alpha_2 + c(\alpha_1 + \alpha_2))$. The opposite inclusion $R(\alpha_1\alpha_2 + c(\alpha_1 + \alpha_2)) \subset R(\alpha_1\alpha_2, \alpha_1 + \alpha_2)$ is obvious.

Thus, $\alpha_1\alpha_2$, $\alpha_1 + \alpha_2 \in R(\alpha_1\alpha_2 + c(\alpha_1 + \alpha_2))$ and $R(\alpha_1\alpha_2 + c(\alpha_1 + \alpha_2)) \subset R(i)$. Therefore α_1 and α_2 are the roots of a second order polynomial with coefficients in $R(i)$. But we have already proved that the roots of any second order polynomial with coefficients from $R(i)$ belong to $R(i)$. Therefore f has a root α_1 which belongs to $R(i)$. This means that $\overline{R} = R(i)$.

The converse statement (if $\overline{R} \neq R$ and $\overline{R} = R(i)$, then R is real closed) is much easier to prove. There are no intermediate fields between R and $\overline{R} = R(i)$, and so it suffices to prove that R is real, i.e., -1 is not the sum of squares of elements of R. By the hypothesis $i \notin R$, i.e., -1 is not a square. Thus, it suffices to prove that in R the sum of squares is a square itself. Since $R(i)$ is algebraically closed, it follows that, if $a, b \in R$, then $\sqrt{a + bi} \in R(i)$, i.e., there are elements c and d in R such that $a + bi = (c + di)^2$. In this case, we also have $a - bi = (c - di)^2$. Therefore

$$a^2 + b^2 = (a + bi)(a - bi) = (c + di)^2(c - di)^2 = (c^2 + d^2)^2.$$

It remains to observe that, if the sum of two squares is a square itself, then the sum of any number of squares is also a square. $\square$

7.2.2 Sylvester's theorem for real closed fields

Let K be an ordered field, and f an irreducible polynomial over K. We call a real closed field $R \supset K$ a *real closure* of K if R is algebraic over K and the ordering of K induced by the ordering of R coincides with the initial ordering of K.

The number of distinct real roots of a real polynomial can be computed without going outside $\mathbb{R}$. This can be done by two methods: with the help of Sturm's theorem or with the help of Sylvester's theorem. Both these theorems can be proved for any ordered field. The most important corollary of this situation is as follows:

if f is a polynomial over an ordered field K, then in any real closure R of K the number of the roots of f is the same.

Initially, Artin's theory was based on Sturm's theorem; for a modernized construction of Artin's theory based on Sturm's theorem, see the book [La2]. Sylvester's theorem, however, is in many ways more convenient. Following [Sc1] we give the solution of Hilbert's seventeenth problem based on Sylvester's theorem.

Over an ordered field K, the signature of a quadratic form φ is defined as follows. Over K, a quadratic form φ can be reduced to the form

$$\varphi(x) = \lambda_1 x_1^2 + \cdots + \lambda_n x_n^2. \tag{1}$$

As for $\mathbb{R}$, the number of positive coefficients λ_i and the number of the negative ones do not depend on the way we reduce the form φ to the canonical form (1). Indeed, suppose that there exist two decompositions

$$V_+ \oplus V_- \oplus V_0 = W_+ \oplus W_- \oplus W_0.$$

Then $V_+ \cap (W_- \oplus W_0) = 0$ so $\dim V_+ + \dim W_- + \dim W_0 \leq 0$. Hence $\dim V_+ \leq \dim W_+$. Similarly, $\dim W_+ \leq \dim V_+$.

Let R be a real closure of K. Theorem 7.2.5 implies that the degree of any irreducible over R polynomial is equal to 1 or 2. Therefore an obvious modification of the arguments used in the proof of Theorem 1.4.5 (see page 33) enables one to prove the following statement.

Theorem 7.2.6. *Let f be a polynomial over K and $\varphi(x, y)$ a bilinear symmetric form on the space $K[x]/(f)$ equal to the trace of the operator of multiplication by xy. Then the signature of φ is equal to the number of distinct roots of f which lie in the real closure R of K.*

In particular, as we have already mentioned, the number of the distinct roots of f is the same for all real closures of K.

The form φ will be called the *trace form* of the space $K[x]/(f)$.

Theorem 7.2.7 (Artin-Schreier). *Let K be an ordered field, and let R and R' be real closures of K. Then there exists precisely one isomorphism $\sigma \colon R \to R'$ over K and this isomorphism preserves the ordering.*

Proof. In the real closed field, the condition $x > y$ is equivalent to the fact that $x - y$ is a square. Therefore any isomorphism $\sigma \colon R \to R'$ preserves the ordering.

The field R is algebraic over K, and so any element $\alpha \in R$ is a root of an irreducible polynomial f over K. Theorem 7.2.6 implies that f has the same number of roots in R and in R'. Let these roots be $\alpha_1 < \cdots < \alpha_n$ and $\alpha_1' < \cdots < \alpha_n'$, respectively. In R, select elements t_i so that $t_i^2 = \alpha_{i+1} - \alpha_i$. By the theorem on a primitive element,

$$K(\alpha_1, \ldots, \alpha_n, t_1, \ldots, t_{n-1}) = K(\theta),$$

where $\theta \in R$ is a root of an irreducible over K polynomial g. In R', the polynomial g has the same number of roots as in R. In particular, it has a root $\theta' \in R'$. Over K, there exists an isomorphism $K(\theta) \to K(\theta')$. This isomorphism is an embedding

$$\sigma\colon K(\alpha_1, \ldots, \alpha_n, t_1, \ldots, t_{n-1}) \to R.$$

It is easy to verify that $\sigma(\alpha_i) = \alpha_i'$. Indeed, σ sends a root of f into a root of f, so that $\sigma(\alpha_{i+1}) - \sigma(\alpha_i) = \sigma(t_i^2) > 0$. On $K(\alpha_1, \ldots, \alpha_n)$, the map σ is uniquely defined. In particular, the image of α is uniquely defined. Now, with the help of Zorn's lemma, we may construct a uniquely defined isomorphism between R and R' over K. $\square$

Let us prove now that every ordered field has a real closure.

Theorem 7.2.8. *Let K be an ordered field, and K' its extension in which there are no relations of the form $-1 = \sum \lambda_i a_i^2$, where λ_i are positive elements of K and $a_i \in K'$. Then the field L obtained from K' by adjoining the quadratic roots of all the positive elements of K is a real field.*

Proof. Suppose on the contrary that the field L is not real. Then $-1 = \sum b_i^2$, where $b_i \in L$. Therefore, in L, we have the relation of the form $-1 = \sum \lambda_i b_i^2$, where the λ_i are positive elements of K and $b_i \in L$. By the hypothesis not all the b_i can simultaneously lie in K'. Therefore there exists a least positive integer r for which a relation of the form indicated holds with $b_i \in K'(\sqrt{\mu_1}, \ldots, \sqrt{\mu_r})$, where $\mu_1, \ldots, \mu_r$ are positive elements of K.

Let us express b_i as $b_i = x_i + y_i \sqrt{\mu_r}$, where $x_i, y_i \in K'(\sqrt{\mu_1}, \ldots, \sqrt{\mu_{r-1}})$. Then

$$-1 = \sum \lambda_i (x_i + y_i \sqrt{\mu_r})^2 = \sum \lambda_i (x_i^2 + y_i^2 \sqrt{\mu_r}) + 2\sqrt{\mu_r} \sum x_i y_i.$$

If $\sum x_i y_i \neq 0$, then $\sqrt{\mu_r} \in K'(\sqrt{\mu_1}, \ldots, \sqrt{\mu_{r-1}})$, which contradicts the assumptions on the minimality of r. Therefore

$$-1 = \sum \lambda_i x_i^2 + \sum \lambda_i \mu_r y_i^2,$$

where λ_i and $\lambda_i \mu_r$ are positive elements of K and $x_i, y_i \in K'(\sqrt{\mu_1}, \ldots, \sqrt{\mu_{r-1}})$. This also contradicts the assumptions on the minimality of r. $\square$

Corollary 1. *Any ordered field K has a real closure.*

Proof. Set $K' = K$. In K, there are no relations of the form $-1 = \sum \lambda_i a_i^2$ with λ_i positive. Therefore the field L obtained from K by adjoining square roots of all the positive elements of K is real. The real closure of L is the required real closure of K. $\square$

Corollary 2. *Let K be an ordered field and let K' be an extension of K. An ordering of K can be extended to K' if and only if there are no relations of the form $-1 = \sum \lambda_i a_i^2$ in K', where the λ_i are positive elements of K and $a_i \in K'$.*

Proof. Clearly, if a relation of the form indicated exists, then it is impossible to extend the ordering from K to K'. Suppose that there are no such relations. Then one can construct a real field $L \supset K'$. Consider its real closure R. The ordering of R induces an ordering of K' required. □

7.2.3 Hilbert's seventeenth problem

In this section we will finally prove that, if a real rational function $r(x_1, \ldots, x_n)$ is non-negative for all real values of $x_1, \ldots, x_n$, then it can be represented as the sum of squares of real rational functions. This proof holds not only for $\mathbb{R}$ but for an arbitrary real closed field R. The proof is easy to deduce from the following rather difficult theorem.

Theorem 7.2.9 (Artin–Lang). *Let R be a real closed field and let $K = R(x_1, \ldots, x_n)$ be an ordered finitely generated extension of R such that the ordering of K is compatible with the ordering of R. Then there exists an R-algebra homomorphism*

$$\varphi \colon R[x_1, \ldots, x_n] \to R$$

identical on R.

Proof. First, consider the case when the transcendence degree of K over R is equal to 1. We may assume that the element $x_1 = x$ is transcendental over R. Then K is a finite algebraic extension of R, and so by Theorem 21.5 (on a primitive element) $K = R(x)[y]$. Precisely the same arguments as in the proof of Noether's lemma on normalization (see Lemma on page 222) show that one can select y which satisfies the equation

$$y^l + c_1(x)y^{l-1} + \cdots + c_l(x) = 0, \text{where } c_1(x), \ldots, c_l(x) \in \mathbb{R}[x].$$

We assume here that l is the least possible.

Consider a polynomial

$$f(X, Y) = Y^l + c_1(X)Y^{l-1} + \cdots + c_l(X)$$

in indeterminates X and Y. To any pair of elements $a, b \in R$ such that $f(a, b) = 0$, there corresponds an R-algebra homomorphism $\sigma \colon R[x, y] \to R$ given by the formulas $\sigma(x) = a$ and $\sigma(y) = b$. Let us show that there are infinitely many such pairs $a, b \in R$.

Let R_K be a real closure of K. The polynomial $\widetilde{f}(Y) = f(x, Y) \in R(x)[Y]$ has a root y in R_K, and so by Theorem 7.2.6 the signature of the trace form φ on the space

$$R(x)[Y]/(\widetilde{f}) \cong R(x)[y]/R(x) = K/R(x)$$

is positive. This form can be reduced to the diagonal form with the elements $h_1(x), \ldots, h_s(x) \in R[x]$ on the diagonal.

Over a real closed field R, any polynomial factorizes into linear and irreducible quadratic factors. The quadratic factors are of the form $(x+\alpha)^2 + \beta^2$, where $\alpha, \beta \in R$. They are therefore positive as elements of $R(x)$, and for all $a \in R$ the elements $(a + \alpha)^2 + \beta^2 \in R$ are also positive.

Let $x - \lambda_1, \ldots, x - \lambda_t$ be all the distinct linear factors in the factorizations of the polynomials $h_1(x), \ldots, h_s(x)$. Let us order the elements $x, \lambda_1, \ldots, \lambda_t \in R(x)$. The following possibilities might occur:

$$\cdots < \lambda_i < x < \lambda_j < \cdots \quad \text{or} \quad \cdots < \lambda_i < x \quad \text{or} \quad x < \lambda_j < \cdots$$

Let a be an arbitrary element of R satisfying, respectively, inequalities $\lambda_i < a < \lambda_j$ or $\lambda_i < a$ or $a < \lambda_j$ (there are infinitely many such elements a). Then the signs of $h_k(a)$ and $h_k(x)$ coincide for all $k = 1, \ldots, s$. The signature of the form φ is therefore equal to the signature of the form with the diagonal elements $h_1(a), \ldots, h_s(a)$.

Let us show that, for almost all a, the trace form φ_a on the space $R[Y]/(f(a, Y))$ can be reduced to the diagonal form with the elements $h_1(a), \ldots, h_s(a)$ on the main diagonal. Indeed, let $A(x) = (a_{ij}(x))$ be the matrix of the form φ in the basis $1, Y, \ldots, Y^{l-1}$ and $B(x) = (b_{ij}(x))$ the matrix such that

$$(B(x))^T A(x) B(x) = \operatorname{diag}(h_1(x), \ldots, h_s(x)).$$

Then, if $\det B(a) \neq 0$ and none of the denominators of the rational functions $b_{ij}(x)$ vanishes at $x = a$, we have

$$(B(a))^T A(a) B(a) = \operatorname{diag}(h_1(a), \ldots, h_s(a)).$$

It remains to observe that $A(a)$ is the matrix of the form φ_a in the basis $1, Y, \ldots, Y^{l-1}$.

Thus, there exist infinitely many elements $a \in R$ for which the signature of the form φ_a is positive. For all such a, the polynomial $f(a, Y)$ has a root $b \in R$, i.e., $f(a, b) = 0$. As we have already observed, to any such pair there corresponds an R-algebra homomorphism $\sigma \colon R[x, y] \to R$. Let us show that almost all such homomorphisms can be extended to

$$R[x, y, x_2, \ldots, x_n] \supset R[x, y].$$

Recall that $x_2, \ldots, x_n \in R[x_1, \ldots, x_n] = K = R(x)[y]$, and therefore $x_i = \dfrac{p_i(x, y)}{q_i(x)}$, where p_i and q_i are polynomials. Let $q = q_1 \cdot \ldots \cdot q_n$. By construction $\sigma(q(x)) = q(a) \neq 0$ for almost all a. In these cases the homomorphism σ can be extended to

$$R[x,y]\left[\frac{1}{q(x)}\right] \supset R[x,y,x_2,\ldots,x_n] \supset R[x_1,\ldots,x_n] = K.$$

The passage from the case when the transcendence degree of K over R is equal to 1 to the case of transcendence degree $m \geq 1$ is performed by a simple induction on m. Suppose that the existence of the homomorphism required is proved for all fields K whose transcendence degree over R is strictly less than m. Consider a field $K = R(x_1,\ldots,x_n)$ whose transcendence degree over R is equal to m. Select an intermediate field $F \colon R \subset F \subset K$ for which the transcendence degree of K over F is equal to 1. Let $R_F \subset R_K$ be real closures of F and K. The transcendence degree of R_K over R_F is equal to 1, and so there exists an R_F-algebra homomorphism $\psi \colon R_F[x_1,\ldots,x_n] \to R_F$.

The transcendence degree of R_F over R is equal to $m-1$, and so the transcendence degree of the field $R(\psi(x_1),\ldots,\psi(x_n)) \subset R_F$ over R does not exceed $m-1$. It is also clear that an ordering of R_F induces an ordering of the field $R(\psi(x_1),\ldots,\psi(x_n))$. Therefore there exists an R-algebra homomorphism $\sigma \colon R[\psi(x_1),\ldots,\psi(x_n)] \to R$. The restriction of the composition of the maps ψ and σ onto the subalgebra

$$R[x_1,\ldots,x_n] \subset R_F[x_1,\ldots,x_n]$$

is the homomorphism required. $\square$

Now it is easy to prove Artin's theorem on non-negative rational functions. Let k be an ordered field. The rational function

$$r(x_1,\ldots,x_n) = \frac{p(x_1,\ldots,x_n)}{q(x_1,\ldots,x_n)}, \quad \text{where} \quad p,q \in k[x_1,\ldots,x_n],$$

is called *non-negative* if $r(a_1,\ldots,a_n) \geq 0$ for all $a_1,\ldots,a_n \in k$ such that $q(a_1,\ldots,a_n) \neq 0$.

Theorem 7.2.10 (Artin). *Let R be a real closed field, let $r \in R(x_1,\ldots,x_n)$ be a non-negative rational function. Then r can be represented as the sum of squares of elements of $R(x_1,\ldots,x_n)$.*

Proof. Suppose on the contrary that r is not the sum of squares of elements of the field $R(x_1,\ldots,x_n)$. Then by Theorem 7.2.4 there exists an ordering of the field $R(x_1,\ldots,x_n)$ for which $r < 0$. Let us express r as an irreducible fraction p/q, where $p,q \in R[x_1,\ldots,x_n]$. Consider an R-algebra

$$R\left[x_1,\ldots,x_n,\frac{1}{q(x_1,\ldots,x_n)}\right]$$

that contains r.

In $\bar{R}_n$, the real closure of the ordered field $R(x_1,\ldots,x_n)$, there exists an element γ such that $\gamma^2 = -r > 0$. The field $R(x_1,\ldots,x_n,\gamma)$ is contained in

$\bar{R}_n$, and therefore the ordering of $\bar{R}_n$ induces an ordering of $R(x_1, \ldots, x_n, \gamma)$. By the Artin–Lang theorem there exists a homomorphism

$$\varphi \colon R\left[x_1, \ldots, x_n, \frac{1}{q(x_1, \ldots, x_n)}\gamma, \frac{1}{\gamma}\right] \to R,$$

identical on R.

Clearly, $\varphi(\gamma)\,\varphi\left(\frac{1}{\gamma}\right) = 1$ and $\varphi(q)\,\varphi\left(\frac{1}{q}\right) = 1$. Hence $\varphi(\gamma) \neq 0$ and $\varphi(q) \neq 0$. Hence also $\varphi(r) = -\varphi(\gamma^2) = -\left(\varphi(\gamma)\right)^2 < 0$. But

$$\varphi(r) = \frac{p(a_1, \ldots, a_n)}{q(a_1, \ldots, a_n)}, \quad \text{where} \quad a_i = \varphi(x_i).$$

Here $q(a_1, \ldots, a_n) = \varphi(q) \neq 0$. The inequality $r(a_1, \ldots, a_n) < 0$ contradicts the hypotheses of the theorem. $\square$

For an arbitrary (i.e., not necessarily real closed) ordered field, Artin's theorem is false (the first example of such a field is given in [Du1]; a simpler proof can be found on page 86 of the book [Pf2]). However, with a slight sharpening of the proof of Theorem 7.2.10, we can prove the following statement for an arbitrary ordered field k.

Theorem 7.2.11. *Let k be an ordered field and R its real closure. If a rational function $r \in k(x_1, \ldots, x_n)$ is such that $r(a_1, \ldots, a_n) \geq 0$ for all $a_1, \ldots, a_n \in R$ (if, of course, the value $r(a_1, \ldots, a_n)$ is defined), then r can be represented as the sum of squares of elements of $k(x_1, \ldots, x_n)$.*

Proof. If r is not the sum of squares of elements of $k(x_1, \ldots, x_n)$, then there exists an ordering of the field $k(x_1, \ldots, x_n)$ for which $r < 0$. Let R' be a real closure of the field $k(x_1, \ldots, x_n)$ with such an ordering. We may assume that $R \subset R'$ (the real closure of the field $k(x_1, \ldots, x_n)$ contains a real closure R_1 of k, and R_1 and R are isomorphic as real closures of the same field). In R', there exists an element γ such that $\gamma^2 = -r > 0$. In the R-algebra

$$R\left[x_1, \ldots, x_n, \frac{1}{q(x_1, \ldots, x_n)}\gamma, \frac{1}{\gamma}\right] \subset R',$$

we introduce an ordering induced by the ordering in R'. The remaining arguments are precisely the same as in the proof of Theorem 7.2.10. $\square$

Corollary. *Let the ordering of the field k be such that the condition "$r(x_1, \ldots, x_n) < 0$ for all $x_1, \ldots, x_n \in k$" implies that $r(x_1, \ldots, x_n) < 0$ for all $x_1, \ldots, x_n \in R$, where R is a real closure of k. Then Artin's theorem for k holds.*

In particular, Artin's theorem holds for the field of rationals $\mathbb{Q}$, i.e., any non-negative rational function with rational coefficients is the sum of squares of rational functions with rational coefficients.

In conclusion we give formulations of two interesting theorems whose proofs are based on Artin's theorem. More precisely, the first of these theorems is deduced from Artin's theorem and the second follows from the first.

Theorem 7.2.12 ([Ri]). *Let I be an ideal in the ring $\mathbb{R}[x_1, \ldots, x_n]$. Then the following conditions are equivalent:*

(1) Any polynomial that vanishes at all the common real zeros of the polynomials from I belongs to I itself;

(2) If the sum of squares of the polynomials from the ring $\mathbb{R}[x_1, \ldots, x_n]$ belongs to I, then these polynomials themselves belong to I.

Theorem 7.2.13 ([St1]). *The homogeneous form F is non-negative if and only if there exists a homogeneous polynomial relation of the form $\varphi(-F) = 0$, where*

$$\varphi(u) = u^{2n+1} + a_1 u^{2n} + \cdots + a_{2n}$$

and the coefficients $a_1, \ldots, a_{2n}$ are the sums of squares of homogeneous forms.

7.3 Pfister's theory

7.3.1 The multiplicative quadratic forms

In this section we consider quadratic forms over an arbitrary field k whose characteristic is distinct from 2. Any quadratic form φ on the space k^n is given by a symmetric $n \times n$ matrix A. Namely, if $x = (x_1, \ldots x_n) \in k^n$, then

$$\varphi(x) = xAx^T = \sum x_i x_j a_{ij}.$$

The quadratic forms φ and ψ are called *equivalent* if there exists a non-singular $n \times n$ matrix P such that $\varphi(x) = \psi(Px)$. Then ψ is given by the matrix PAP^T. For equivalent forms we write $\varphi \cong \psi$.

We say that the form φ *represents* an element $a \in k$ if $a = \varphi(x)$ for some $x \in k^n$. For example, the form $\varphi(x) = x_1^2 + \cdots + x_n^2$ represents a if and only if a can be represented as the sum of n squares.

The main tools of Pfister's theory are multiplicative forms of a particular type he introduced. Recall that the quadratic form φ is *multiplicative* if the forms φ and $a\varphi$ are equivalent for any nonzero element a representable by φ.

Any multiplicative form φ possesses the following remarkable property: if a and b are representable by φ, then ab is also representable by φ. Indeed, equivalent forms represent the same set of elements of the field k. Now if a multiplicative form φ represents a and b, then the form $a\varphi$ equivalent to φ represents ab.

We introduce the following notations. Denote

$$\langle a_1, \ldots, a_n \rangle = a_1 x_1^2 + \cdots + a_n x_n^2.$$

For the forms $\varphi = \langle a_1, \ldots, a_m \rangle$ and $\psi = \langle b_1, \ldots, b_n \rangle$, we define

$$\varphi \oplus \psi = \langle a_1, \ldots, a_m, b_1, \ldots, b_n \rangle,$$
$$\varphi \otimes \psi = \langle a_1 b_1, \ldots, a_m b_1, a_1 b_2, \ldots, a_m b_2, \ldots, a_1 b_n, \ldots, a_m b_n \rangle.$$

Pfister introduced the forms

$$\langle\langle a_1, \ldots, a_n \rangle\rangle = \langle 1, a_1 \rangle \otimes \langle 1, a_2 \rangle \otimes \cdots \otimes \langle 1, a_n \rangle, \text{ where } a_1 \cdots a_n \neq 0,$$

which play the main role in his theory. The most interesting for us is the form $\langle\langle \underbrace{1, \ldots, 1}_{n} \rangle\rangle$, i.e., the sum of 2^n squares. However, in the proof by induction on n we cannot avoid considering the general case $\langle\langle a_1, \ldots, a_n \rangle\rangle$.

Theorem 7.3.1 (Pfister). *The form $\langle\langle a_1, \ldots, a_n \rangle\rangle$ is multiplicative.*

Proof. We consider first the case $n = 1$. Let the form $\langle\langle a_1 \rangle\rangle = x_1^2 + a_1 x_2^2$ represent $b \neq 0$, i.e., $b = c_1^2 + a_1 c_2^2$. Set $A = \begin{pmatrix} 1 & 0 \\ 0 & a_1 \end{pmatrix}$ and $P = \begin{pmatrix} c_1 & c_2 \\ -a_1 c_2 & c_1 \end{pmatrix}$. It is easy to verify that $PAP^T = bA$ and $\det P \neq 0$. This means that $\langle\langle a_1 \rangle\rangle \cong b\langle\langle a_1 \rangle\rangle$.

The form $\langle\langle a_1, \ldots, a_n \rangle\rangle$ is of the type $\varphi \oplus a_n \varphi$, where $\varphi = \langle\langle a_1, \ldots, a_{n-1} \rangle\rangle$. It therefore suffices to prove that if φ is multiplicative and $a \neq 0$, then the form $\varphi \oplus a\varphi$ is also multiplicative. Let $b = \varphi(x) + a\varphi(y) = \xi + a\eta \neq 0$, where ξ and η are representable by φ. If $\xi = 0$, then $\eta \neq 0$, and hence, $\eta\varphi \cong \varphi$. Hence again

$$b(\varphi \oplus a\varphi) = a\eta(\varphi \oplus a\varphi) \cong a\varphi \oplus a^2\varphi \cong \varphi \oplus a\varphi,$$

since $a^2\varphi \cong \varphi$. The case $\eta = 0$ is similarly considered. It remains to consider the case $\xi\eta \neq 0$. In this case $\varphi \cong \xi\varphi$ and $\varphi \cong \xi^{-1}\varphi \cong (\eta\xi^{-1})\varphi$. Hence

$$b(\varphi \oplus a\varphi) = (\xi + a\eta)(\varphi \oplus a\varphi) \cong (1 + a\eta\xi^{-1})(\varphi \oplus (a\eta\xi^{-1})\varphi)$$
$$\cong (1 + a\eta\xi^{-1})\langle\langle a\eta\xi^{-1} \rangle\rangle \otimes \varphi.$$

When we considered the case $n = 1$ we showed that the form $\langle\langle a\eta\xi^{-1} \rangle\rangle$ is multiplicative. This form represents an element $1 + a\eta\xi^{-1} = b\xi^{-1} \neq 0$. Hence $(1 + a\eta\xi^{-1})\langle\langle a\eta\xi^{-1} \rangle\rangle \cong \langle\langle a\eta\xi^{-1} \rangle\rangle$, and therefore

$$b(\varphi \oplus a\varphi) \cong \langle\langle a\eta\xi^{-1} \rangle\rangle \otimes \varphi = \varphi \oplus (a\eta\xi^{-1})\varphi \cong \varphi \oplus a\varphi,$$

as was required. $\square$

In the proof of Pfister's theorem we did not use the fact that char $k \neq 2$, and therefore it is true for any field. The case where $a_1 = \cdots = a_n = 1$ is particularly interesting. In this case Pfister's theorem implies that

in any field the product of elements each representable as the sum of 2^n squares is also representable as the sum of 2^n squares.

For $n = 1, 2, 3$, there are explicit general formulas for the representation. For example, when $n = 1$ and we have the sum of two squares, the corresponding identity is

$$(x_1^2 + x_2^2)(y_1^2 + y_2^2) = (x_1y_1 + x_2y_2)^2 + (x_1y_2 - x_2y_1)^2.$$

But for $n > 3$ there are no identities of similar form.

We will need the following auxiliary statement.

Lemma 7.3.2. *Let us represent the form*

$$\varphi = \langle\langle a_1, \ldots, a_n \rangle\rangle, \quad \text{where } a_1 \cdots a_n \neq 0,$$

as $\varphi = \langle 1 \rangle \oplus \varphi'$. Let φ' represent an element $b_1 \neq 0$. Then $\varphi \cong \langle\langle b_1, \ldots, b_n \rangle\rangle$ for some nonzero $b_2, \ldots, b_n$.

Proof. We use induction on n. For $n = 1$, the form φ' is of the type $a_1 x_1^2$. If $b_1 = a_1 c^2$, then $b_1 x_1^2 = a_1 (cx_1)^2 \cong a_1 x_1^2$. Thus $\langle a_1 \rangle \cong \langle b_1 \rangle$, and therefore $\langle\langle a_1 \rangle\rangle \cong \langle\langle b_1 \rangle\rangle$.

Now suppose that the statement required is proved for all the forms of the type $\langle\langle c_1, \ldots, c_{n-1} \rangle\rangle$, where $c_1 \cdots c_{n-1} \neq 0$. To prove the statement required for the form $\varphi = \langle\langle a_1, \ldots, a_n \rangle\rangle$, consider the form

$$\psi = \langle\langle a_1, \ldots, a_{n-1} \rangle\rangle = \langle 1 \rangle \oplus \psi'.$$

Clearly, $\varphi = \psi \oplus a_n \psi$ and $\varphi' = \psi' \oplus a_n \psi$. Therefore the element b_1 representable by the form φ' can be expressed as $b_1 = b_1' + a_n b$, where the elements b_1' and b are represented by the forms ψ' and ψ, respectively.

First, let $b = 0$. In this case, $b_1' = b_1$. By the induction hypothesis $\psi \cong \langle\langle b_1, \ldots, b_{n-1} \rangle\rangle$, and therefore $\varphi \cong \langle\langle b_1, \ldots, b_{n-1}, a_n \rangle\rangle$.

Now consider the case $b \neq 0$. In this case, the form ψ represents a nonzero element b. By Pfister's theorem, the form ψ is multiplicative, and so $\psi \cong b\psi$. Hence

$$\varphi' = \psi' \oplus (a_n b)(b^{-1}\psi) \cong \psi' \oplus c_n \psi,$$

where $c_n = a_n b$. Here $b_1 = b_1' + c_n$. Let $b_1' = 0$. Then $b_1 = c_n$ and

$$\varphi \cong \langle\langle c_n \rangle\rangle \otimes \psi = \langle\langle c_n, a_1, \ldots, a_{n-1} \rangle\rangle = \langle\langle b_1, a_1, \ldots, a_{n-1} \rangle\rangle.$$

It remains to consider the case when $b_1 = b_1' + c_n$ and $b_1' c_n \neq 0$. The form ψ' represents b_1', and so by the induction hypothesis $\psi \cong \langle\langle b_1', b_2, \ldots, b_{n-1} \rangle\rangle$. Hence,

$$\varphi \cong \langle\langle b_1', b_2, \ldots, b_{n-1}, c_n \rangle\rangle \cong \langle\langle b_1', c_n, b_2, \ldots, b_{n-1} \rangle\rangle$$
$$\cong \langle\langle b_1', c_n \rangle\rangle \otimes \langle\langle b_2, \ldots, b_{n-1} \rangle\rangle.$$

Let $\lambda = b'_1$ and $\mu = c_n$. Then $\lambda + \mu = b_1 \neq 0$. The identity

$$\begin{pmatrix} 1 & 1 \\ mu & -\lambda \end{pmatrix} \begin{pmatrix} \lambda & 0 \\ 0 & \mu \end{pmatrix} \begin{pmatrix} 1 & \mu \\ 1 & -\lambda \end{pmatrix} = \begin{pmatrix} \lambda + \mu & 0 \\ 0 & (\lambda + \mu)\lambda\mu \end{pmatrix}$$

shows that $\langle b'_1, c_n \rangle \cong \langle b_1, b_1 b'_1 c_n \rangle$. Hence

$$\langle\langle b'_1, c_n \rangle\rangle = \langle 1, b'_1, c_n, b'_1 c_n \rangle \cong \langle 1, b_1, b'_1 c_n, b_1 b'_1 c_n \rangle = \langle\langle b_1, b'_1 c_n \rangle\rangle.$$

Set $b_n = b'_1 c_n$. Then

$$\varphi \cong \langle\langle b_1, b_n \rangle\rangle \otimes \langle\langle b_2, \ldots, b_n \rangle\rangle \cong \langle\langle b_1, \ldots, b_n \rangle\rangle. \quad \square$$

7.3.2 C_i-fields

In the preceding section we met one of the tools of Pfister's theory — multiplicative forms. The other tool of this theory — C_i-fields — was developed in 1933–1936 by the Chinese mathematician Tsen Chiungze and rediscovered in 1952 by Lang.

The field K is called a *C_i-field* if any system of homogeneous polynomials

$$f_1, \ldots, f_r \in K[x_1, \ldots, x_n]$$

such that $d_1^i + \cdots + d_r^i < n$, where $d_s = \deg f_s$, has a common non-trivial zero.

Theorem 7.3.3 (Tsen–Lang). *If the field L is algebraically closed, then the field $K = L(t_1, \ldots, t_i)$, i.e., the field of rational functions in i indeterminates over L, is a C_i-field.*

Proof. We use induction on i. For $i = 0$, the field K coincides with L, i.e., K is algebraically closed and the condition $d_1^i + \cdots + d_r^i < n$ means that $r < n$. In this case the statement required coincides with the homogeneous Hilbert's Nullstellensatz (Theorem 6.1.10 on page 231).

The inductive step consists in the proof of the fact that, if L is a C_i-field, then $K = L(t)$ is a C_{i+1}-field. Let $f_1, \ldots, f_r \in K[x_1, \ldots, x_n]$ be homogeneous polynomials such that $d_1^{i+1} + \cdots + d_r^{i+1} < n$, where $d_s = \deg f_s$. We have to prove that $f_1, \ldots, f_r$ have a common non-trivial zero in K. The coefficients of the polynomials $f_1, \ldots, f_r$ belong to $K = L(t)$, i.e., they are rational functions in t over L. Having multiplied all these coefficients by an appropriate polynomial from $L[t]$ we may assume that the coefficients of the polynomials $f_1, \ldots, f_r$ belong to $L[t]$, i.e., these coefficients are polynomials in t over L. The degrees of these polynomials are bounded by a number m, and so any coefficient α can be expressed as

$$\alpha = a_0 + a_1 t + \cdots + a_m t^m, \quad \text{where} \quad a_0, \ldots, a_m \in L. \tag{1}$$

For $p = 1, \ldots, n$, define

$$x_p = x_{p0} + x_{p1}t + \cdots + x_{ps}t^s, \tag{2}$$

where $x_{p0}, \ldots, x_{ps}$ are independent variables over L and s is sufficiently large (we will determine this number shortly). In every homogeneous form $f_1, \ldots, f_r$, replace the coefficients and indeterminates by expressions (1) and (2), respectively. As a result, the form f_j takes the form

$$g_0 + g_1 t + \cdots + g_N t^N,$$

where $N = sd_j + m$ and $g_0, \ldots, g_N$ are the forms of degree d_j in $(s+1)n$ variables x_{pq} over L. By the induction hypothesis all the forms g for the polynomials $f_1, \ldots, f_r$ have a common non-trivial zero if

$$\sum_{j=1}^{r} (sd_j + m + 1)d_j^i < (s+1)n,$$

i.e.,

$$(m+1) \sum d_j^i - n < s(n - \sum d_j^{i+1}).$$

By the hypothesis we have $n - \sum d_j^{i+1} > 0$, and so the inequality required holds for sufficiently large values of s, such as $s > (m+1) \sum d_j^i$. $\square$

7.3.3 Pfister's theorem on the sums of squares of rational functions

In the two preceding sections we have prepared the main tools for the proof of Pfister's theorem. We will also need the following property of quadratic forms: if the non-degenerate quadratic form φ over K is isotropic (i.e., $\varphi(u) = 0$ for some $u \neq 0$), then it is *universal*, i.e., it represents all the elements of the field K. Indeed, to any non-degenerate quadratic form φ there corresponds a non-degenerate bilinear symmetric form

$$f(x, y) = \frac{1}{2} \left(\varphi(x+y) - \varphi(x) - \varphi(y) \right).$$

Therefore there exists a vector v such that $f(u, v) = 1$. In this case

$$\varphi(v + \lambda u) = \varphi(v) + \varphi(\lambda u) + 2f(v, \lambda u) = \varphi(v) + 2\lambda.$$

For any $b \in K$, we can select a λ such that $\varphi(v) + 2\lambda = b$.

Theorem 7.3.3 implies that, if L is algebraically closed, then any non-degenerate quadratic form φ in 2^n variables over $K = L(t_1, \ldots, t_n)$ is universal. Indeed, let $r \in K$. Consider an auxiliary quadratic form

$$\tilde{\varphi}(u, t) = \varphi(u) - rt^2$$

in $2^n + 1$ variables. By Theorem 7.3.3 the form $\tilde{\varphi}$ has a non-trivial zero (u_0, t_0). If $t_0 = 0$, then $\varphi(u_0) = 0$ for some $u_0 \neq 0$. This means that φ is isotropic, and therefore universal. If $t_0 \neq 0$, then $\varphi\left(\dfrac{u_0}{t_0}\right) = r$, i.e., φ represents r.

Theorem 7.3.4 (Pfister). *Let R be a real closed field; let $\varphi = \langle\langle a_1, \ldots, a_n \rangle\rangle$ be a non-degenerate quadratic form in 2^n variables over $R(x_1, \ldots, x_n)$. If b is the sum of squares of elements $R(x_1, \ldots, x_n)$, then φ represents b.*

Proof. If the form φ is isotropic, then it is universal. We may therefore assume that φ is anisotropic, i.e., $\varphi(u) \neq 0$ for $u \neq 0$. By the hypothesis $b = b_1^2 + \cdots + b_m^2$. Let us use induction on m. For $m = 1$, the statement is obvious, since any multiplicative form represents 1, and hence it represents $b_1^2 \cdot 1$ as well.

Now let $m = 2$, i.e., $b = b_1^2 + b_2^2$, where $b_1 b_2 \neq 0$. Let $L = R(i)$ be the algebraic closure of R. Let $\beta = b_1 + i b_2$ generate the field $L(x_1, \ldots, x_n)$ over the field $R(x_1, \ldots, x_n)$, i.e.,

$$L(x_1, \ldots, x_n) = R(x_1, \ldots, x_n)(\beta).$$

The field R is real, and so $i \notin R(x_1, \ldots, x_n)$ and $\beta \notin R(x_1, \ldots, x_n)$.

The form φ can be also considered as a form over the field $L(x_1, \ldots, x_n) \supset R(x_1, \ldots, x_n)$. Over this field φ is universal since L is algebraically closed. Therefore there exist 2^n-dimensional vectors u, v over $R(x_1, \ldots, x_n)$ for which $\varphi(u + \beta v) = \beta$, i.e.,

$$\varphi(u) + 2\beta f(u, v) + \beta^2 \varphi(v) = \beta, \tag{1}$$

where f is a bilinear symmetric form corresponding to φ. Here $v \neq 0$ since otherwise $\beta = \varphi(u) \in R(x_1, \ldots, x_n)$. Since φ is anisotropic, it follows that $\varphi(v) \neq 0$.

An irreducible over $R(x_1, \ldots, x_n)$ equation for β is of the form $(\beta - b_1)^2 + b_2^2 = 0$, i.e.,

$$\beta^2 - 2b_1\beta + b = 0. \tag{2}$$

Comparing (1) and (2) we deduce that $b = \varphi(u)/\varphi(v)$. Since φ is multiplicative, it follows that φ represents both $\dfrac{1}{\varphi(v)}$ and the product of $\varphi(u)$ and $\dfrac{1}{\varphi(v)}$, i.e., b.

Now suppose that the statement required is proved for some $m \geq 2$, i.e., any form φ of the type $\varphi = \langle\langle a_1, \ldots, a_n \rangle\rangle$ represents any element of the type $b_1^2 + \cdots + b_m^2$. We have to prove that the form φ represents any element of the type $b_1^2 + \cdots + b_m^2 + b_{m+1}^2$, where $b_{m+1} \neq 0$. Let us express this element in the form $b_{m+1}^2(b + 1)$, where b is the sum of m squares. It suffices to prove that the form φ represents an element $c = b + 1$. We may assume that $c \neq 0$.

To make use of Lemma 7.3.2 (see page 272), we express the form φ as $\varphi = \langle 1 \rangle \oplus \varphi'$. By the induction hypothesis φ represents b, i.e., $b = b_0^2 + b'$, where b' is represented by φ'. Consider the multiplicative form $\psi = \varphi \otimes \langle 1, -c \rangle$ in 2^{n+1} variables. Clearly,

$$\psi = \langle 1 \rangle \oplus \varphi' \oplus (-c)\varphi = \langle 1 \rangle \oplus \psi',$$

where the form $\psi' = \varphi' \oplus (-c\varphi)$ represents the element

$$b' - c = (b - b_0^2) - (1 + b) = -1 - b_0^2.$$

Here $-1 - b_0^2 \neq 0$ since $i \notin R(x_1, \ldots, x_n)$. In this case we may apply Lemma 7.3.2 to the form ψ and the element $-1 - b_0^2$.

As a result, we find that there exist non-zero elements $c_1, \ldots, c_n \in R(x_1, \ldots, x_n)$ such that

$$\psi \cong \langle\langle -1 - b_0^2, c_1, \ldots, c_n \rangle\rangle,$$

i.e., $\psi \cong \langle -1 - b_0^2 \rangle \otimes \chi = \chi \oplus (-1 - b_0^2)\chi$, where $\chi \cong \langle\langle c_1, \ldots, c_n \rangle\rangle$.

Let us now apply the induction hypothesis again, this time to the form χ. The element $1 + b_0^2$ is the sum of not more than m squares, and so the form χ represents it. It then follows from the multiplicativity of χ that $\chi \cong (1 + b_0^2)\chi$, and therefore

$$\varphi \oplus (-c\varphi) = \psi \cong \chi \oplus (-1 - b_0^2)\chi \cong (1 + b_0^2)\chi \oplus (-1 - b_0^2)\chi.$$

Let $\xi = (1 + b_0^2)\chi$. Then

$$\varphi(Px + Qy) - c\varphi(Rx + Sy) = \xi(x) - \xi(y),$$

where $\begin{pmatrix} P & Q \\ R & S \end{pmatrix} = U$ is a non-singular $2^n \times 2^n$ matrix. By setting $x = y$ we find that

$$\varphi(Ax) = c\varphi(Bx), \quad \text{where } A = P + Q \text{ and } B = R + S.$$

Since U is non-singular, it follows that, if $x \neq 0$, then $(Ax, Bx) \neq (0, 0)$. If at least one of the matrices A and B were singular, the form φ would be isotropic, whereas we are considering an anisotropic form. Since therefore both the matrices are invertible, $\varphi \cong c\varphi$, and therefore the multiplicative form φ represents the element c. $\square$

By Artin's theorem any non-negative element of the field $R(x_1, \ldots, x_n)$, where R is a real closed field, is the sum of squares. The multiplicative form $\langle\langle \underbrace{1, \ldots, 1}_{n} \rangle\rangle$ represents, therefore, any non-negative element of $R(x_1, \ldots, x_n)$,

i.e., any non-negative element of this field can be represented as the sum of 2^n squares.

On page 251 we gave example of an element of $\mathbb{R}(x_1, \ldots, x_n)$ that cannot be represented as the sum of n squares. Therefore, if N is the least number for which any element of the field $\mathbb{R}(x_1, \ldots, x_n)$ can be represented as the sum of N squares, then $n + 1 \leq N \leq 2^n$.

In the general case no other estimates on N are known. Only for $n = 2$ it is known that $N = 4$. This statement is proved in two distinct ways but both the proofs are rather complicated. One of the proofs uses the theory of elliptic curves over the field $\mathbb{C}(x)$ (see [Ca6] and [Ch2]). The other proof is based on the Noether–Lefshetz theorem which states that on any generic surface of degree $d \geq 4$ in three-dimensional projective space, any curve is singled out as the intersection with certain other surface (see [Co1]).

In the proof of Pfister's theorem one essentially uses the fact that R is real closed. For $\mathbb{Q}$, Pfister's theorem is false. This is clear already for $n = 0$. Indeed, not every positive rational number is the square of a rational number. But for $n = 0$ the estimate required is known:

every positive rational number is the sum of squares of four rational numbers.

Indeed, by Meier's theorem (see [BS], [Ca7] or [Se1]) any non-degenerate quadratic form over $\mathbb{Q}$ in $n \geq 5$ indeterminates non-trivially represents 0 if it represents 0 over $\mathbb{R}$. In particular, if $r > 0$, the form

$$x_1^2 + x_2^2 + x_3^2 + x_4^2 - rx_5^2$$

represents 0 over $\mathbb{Q}$.

For $n = 1$, i.e., for the polynomials over $\mathbb{Q}$ in one indeterminate, the first to obtain the estimate was E. Landau in the 1906 paper [La1]. He showed that any non-negative polynomial (in one indeterminate) with rational coefficients can be represented as the sum of 8 squares of polynomials with rational coefficients (a simple proof of Landau's theorem is given in Chapter 7 of the book [Pf2]). But this estimate is not an exact one. The exact estimate was obtained by Pourchet [Po3]. Every non-negative polynomial over $\mathbb{Q}$ can be represented as the sum of squares of 5 polynomials over $\mathbb{Q}$. Chapter 17 of the book [Ra3] is devoted to the detailed exposition of Pourchet's theorem.

For polynomials with certain special properties more precise results can be obtained. For example, in [Da3] it is shown that, if the values of a polynomial $f(x)$ at all integer, x are the sums of squares of two integers, then $f(x)$ is the sum of squares of two polynomials with integer coefficients.

An example of a non-negative polynomial which cannot be represented over $\mathbb{Q}$ as the sum of squares of four polynomials is sufficiently easy to construct. To see this, suppose that

$$ax^2 + bx + c = \sum_{s=1}^{4}(a_s x + b_s)^2.$$

Then $4ac - b^2$ is the sum of squares of three rational numbers. Indeed,

$$4ac - b^2 = 4\left(\sum a_s^2\right)\left(\sum b_t^2\right) - 4\left(\sum a_s b_s\right)^2$$

and, if we consider the product of quaternions $a_1 + a_2 i + a_3 j + a_4 k$ and $b_1 - b_2 i - b_2 j - b_2 k$, we see that

$$\left(\sum a_s^2\right)\left(\sum b_t^2\right) = \left(\sum a_s b_s\right)^2 \quad \text{plus the sum of three squares.}$$

Now it is easy to show that the quadratic $x^2 + x + 4$ cannot be represented as the sum of squares of four polynomials. In this case $4ac - b^2 = 15$ and we have to prove that 15 cannot be represented as the sum of squares of three rational numbers. If 15 were equal to $p^2 + q^2 + r^2$ then, after reducing to the common denominator, we would have obtained the congruence

$$a^2 + b^2 + c^2 \equiv 15d^2 \pmod 8 \equiv -d^2 \pmod 8,$$

where at least one of the numbers is odd. But such a congruence is impossible.

8

Appendix

8.1 The Lenstra-Lenstra-Lovász algorithm

In 1982, in the paper [Le2], there was suggested an algorithm called The *Lenstra-Lenstra-Lovász algorithm*, or the *LLL-algorithm* for short. It enables one to factorize any polynomial over $\mathbb{Z}$ into irreducible factors over a polynomial time. More precisely, if $f(x) = \sum_{i=0}^{n} a_i x^i \in \mathbb{Z}[x]$ and $|f| = \sqrt{\sum_{i=0}^{n} a_i^2}$, then, to factorize f into irreducible factors with this algorithm, one needs not more than $O\left(n^{12} + n^9 (\ln |f|)^3\right)$ operations.

The LLL-algorithm is of huge theoretical interest, but practically it is no more effective than a comparatively simple algorithm we described in section 2.5.2. At the outset both algorithms work similarly: we factorize f into irreducible factors modulo a prime p with the help of Berlekamp's algorithm, and then with the help of Henzel's lemma we compute, with certain accuracy, a p-adic irreducible factor h of f. After that the LLL-algorithm proceeds differently: for h, we seek an irreducible divisor h_0 of f in $\mathbb{Z}[x]$ divisible by h modulo p. The condition of divisibility of h_0 by h means that the coefficients of the polynomial h_0 are the coordinates of points on a lattice, and the condition of divisibility of f by h_0 means that the coefficients of h_0 are not too high.

To determine h_0, we use an algorithm for constructing a reduced basis of a lattice. Observe that the latter algorithm also has numerous applications that go beyond the problem of factorization of polynomials.

8.1.1 The general description of the algorithm

Let us pass directly to the description of the factorization algorithm. We assume that the polynomials f and f' are relatively prime and $\mathrm{cont}(f) = 1$. Let us start by computing the resultant $R(f, f') \in \mathbb{Z}$. Let p be the least prime which does not divide $R(f, f')$. Then the degree of $f \pmod{p}$ is equal to n

and the polynomials f (mod p) and f' (mod p) are relatively prime. To prove this, it suffices to observe that $R(f, f') = \pm a_n D(f)$ (and therefore a_n is not divisible by p) and $R(f, f') = \varphi f + \psi f'$, where $\varphi, \psi \in \mathbb{Z}[x]$.

The polynomial f (mod p) has no multiple divisors, and so it can be factorized into irreducible factors using Berlekamp's algorithm.

Let h (mod p) be one of the irreducible factors of f (mod p). In what follows we assume that h is a monic polynomial and the coefficients of h are reduced modulo p, i.e., lie between 0 and $p - 1$.

Consider the factorization of f into irreducible factors over $\mathbb{Z}$ and pass from this factorization to a factorization modulo p. The polynomial f (mod p) has no multiple irreducible factors, and so the polynomial h (mod p) corresponds to precisely one irreducible factor h_0 of f over $\mathbb{Z}$. Therefore there exists precisely one irreducible factor h_0 of f over $\mathbb{Z}$ for which h_0 (mod p) is divisible by h (mod p).

If one knows an algorithm which recovers h_0 from h, it is easy to factorize f. Indeed, let the factorization $f = f_1 f_2$ over $\mathbb{Z}$ be given, where the complete factorization of f_1 over $\mathbb{Z}$ is known and for f_2 the complete factorization modulo p is known (at the first step $f_1 = 1$ and $f_2 = f$). Take an irreducible divisor h (mod p) of f_2 (mod p) and find the irreducible divisor h_0 of f_2 over $\mathbb{Z}$ for which h_0 (mod p) is divisible by h (mod p). Replace f_1 by $f_1 h_0$ and f_2 by f_2/h_0. After that we repeat this operation until we obtain a complete factorization of f.

The algorithm which computes h_0 for a given h (mod p) will be described later in Section 8.1.3. This algorithm uses the algorithm that computes a *reduced basis* of the lattice. We therefore first discuss the notion of a reduced basis of a lattice and the algorithm of calculating the reduced basis (this algorithm was also suggested in [Le2], and therefore is also called an *LLL-algorithm*).

8.1.2 A reduced basis of the lattice

A subset $L \subset \mathbb{R}^n$ is called a *lattice of rank n* if there exists a basis $b_1, \ldots, b_n$ in $\mathbb{R}^n$ such that

$$L = \sum_{i=1}^n \mathbb{Z} \cdot b_i = \left\{ \sum_{i=1}^n r_i b_i \mid r_i \in \mathbb{Z} \right\}.$$

The *determinant* of the lattice L is the number

$$d(L) = \left| \det(b_1, \ldots, b_n) \right|,$$

where b_i denotes the column-vector of coordinates of the vector b_i. The determinant of the lattice is equal to the volume of the parallelepiped spanned by $b_1, \ldots, b_n$. There are several bases of the lattice but the determinant of transition from one basis to another is equal to ± 1, and so the number $d(L) > 0$ does not depend on the choice of a basis.

Let $b_1, \ldots, b_n$ be a basis of L. The Gram–Schmidt orthogonalization yields an orthogonal (not necessarily orthonormal) basis

$$b_i^* = b_i - \sum_{j=1}^{i-1} \mu_{ij} b_j^*, \quad \text{where} \quad \mu_{ij} = \frac{(b_i, b_j^*)}{(b_j^*, b_j^*)}, \; 1 \leq j < i \leq n.$$

A basis $b_1, \ldots, b_n$ of the lattice L is called *reduced* if

$$|\mu_{ij}| \leq \frac{1}{2} \text{ for } 1 \leq j < i \leq n \text{ and } |b_i^* + \mu_{i,i-1} b_{i-1}^*|^2 \geq \frac{3}{4} |b_{i-1}^*|^2 \text{ for } 1 < i \leq n.$$

Theorem 8.1.1 (Hadamard's inequality). *The volume of a given paral-lelepiped does not exceed the product of the length of its edges, i.e.,*

$$d(L) \leq \prod_{i=1}^{n} |b_i|.$$

Proof. The vectors b_i^* are orthogonal, and so

$$|b_i|^2 = |b_i^*|^2 + \sum_{j=1}^{i-1} \mu_{ij}^2 |b_j^*|^2 \leq |b_i^*|^2.$$

Moreover, $d(L) = \prod_{i=1}^{n} |b_i^*|.$ $\square$

In the following theorem we have collected the main inequalities for the vectors of a reduced basis.

Theorem 8.1.2. *Let $b_1, \ldots, b_n$ be a reduced basis of the lattice L. Then:*
   a) $d(L) \leq \prod_{i=1}^{n} |b_i| \leq 2^{n(n-1)/4} d(L)$.
   b) $|b_j| \leq 2^{(i-1)/2} |b_i^*|$ *for* $1 \leq j \leq i \leq n$.
   c) $|b_1| \leq 2^{(n-1)/4} d(L)^{1/n}$.
   d) *If* $x \in L$ *and* $x \neq 0$, *then* $|b_1| \leq 2^{(n-1)/2} |x|$.
   e) *If the vectors* $x_1, \ldots, x_t \in L$ *are linearly independent, then*

$$|b_j| \leq 2^{(n-1)/2} \max\{|x_1|, \ldots, |x_t|\} \text{ for } 1 \leq j \leq t.$$

Proof. a) From the definition of a reduced basis it follows that

$$|b_i^*|^2 \geq \left(\frac{3}{4} - \mu_{i,i-1}^2 \right) |b_{i-1}^*|^2 \geq \frac{1}{2} |b_{i-1}^*|^2,$$

since $|\mu_{i,i-1}| \leq \frac{1}{2}$.

A simple induction shows that $|b_j^*|^2 \leq 2^{i-j}|b_i^*|^2$ for $i \geq j$. Therefore

$$|b_i|^2 = |b_i^*|^2 + \sum_{j=1}^{i-1} \mu_{ij}^2 |b_j^*|^2 \leq |b_i^*|^2 \left(1 + \frac{1+2+\cdots+2^{i-2}}{2}\right) = |b_i^*|^2 \left(\frac{1+2^{i-1}}{2}\right).$$

Hence

$$\prod_{i=1}^{n} |b_i|^2 \leq \frac{1+1}{2} \cdot \frac{1+2}{2} \cdot \ldots \cdot \frac{1+2^{n-1}}{2} \prod_{i=1}^{n} |b_i^*|^2 \leq 1 \cdot 2 \cdot \ldots \cdot 2^{n-1} d(L)^2 = 2^{n(n-1)} d(L)^2.$$

b) From the inequalities

$$|b_j^*|^2 \leq 2^{i-j}|b_i^*|^2 \quad \text{for } i \geq j$$

$$|b_j|^2 \leq \frac{1+2^{j-1}}{2}|b_j^*|^2 \leq 2^{j-1}|b_j^*|^2$$

we deduce that

$$|b_j|^2 \leq 2^{i-j} \cdot 2^{j-1}|b_j^*|^2 = 2^{i-1}|b_i^*|^2 \quad \text{for } i \geq j.$$

c) Thanks to b) we have

$$|b_1|^2 \leq |b_1^*|^2,$$
$$|b_1|^2 \leq 2|b_2^*|^2,$$
$$\ldots\ldots\ldots\ldots\ldots$$
$$|b_1|^2 \leq 2^{n-1}|b_n^*|^2.$$

The product of these inequalities yields

$$|b_1|^{2n} \leq 2^{n(n-1)/2} \prod_{i=1}^{n} |b_i^*|^2 = 2^{(n-1)/2} d(L)^2.$$

d) Express x in the form $x = \sum_{i=1}^{n} r_i b_i = \sum_{i=1}^{n} s_i b_i^*$, where $r_i \in \mathbb{Z}$ and $s_i \in \mathbb{R}$. Let i_0 be the greatest index for which $r_i \neq 0$. Then $r_{i_0} = s_{i_0}$. Hence

$$|x|^2 \geq s_{i_0}^2 |b_{i_0}^*|^2 = r_{i_0}^2 |b_{i_0}^*|^2 \geq |b_{i_0}^*|^2 \geq 2^{1-i_0}|b_1|^2 \geq 2^{1-n}|b_1|^2.$$

e) The vectors $x_1, \ldots, x_t$ cannot all belong to the subspace spanned by $b_1, \ldots, b_{t-1}$. Hence, for a vector x_s, we have $|x_s| \geq |b_i^*|^2$, where $i \geq t$ (see d)). Therefore, for $j \leq i$, we obtain

$$|x_s|^2 \geq |b_i^*|^2 \geq 2^{j-i}|b_j^*|^2 \geq 2^{j-i} \cdot 2^{1-j}|b_j|^2 = 2^{1-i}|b_j|^2 \geq 2^{1-n}|b_j|^2. \quad \square$$

Let us now describe the algorithm for calculating a reduced basis of the lattice L. Suppose that the vectors $b_1, \ldots, b_{k-1}$ form a reduced basis of the lattice they span (we start with one vector, i.e., with $k = 2$). We adjoin to them a vector b_k which belongs to L but does not lie in the subspace spanned by $b_1, \ldots, b_{k-1}$. The construction of the basis of the lattice generated by $b_1, \ldots, b_k$ is performed as follows.

Step 1 (Fulfilling the condition $|\mu_{kj}| \leq \frac{1}{2}$.) Recall that $\mu_{ij} = \frac{(b_i, b_j^*)}{(b_j^*, b_j^*)}$. Start with $l = k$ and suppose that $|\mu_{kj}| \leq \frac{1}{2}$ for $l < j < k$. Replace b_k by $b_k - q b_l$, where q is the integer nearest to μ_{kl}. This transformation preserves μ_{kj} for $j > l$ (since $b_j^* \perp b_l$ for $l < j$) and replaces μ_{kl} by $\mu_{kl} - q$ since $(b_l, b_l^*) = (b_l^*, b_l^*)$. Clearly, $|\mu_{kl} - q| \leq \frac{1}{2}$, and so after such a modification of the b_i the condition $|\mu_{kj}| \leq \frac{1}{2}$ will be satisfied for $l - 1 < j < k$. Now repeat the operation.

Step 2 (Fulfilling the condition $|b_k^*|^2 \geq (\frac{3}{4} - \mu_{k,k-1}^2)|b_{k-1}^*|^2$.) Suppose that

$$|b_k^*|^2 < (\frac{3}{4} - \mu_{k,k-1}^2)|b_{k-1}^*|^2.$$

Then we replace the ordered set $(b_1, \ldots, b_{k-2}, b_{k-1}, b_k)$ by the ordered set $(b_1, \ldots, b_{k-2}, b_k, b_{k-1})$. Under this change b_{k-1}^* will be replaced by $b_k^* + \mu_{k,k-1}b_{k-1}^*$, and so $|b_{k-1}^*|^2$ will be replaced by

$$|b_k^*|^2 + \mu_{k,k-1}^2|b_{k-1}^*|^2 < \left(\frac{3}{4} - \mu_{k,k-1}^2\right)|b_{k-1}^*|^2 + \mu_{k,k-1}^2|b_{k-1}^*|^2 = \frac{3}{4}|b_{k-1}^*|^2.$$

Let us consider the reduced basis $b_1, \ldots, b_{k-2}$ and apply the first step of the algorithm to it. Since $|b_{k-1}^*|^2$ decreases it follows that the algorithm converges (for the rigorous proof of the convergence of the algorithm, see [Le2] and [Co3]).

8.1.3 The lattices and factorization of polynomials

We recall that it remains to construct an algorithm which, for an irreducible divisor h (mod p) of the polynomial f (mod p) without multiple divisors, computes an irreducible divisor h_0 of f for which h_0 (mod p) is divisible by h (mod p). In doing so, we may assume that h is a monic polynomial.

In intermediate calculations we have to consider divisibility modulo p^k. We therefore consider a more general case: let us assume that h (mod p^k) is an irreducible divisor of the polynomial f (mod p^k) without multiple divisors, the polynomial h is monic and h_0 is an irreducible divisor of f for which h_0 (mod p) is divisible by h (mod p). It is easy to verify that in this case h_0 (mod p^k) is divisible by h (mod p^k). Indeed, the polynomial $\frac{f}{h_0}$ (mod p) is not divisible by the irreducible polynomial h (mod p), i.e., these polynomials are relatively prime. Therefore

$$\lambda h + \mu \frac{f}{h_0} = 1 - p\nu \quad \text{for some } \lambda, \mu, \nu \in \mathbb{Z}[x].$$

Let us multiply both sides of this equality by $(1 + p\nu + \cdots + p^{k-1}\nu^{k-1})h_0$. As a result, we obtain $\lambda_1 h + \mu_1 f \equiv h_0 \pmod{p^k}$. But $f \pmod{p^k}$ is divisible by $h \pmod{p^k}$, and so $h_0 \pmod{p^k}$ is divisible by $h \pmod{p^k}$.

Let $n = \deg f$ and $l = \deg h$. Fix an integer $m \geq l$ and consider the set of all polynomials with integer coefficients of degree not greater than m which are divisible by p^k modulo h. As we have just shown, h_0 belongs to this set if $\deg h_0 \leq m$.

To $g(x) = a_0 + \cdots + a_m x^m$, we assign the point $(a_0, \ldots, a_m) \in \mathbb{R}^{m+1}$. Under this map the polynomials considered form a lattice L. Clearly, the norm $|g| = \sqrt{\sum a_i^2}$ is just the Euclidean length in $\mathbb{R}^{m+1}$.

The basis of the lattice L consists of the polynomials $p^k x^i$, where $0 \leq i < l$, and the polynomials $h(x)x^j$, where $0 \leq j \leq m - l$. The coordinates of these vectors (polynomials) relative to the basis $1, x, \ldots, x^m$ constitute a matrix of the form $\begin{pmatrix} p^k I_l & * \\ 0 & I' \end{pmatrix}$, where I_l is the $l \times l$ unit matrix and I' is an upper triangular matrix with units on the main diagonal. Hence $d(L) = p^{kl}$.

Before we start describing the algorithm for calculating the polynomial h_0, we prove two theorems which provide the estimates we need.

Theorem 8.1.3. *If a polynomial $b \in L$ is such that $|b|^n \cdot |f|^m < p^{kl}$, then b is divisible by h_0. (In particular, $(f, b) \neq 1$, where (f, b) is the greatest common divisor of f and b.)*

Proof. We may assume that $b \neq 0$. Set $g = (f, b)$. We would like to prove that g is divisible by h_0. To this end, it suffices to prove that $g \pmod p$ is divisible by $h \pmod p$. Indeed, if $g \pmod p$ is divisible by $h \pmod p$ and $\dfrac{f}{g} \in \mathbb{Z}[x]$, then $\dfrac{f}{g}$ is not divisible by h_0 because $h \pmod p$ is a simple (of multiplicity 1) divisor of $f \pmod p$. Hence g is divisible by h_0.

Suppose that $g \pmod p$ is not divisible by $h \pmod p$. The polynomial $h \pmod p$ is irreducible, and so $g \pmod p$ and $h \pmod p$ are relatively prime, i.e., there exist polynomials $\lambda_1, \mu_1, \nu_1 \in \mathbb{Z}[x]$ such that

$$\lambda_1 h + \mu_1 g = 1 - p\nu_1. \tag{1}$$

Let $e = \deg g$ and $m' = \deg b$. Clearly, $0 \leq e \leq m' \leq m$. Set

$$M = \{\lambda f + \mu b \mid \lambda, \mu \in \mathbb{Z}[x], \deg \lambda < m' - e, \deg \mu < n - e\}$$
$$\subset \mathbb{Z} + \mathbb{Z} \cdot x + \cdots + \mathbb{Z} \cdot x^{n+m'-e-1}.$$

Denote by M' the image of M under the natural projection onto

$$\mathbb{Z} \cdot x^e + \cdots + \mathbb{Z} \cdot x^{n+m'-e-1}.$$

First we show that, if the image of $\lambda f + \mu b \in M$ is equal to $0 \in M'$, then $\lambda = \mu = 0$. Indeed, in this case $\deg(\lambda f + \mu b) < e$, but $\lambda f + \mu b$ is divisible by g

and $\deg g = e$. Hence $\lambda f + \mu b = 0$, i.e., $\lambda\left(\dfrac{f}{g}\right) = -\mu\left(\dfrac{b}{g}\right)$. The polynomials $\dfrac{f}{g}$ and $\dfrac{b}{g}$ are relatively prime, and so μ is divisible by $\dfrac{f}{g}$. But $\deg \mu < n - e = \deg \dfrac{f}{g}$. Hence $\mu = 0$, and therefore $\lambda = 0$.

Thus, the projections of the sets $\{x^i f \mid 0 \le i < m' - e\}$ and $\{x^j b \mid 0 \le j < n - e\}$ onto M' are, on the one hand, linearly independent and, on the other hand, generate M'. This means that the projections of these two sets form a basis of the lattice M'. In particular, the rank of the lattice M' is equal to $n + m' - 2e$. Moreover, by Hadamard's inequality $d(M') \le |f|^{m'-e}|b|^{n-e}$.

By the hypothesis, $|f|^m|b|^n < p^{kl}$. Since $m' \le m$, we have

$$d(M') \le |f|^m|b|^n < p^{kl}. \tag{2}$$

Let us show that, if $\nu \in M$ and $\deg \nu < e + l$, then $p^{-k}\nu \in \mathbb{Z}[x]$. Indeed, the polynomial $\nu = \lambda f + \mu b$ is divisible by $g = (f, b)$, and so having multiplied (1) by $(1 + p\nu_1 + \cdots + p^{k-1}\nu_1^{k-1})\dfrac{\nu}{g}$ we obtain

$$\lambda_2 h + \mu_2 h \equiv \frac{\nu}{g} \pmod{p^k}. \tag{3}$$

We consider the situation when $f \pmod{p^k}$ is divisible by $h \pmod{p^k}$. Further, $b \in L$, and so $b \pmod{p^k}$ is divisible by $h \pmod{p^k}$. Since $\nu \in M$, it follows that $\nu = \lambda f + \mu b$, so $\nu \pmod{p^k}$ is divisible by $h \pmod{p^k}$. Therefore (3) implies that $\nu/g \pmod{p^k}$ is also divisible by $h \pmod{p^k}$. But

$$\deg\left(\frac{\nu}{g} \pmod{p^k}\right) < e + l - e = l$$

and $h \pmod{p^k}$ is a monic polynomial of degree l. Hence $\dfrac{\nu}{g} \equiv 0 \pmod{p^k}$, and therefore $\nu \equiv 0 \pmod{p^k}$.

In M', we can select a basis $b_e, b_{e+1}, \ldots, b_{n+m'-e-1}$ such that $\deg b_j = j$. It is easy to verify that $e + l - 1 \le n + m' - e - 1$. Indeed, the polynomial b is divisible by g, and so $e = \deg g \le \deg b = m'$; also $\dfrac{f}{g} \pmod{p}$ is divisible by $h \pmod{p}$, and so

$$l = \deg h \le \deg f - \deg g = n - e.$$

The elements $b_e, \ldots, b_{e+l-1} \in M'$ are obtained from the polynomials that lie in M and are divisible by p^k under the projection that annihilates terms of degree lower than e. Therefore all the coefficients of the polynomials $b_e, \ldots, b_{e+l-1}$ (in particular, their highest coefficients) are divisible by p^k.

Clearly, the discriminant $d(M')$ is equal to the absolute value of the product of the highest coefficients of the polynomials $b_e, \ldots, b_{n+m'-e-1}$, and so it is not less than the product of the absolute values of the highest coefficients of $b_e, \ldots, b_{e+l-1}$. Hence $d(M') \ge p^{kl}$. This contradicts inequality (2). $\square$

In the next theorem we, as earlier, assume that h is a monic polynomial and $f \pmod{p^k}$ is divisible by $h \pmod{p^k}$; L is the lattice of polynomials of degree not higher than m and divisible by h modulo p^k; $l = \deg h$ and $n = \deg f$; h_0 is an irreducible divisor of f for which $h_0 \pmod{p}$ is divisible by $h \pmod{p}$, i.e., $h_0 \pmod{p^k}$ is divisible by $h \pmod{p^k}$.

Theorem 8.1.4. *Let $b_1, b_2, \ldots, b_{m+1}$ be a reduced basis of the lattice L. Suppose that*

$$p^{kl} > 2^{mn/2} \binom{2m}{m}^{n/2} |f|^{m+n}.$$

a) *Then $\deg h_0 \leq m$ if and only if*

$$|b_1| < \sqrt[n]{\frac{p^{kl}}{|f|^m}}.$$

b) *Suppose that, for a basis vector b_j, we have*

$$|b_j| < \sqrt[n]{\frac{p^{kl}}{|f|^m}}. \tag{$*$}$$

Let t be the greatest of all such indices j. Then

$$\deg h_0 = m + 1 - t; \qquad h_0 = \mathrm{GCD}(b_1, \ldots, b_t);$$

and inequality $(*)$ holds for $j = 1, \ldots, t$.

Proof. a) First, suppose that $|b_1| < \sqrt[n]{\dfrac{p^{kl}}{|f|^m}}$, i.e., $|b_1|^n \cdot |f|^m < p^{kl}$. Then by Theorem 8.1.3, the polynomial $b_1 \in L$ is divisible by h_0. On the other hand, the condition $b_1 \in L$ implies that $\deg b_1 \leq m$. Thus $\deg h_0 \leq m$.

Now suppose that $\deg h_0 \leq m$, i.e., $h_0 \in L$. By Corollary of Mignotte's theorem (see page 154), $|h_0| \leq \sqrt{\binom{2m}{m}} |f|$. By applying Theorem 8.1.2 (d) for $x = h_0$, we obtain

$$b_1 \leq 2^{m/2} |h_0| \leq 2^{m/2} \sqrt{\binom{2m}{m}} |f|. \tag{4}$$

By the hypothesis, $2^{mn/2} \binom{2m}{m}^{n/2} |f|^n < \frac{p^{kl}}{|f|^m}$, i.e.,

$$2^{m/2} \sqrt{\binom{2m}{m}} |f| < \sqrt[n]{\frac{p^{kl}}{|f|^m}}. \tag{5}$$

Formulas (1) and (5) yield the inequality desired: $|b_1| < \sqrt[n]{\dfrac{p^{kl}}{|f|^m}}$.

b) Let J be the set of all the indices j for which $(*)$ holds. By Theorem 8.1.3, if $j \in J$, then b_j is divisible by h_0. Therefore $h_1 = \text{GCD}(b_j \mid j \in J)$ is divisible by h_0. Here if $j \in J$, then b_j is divisible by h_1 and $\deg b_j \leq m$, i.e., b_j belongs to the lattice

$$\mathbb{Z} \cdot h_1 + \mathbb{Z} \cdot h_1 x + \cdots + \mathbb{Z} \cdot h_1 x^{m - \deg h_1}.$$

Since the vectors $b_1, \ldots, b_m$ are linearly independent, it follows that

$$|J| \leq m + 1 - \deg h_1, \tag{6}$$

where $|J|$ is the cardinality of J.

By Corollary of Mignotte's theorem (see page 154), we have

$$|h_0 x^i| = |h_0| \leq \sqrt{\binom{2m}{m}} |f| \quad \text{for any } i \geq 0.$$

By definition, $i = 0, 1, \ldots, m - \deg h_0$ for $h_0 x^i \in L$ and these vectors are linearly independent. Theorem 8.1.2 (e) is therefore applicable to them:

$$|b_j| \leq 2^{m/2} |h_0 x^i| \leq 2^{m/2} \sqrt{\binom{2m}{m}} |f| \quad \text{for} \quad 1 \leq j \leq m + 1 - \deg h_0.$$

By the hypothesis, $2^{m/2} \sqrt{\binom{2m}{m}} |f| < \sqrt[n]{\dfrac{p^{kl}}{|f|^m}}$, and so

$$\{1, 2, \ldots, m + 1 - \deg h_0\} \subset J.$$

Since h_1 is divisible by h_0, it follows that $\deg h_1 \leq \deg h_0$, and therefore

$$|J| \geq m + 1 - \deg h_0 \geq m + 1 - \deg h_1. \tag{7}$$

By comparing the inequalities (6) and (7) we see that $\deg h_0 = \deg h_1 = t$ and $J = \{1, 2, \ldots, t\}$.

It remains to verify that $h_0 = \pm h_1$. Since h_0 is a divisor of the polynomial f with content 1, it follows that $\text{cont}(h_0) = 1$. Let $j \in J$ and $d_j = \text{cont}(b_j)$. By Theorem 8.1.3, the polynomial b_j is divisible by h_0. Therefore $\dfrac{b_j}{d_j}$ is also divisible by h_0. Since $h_0 \in L$, we deduce that $\dfrac{b_j}{d_j} \in L$. But b_j is a basis element of the lattice L, and so $d_j = 1$. This means that $\text{cont}(h_1) = 1$, since b_j is divisible by h_1. Thus h_1 is divisible by h_0 and $\text{cont}(h_1) = 1$, and so $h_0 = \pm h_1$. $\square$

Now we are able to describe the algorithm for calculating the polynomial h_0.

An auxiliary algorithm. (For a fixed m, this algorithm verifies whether or not $\deg h_0 \leq m$. If this is true, then it calculates the polynomial h_0.)

INITIAL DATA:

- polynomial f of degree n;
- a prime p;
- a positive integer k;
- a monic polynomial h for which f (mod p^k) is divisible by h (mod p^k); the polynomial h (mod p) is irreducible and f (mod p) is not divisible by h^2 (mod p).

We also assume that the coefficients of h are reduced modulo p^k, in other words, lie between 0 and $p^k - 1$ (here $|h|^2 \leq 1 + lp^{2k}$.)

- a positive integer $m \geq l = \deg h$ such that

$$p^{kl} > 2^{mn/2} \binom{2m}{m}^{n/2} |f|^{m+n}. \tag{8}$$

ALGORITHM'S PERFORMANCE. For a lattice L with basis

$$\{p^k x^i \mid 0 \leq i < l\} \cup \{h^k x^j \mid 0 \leq j < m - l\},$$

we find a reduced basis $b_1, \ldots, b_{m+1}$.

If $|b_1| \geq \sqrt[n]{\dfrac{p^{kl}}{|f|^m}}$, then $\deg h_0 > m$ and the algorithm stops.

If $|b_1| < \sqrt[n]{\dfrac{p^{kl}}{|f|^m}}$, then $\deg h_0 \leq m$ and $h_0 = \mathrm{GCD}(b_1, \ldots, b_t)$, where t is determined in the formulation of Theorem 8.1.4 (b).

Main algorithm. (The algorithm calculates an irreducible divisor h_0 of f such that h_0 (mod p) is divisible by h (mod p)).

We may assume that $l = \deg h < \deg f = n$ and the coefficients of h are reduced modulo p, i.e., lie between 0 and $p - 1$.

ALGORITHM'S PERFORMANCE. First, we compute the least positive integer k for which inequality (8) holds for $m = n - 1$:

$$p^{kl} > 2^{n(n-1)/2} \left(\frac{2(n-1)}{n-1}\right)^{n/2} |f|^{2n-1}.$$

Next, for the factorization $f \equiv hg$ (mod p), we calculate its Henzel's lift $f \equiv \overline{h}\overline{g}$ (mod p^k), where k is the positive integer just computed. Here $\overline{h} \equiv h$ (mod p) and the coefficients of $\overline{h}$ are assumed to be reduced modulo p^k.

Let u be the largest integer for which $l \leq \frac{n-1}{2^u}$. We consecutively execute the auxiliary algorithm for $m = \left[\frac{n-1}{2^u}\right], \left[\frac{n-1}{2^{u-1}}\right], \ldots, \left[\frac{n-1}{2}\right], n - 1$ until we compute h_0. If we are unable to compute h_0, then $\deg h_0 > n - 1$ and $h_0 = f$, i.e., f is irreducible.

References

[Ad]   Adams W. W., Loustaunau Ph., *An introduction to Gröbner bases*, AMS, 1994.

[Al]   Cassels J. W. S., Fröhlich A. (eds.), *Algebraic number theory*, Academic Press, 1967.

[An1] Anderson B., *Polynomial root dragging*, Amer. Math. Monthly **100** (1993), 864–866.

[An2] Anderson N., Saff E. B., Varga R. S., *On the Eneström–Kakeya theorem and its sharpness*, Linear Algebra and Appl. **28** (1979), 5–16.

[An3] Andrushkiw J. W., *Polynomials with prescribed values at critical points*, Bull. Amer. Math. Soc. **62** (1956), 243.

[Ar1] Artin E., *Über die Zerlegung definiter Funktionen in Quadrate*, Abh. Math. Sem. Hamburg **5** (1927), 100–115.

[Ar2] Artin E., Schreier O., *Algebraische Konstruktion reeler Körper*, Abh. Math. Sem. Hamburg **5** (1927), 85–99.

[As]   Askey R., *An inequality for Tchebycheff polynomials and extensions*, J. Approx. Theory **14** (1975), 1–11.

[Ay]   Ayoub R. G., *On the nonsolvability of the general polynomial*, Amer. Math. Monthly **89** (1982), 397–401.

[Az]   Aziz A., *On the zeros of composite polynomials*, Pacific J. Math. **103** (1982), 1–7.

[Be1] Beckenbach E. F., Bellman R. *Inequalities*, Springer, Berlin, 1961.

[Be2] van den Berg F. J., *Nogmaals over afgeleide Wortelpunten*, Nieuw Archiev voor Wiskunde **15** (1888), 100–164.

[Be3] Berg L., *Abschätzung von Nullstellen eines Polynoms*, Z. angew. Math. Mech. **67** (1987), 57–58.

[BR1] Bergkvist T., Rullgård H., On polynomial eigenfunctions for a class of differential operators. Math. Res. Lett. 9 (2002), no. 2-3, 153–171

[BR2] Bergkvist T., Rullgård H., and Shapiro B., On Bochner-Krall polynomial systems (accepted to Math. Scand.), arXive: math.SP/0209064

[Be4] Berlekamp E. R., *Factoring polynomials over finite fields*, Bell System Tech. J. **46** (1967), 1853–1859.

[Be5] Berlekamp E. R., *Factoring polynomials over large finite fields*, Math. Comp. **24** (1970), 713–735.

[Be6] Bernoulli J., *Ars conjectandi*, Basileae, 1713.

[Be7]   Beukers F., *A note on the irrationality of* $\zeta(2)$ *and* $\zeta(3)$, Bull. London Math.
         Soc. **11** (1979), 268–272.

[BCS]   Bochner S. Über Sturm-Liouvillesche Polynomsysteme, Math. Z., 29 (1929)
         730–736. For a modern exposition see, e.g., Chihara T. S., *An introduction
         to orthogonal polynomials*, Gordon and Breach, New York, 1977 and Niki-
         forov A. F., Suslov S. K., Uvarov V. B. *Classical orthogonal polynomials
         of a discrete variable.* Springer-Verlag, Berlin, 1991. xvi+374 pp.

[BS]   Borevich Z. I., Shafarevich I. R. Теория чисел. (Russian) [The theory
         of numbers] Third edition. Nauka, Moscow, 1985. 504 pp. MR 88f:11001;
         Borevich Z. I., Shafarevich I. R. *Number theory.* Translated from the Rus-
         sian by Newcomb Greenleaf. Pure and Applied Mathematics, Vol. 20, Aca-
         demic Press, New York-London, 1966, x+435 pp.

[JB6]   Borcea J. On the Sendov conjecture for polynomials with at most six distinct
         roots, J. Math. Anal. Appl., 200 (1996), 182–206.

[JB7]   Borcea J. The Sendov conjecture for polynomials with at most seven distinct
         zeros, Analysis, 16 (1996), 137–159

[JB8]   Borcea J. Maximal and inextensible polynomials, and the geometry of the
         spectra of normal operators, arXive: math.CV/0309233

[BX]   Brown J. E., Xiang G. Proof of the Sendov conjecture for polynomials of
         degree at most eight, J. Math. Anal. Appl., 232 (1999), 272–292.

[Bo]   Boyd D. W., *Sharp inequalities for the product of polynomials*, Bull. London
         Math. Soc. **26** (1994), 449–454.

[Br]   Brauer A., *On algebraic equations with all but one root in the interior of the
         unit circle*, Math. Nachrichten **4** (1950/51), 250–257.

[Bu1]   Buchberger B., *Ein Algorithmus zum Auffinden der Basiselemente des
         Restklassenringes nach einem nulldimensionalen Polynomideal*, Ph.D. The-
         sis, Inst. University of Innsbruck, Austria, 1965.

[Bu2]   Buchberger B., *Gröbner bases: An algorithmic method in polynomial ideal
         theory*, in: N. K. Bose (ed.), *Multidimensional systems theory*, Reidel, Dor-
         drecht, 1985, 184–232.

[Bu3]   Buchberger B., Winkled F. (Eds.), *Gröbner bases and applications*, London
         Math. Soc. Lecture Notes Series, V. 251, 1998.

[Bu4]   Burnside W. S., Panton A. W., *The theory of equations*, Dublin Univ. Press,
         Dublin-London, 1928.

[Ca1]   Cahen P.-J., Chabert J.-L., *Integer-valued polynomials*, AMS, 1997.

[Ca2]   Cantor D. G., Zassenhaus H., *A new algorithm for factoring polynomials
         over finite fields*, Math. Comp. **36** (1981), 587–592.

[Ca3]   Cartier P.,Tate J., *A simple proof of the main theorem of eliminating theory
         in algebraic geometry*, L'Enseignement Math. **24** (1978), 311–317.

[Ca4]   Cassels J. W. S., *An introduction to Diophantine approximation*, Cambridge
         Univ. Press, Cambridge, 1957.

[Ca5]   Cassels J. W. S., *On the representation of rational functions as sums of
         squares*, Acta Arith. **9** (1964), 79–82.

[Ca6]   Cassels J. W. S., Ellison W. J., Pfister A., *On sums of squares and on elliptic
         curves over function fields*, J. Number Theory **3** (1971), 125–149.

[Ca7]   Cassels J. W. S., *Rational quadratic forms*, Academic Press, 1978.

[Ch1]   Chandrasekharan K., *Introduction to analytic number theory*, Springer,
         Berlin, 1968.

[Ch]   Chebotarev N.G., *Foundations of Galois theory*, ONTI, Moscow-Leningrad,
         1937 (in Russsian).

[Ch2]   Christie M. R., *Positive definite rational functions of two variables which are not the sum of three squares*, J. Number Theory **8** (1976), 224–232.

[Co1]   Colliot-Thélène J.-L., *The Noether–Lefschetz theorem and sums of 4 squares in the rational function field* $\mathbb{R}(x, y)$, Compositio Math. **86** (1993), 235–243.

[Co2]   Cohen G. L., Smith G. H., *A simple verification of Ilieff's conjecture for polynomials with three zeros*, Amer. Math. Monthly **95** (1988), 734–737.

[Co3]   Cohen H. *A course in computational algebraic number theory*, Springer, Berlin, 1993.

[Cr]   Cross G. E., Kannappan Pl. *A functional identity characterizing polynomials*, Aequat. Math. **34** (1987), 147–152.

[Da1]   Das M., *Sur un déterminant á permutation circulaire pour les polynômes de Tchebychev de première espeèe*, C. R. Acad. Sci. Paris **268** (1969), A385–A386.

[Da2]   Davenport H., *Multiplicative number theory*, Markham Publ. Co., Chicago, 1967.

[Da3]   Davenport H., Lewis D. J., Schinzel A., *Polynomials of certain special types*, Acta Arithm. **9** (1964), 108–116.

[Do1]   Dobbertin H., Schmieder G., *Zur Charakterisierung von Polynomen durch ihre Null- und Einsstellen*, Arch. Math. **48** (1987), 337–342.

[Do2]   Dörge K., *Einfacher Beweis des Hilbertschen Irreduzibilitätssatzes*, Math. Ann. **96** (1927), 176–182.

[Du1]   Dubois D. W., *Note on Artin's solution of 17th Hilbert's problem*, Bull. Amer. Math. Soc. **73** (1967), 540–541.

[Du2]   Dumas G., *Sur quelques cas d'irréductibilité des polynômes à coefficients rationels*, J. Math. Pures Appl. **2** (1906), 191–258.

[Ed]   Edwards H. M., *Galois theory*, Springer, New York, 1984.

[EN]   *Encyclopedia of elementary mathematics*, v. 2 (Algebra), GITL, Moscow-Leningrad, 1951 (in Russian).

[EG]   Eremenko A., Gabrielov A., Rational functions with real critical points and the B. and M. Shapiro conjecture in real enumerative geometry. Ann. of Math. (2) 155 (2002), no. 1, 105–129

[Ev]   Evyatar A., *On polynomial equations*, Israel J. Math., **10** (1971), 321–326.

[Fa]   Faulhaber J., *Academia algebrae*, Augsburg, 1631.

[Fr]   Fried M., *On Hilbert's irreducibility theorem*, J. Number Theory, **6** (1974), 211–231.

[GNS]   Galochkin A. I., Nesterenko Yu. V., Shidlovskii A. B. Введение в теорию чисел [Introduction to number theory]. Second edition. Izdatelstvo Moskovskogo Universiteta imeni M. V. Lomonosova [Moscow University Press], Moscow, 1995. 160 pp. (Russian).

[Gal]   Galois É., *Works*, ONTI, Moscow-Leningrad, 1936. (in Russian).
Galois É., *Œuvres mathématiques*. [Mathematical works] Publiées en 1846 dans le Journal de Liouville, suivies d'une étude par Sophus Lie, "Influence de Galois sur le développement des mathématiques" (1895). [Published in 1846 in the Journal de Liouville, followed by a study by Sophus Lie, "Influence of Galois on the development of mathematics" (1895)] With a foreword by J. Liouville. Éditions Jacques Gabay, Sceaux, 1989. 78 pp. (in French)

[Ga1]   Gardner R. B., Govil N. K., *On the location of zeros of a polynomial*, J. Approx. Theory **76** (1994), 286–292.

[Ga2]   Garling D. J. H., *A course in Galois theory*, Cambridge Univ. Press, Cambridge, 1986.

[Ga3]   Gasper G., Rahman M., *Basic hypergeometric series*, Cambridge Univ. Press, Cambridge, 1990.

[Ge1]   Gelfond A. O. *Transcendental and algebraic numbers*. Translated from the first Russian edition by Leo F. Boron Dover Publications, Inc., New York, 1960, vii+190 pp.

[Ge2]   Gelfond A. O. *Calculus of finite differences*. Translated from the Russian. International Monographs on Advanced Mathematics and Physics. Hindustan Publishing Corp., Delhi, 1971. vi+451 pp.

[Gr]   Grosswald E., *Recent applications of some old work of Laguerre*, Amer. Math. Monthly **86** (1979), 648–658.

[Ha]   Haruki Sh., *A property of quadratic polynomials*, Amer. Math. Monthly **86** (1979), 577–578.

[Hi1]   Hilbert D., *Über die Darstellung definierter Formen als Summe von Formenquadraten*, Math. Ann. **32** (1888), 342–350.

[Hi2]   Hilbert D., *Über die Theorie der algebraische Formen*, Math. Ann. **36** (1890), 473–534.

[Hi3]   Hilbert D., *Über die Irreduzibilität ganzer rationaler Funktionen mit ganzzahligen Koeffizienten*, J. reine angew. Math., **110** (1892), 104–129.

[Hi4]   Hilbert D., *Über die vollen Invariantensysteme*, Math. Ann. **42** (1893), 313–373.

[Ho]   Hornfeck B., *Primteiler von Polynomen*, J. reine angew. Math., **243** (1970), 120.

[Hu]   Hurwitz A., *Über der Vergleich des arithmetischen und des geometrischeh Mittels*, J. fur Math., **108** (1891), 266–268.

[J]   Janusz G. J., *Algebraic number fields*, AMS, 1996.

[Kh]   Khovanskiĭ A. G., *Fewnomials*, AMS, 1991.

[Ki]   Kirillov A. A. Что такое число? (Russian) [What is a number?] Современная Математика для Студентов [Contemporary Mathematics for Students], 4. Nauka, Moscow, 1993. 80 pp. (in Russian).

[KF]   Klein F. Лекции об икосаедре и решении уравнений пятой степени. (Russian) [Lectures on the icosahedron, and solutions of equations of the fifth degree] Translated from the German by A. L. Gorodentsev and A. A. Kirillov. Translation edited and with a preface by A. N. Tyurin. With appendices by V. I. Arnold, J.-P. Serre and A. N. Tyurin. Nauka, Moscow, 1989. 336 pp. (in Russian).
        Klein F., *Lectures on the icosahedron and the solution of equations of the fifth degree*. Translated into English by George Gavin Morrice. Second and revised edition. Dover Publications, Inc., New York, 1956. xvi+289 pp.

[Kl]   Kleiman H., *Irreducibility criteria*, J. London Math. Soc. (2) **5** (1972), 133–138.

[KS]   Kostov V., Shapiro B., On arrangements of roots for a real hyperbolic polynomial and its derivatives, Bulletin des sciences mathématiques, v. 126, 2002, no. 1, 45–60

[Kr1]   Kronecker L., *Zwei Sätze ueber Gleichungen mit ganzzahligen Coefficienten*, J. reine angew. Math. **53** (1857), 173–175.

[Kr2]   Krusemeyer M., *Why does the Wronskian work?* Amer. Math. Monthly **95** (1988), 46–49.

[Ku]   Kubert D., *The universal ordinary distribution*, Bull. Soc. Math. France **107** (1979), 179–202.

[La1]   Landau E., *Über die Darstellung definiter Funktionen als Summe von Quadraten*, Math. Ann. **62** (1906), 290–329.

[La2]   Lang S., *Algebra*. Revised third edition. Graduate Texts in Mathematics, 211. Springer-Verlag, New York, 2002. xvi+914 pp.

[La3]   Lang S., *Fundamentals of Diophantine geometry*, Springer, 1983.

[La4]   Lang S., *Old and new conjectured diophantine inequalities*, Bull. Amer. Math. Soc. **23** (1990), 37–83.

[Le1]   Lehmer D. H., *A new approach to Bernoulli polynomials*, Amer. Math. Monthly **95** (1988), 905–911.

[Le2]   Lenstra A. K., Lenstra H. W., Lovàsz L., *Factoring polynomials with rational coefficients*, Math. Ann. **261** (1982), 515–534.

[Lj]   Ljunggren W., *On the irreducibility of certain trinomials and quadrinomials*, Math. Scand. **8** (1960), 65–70.

[Ma1]   Macdonald I. G., *Symmetric functions and Hall polynomials*, second edition, Oxford Univercity press, 1995.

[Ma2]   Mahler K., *An application of Jensen formula to polynomials*, Mathematika **7** (1960), 98–100.

[Ma3]   Mahler K., *On some inequalities for polynomials in several variables*, J. London Math. Soc. **37** (1962), 341–344.

[Ma4]   Mahler K., *An inequality for a pair of polynomials that are relatively prime*, J. Austral. Math. Soc. **4** (1964), 418–420.

[Mal]   Malamud, S. M., Inverse spectral problem for normal matrices and a generalization of the Gauss-Lucas theorem, arXiv:math.CV/0304158v1-v3.

[Ma5]   Marcus M., Lopes L., *Inequalities for symmetric functions and Hermitian matrices*, Canad. J, Math. **8** (1956), 524–531.

[Ma6]   Marden M., *Geometry of polynomials*, AMS, 1966.

[Ma7]   Marden M., *The search for a Rolle's theorem in the complex domain*, Amer. Math. Monthly **92** (1985), 643–650.

[MS]   Másson G., Shapiro B., *On polynomial eigenfunctions of a hypergeometric-type operator*, Experiment. Math. 10 (2001), no. 4, 609–618.

[Mi1]   Mignotte M., *An inequality about factors of polynomials*, Math. Comput. **28** (1974), 1153–1157.

[Mi2]   Mikusiński J., Schinzel A., *Sur la rèductibilitè de certains trinômes*, Acta Arithm. **9** (1964), 91–95.

[Mi3]   Milnor J., *On polylogarithms, Hurwitz zeta functions, and the Kubert identities*, L'Enseignement Math. **29** (1983), 281–322.

[Mo]   Motzkin T. S., *Algebraic Inequalities*, in: O. Shisha (ed.), *Inequalities*, Academic Press, New York, 1967, 199–203.

[Mu1]   Muirhead R. F., *Some methods applicable to identities and inequalities of symmetric algebraic functions of n letters*, Proc. Edinburgh Math. Soc. **21** (1903), 144–157.

[Mu2]   Mumford D., *Algebraic geometry, I, Complex projective varieties*, Springer, Berlin, 1976.

[My]   Mycielski Y., *Polynomials with pre-assigned values at their branching points*, Amer. Math. Monthly **77** (1970), 853–855.

[Na]   Nathanson M. B., *Catalan's equation in $K(t)$*, Amer. Math. Monthly **81** (1974), 371–373.

[Ne]   Newman D. J., Slater M. *Waring's problem for the ring of polynomials*, J. Number Theory **11** (1979), 477–487.

[O]   O'Hara P. J., *Another proof of Bernstein's theorem*, Amer. Math. Monthly **80** (1973), 673–674.

[Os1]   Osada H., *The Galois groups of the polynomials $x^n + ax^l + b$*, J. Number Theory **25** (1987), 230–238.

[Os2]   Ostrowski A., *Über ganzwertige Polynome in algebraischen Zahlkörpern*, J. reine angew. Math. **149** (1919), 117–124.

[Os3]   Ostrowski A. M. *Solution of equations and system of equations*, Academic Press, N.Y.-London, 1960.

[Oz]   Ozeki K., *A certain property of polynomials*, Aequat. Math. **25** (1982), 247–252.

[Per]   Pereira, R., Differentiators and the geometry of polynomials, J. Math. Anal. Appl. **285** (2003), 336–348.

[Pe]   Perron O., *Algebra II*, de Gruyter, Leipzig, 1933.

[Pf1]   Pfister A., *Multiplikative quadratische Formen*, Arch. Math. **16** (1965), 363–370.

[Pf2]   Pfister A., *Quadratic forms with applications to algebra, geometry and topology*, London Math. Soc. Lecture Notes Series, V. 217, 1995.

[Po1]   Pólya G., *Über ganzwertige ganze Funktionen*, Rend. Circ. Matem. Palermo **40** (1915), 1–16.

[Pol]   Pólya G., Szegö G. *Problems and theorems in analysis. I. Series, integral calculus, theory of functions.* Translated from the German by D. Aeppli. Corrected printing of the revised translation of the fourth German edition. Grundlehren der Mathematischen Wissenschaften [Fundamental Principles of Mathematical Sciences], 193. Springer-Verlag, Berlin–New York, 1978; Vol. II. *Theory of functions, zeros, polynomials, determinants, number theory, geometry.* Revised and enlarged translation by C. E. Billigheimer of the fourth German edition. Springer Study Edition. Springer-Verlag, New York–Heidelberg, 1976.

[Po1]   Postnikov M. M. *Foundations of Galois theory.* Translated by Ann Swinfen with introductory foreword by translation editor P. J. Hilton. International Series of Monographs on Pure and Applied Mathem atcs, Vol. 29, Pergamon Press, Oxford-London-New York-Paris, 1962, x+109 pp.; Postnikov M. M., Fundamentals of Galois theory.Translated from the first Russian edition by Leo F. Boron, with the editorial collaboration of Robert A. Moore. P. Noordhoff, Ltd., Groningen, 1962, 142 pp.; Postnikov M. M., *Fundamentals of Galois theory.* Translated from the first Russian edition by Leo F. Boron, with the editorial collaboration of Robert A. Moore. P. Noordhoff, Ltd., Groningen, 1962, 142 pp.

[Po2]   Postnikov M. M. Устойчивые многочлены. [Stable polynomials], Moscow, Nauka, 1981. 176 pp. (in Russian).

[Po3]   Pourchet Y., *Sur la représentation en somme de carres des polynomes a une indeterminee sur un corps de nombres algèbriques*, Acta Arith. **19** (1971), 89–104.

[Pr1]   Prasolov V. V., *Problems and theorems in linear algebra*, AMS, 1994

[Pr2]   Prasolov V. V., *Intuitive topology*, AMS, 1995.

[Pr3]   Prasolov V., Solovyev Yu., *Elliptic functions and elliptic integrals*, AMS, 1997.

[Pr4]   Prasolov V., Shvartsman O., Азбука римановых поверхностеи [*The alphabet of Riemannian surphaces*], Moscow, Fazis, 1998 (in Russian).

[Ra1]   Rabinowitsch S., *Zum Hilbertschen Nullstellensatz*, Math. Ann. **102** (1929), 520.

[Ra2]   Rabinowitz S., *The factorization of $x^5 \pm x + n$*, Math. Mag. **61** (1988), 191–193.

[Ra3]   Rajwade A. R.,*Squares*, London Math. Soc. Lecture Notes Series, v. 171, 1993.

[Ri]   Risler J.-J., *Une caractérisation des idéaux des variétés algébriques réeles*, C. R. Acad. Sci. **271** (1970), Serie A, 1171–1173.

[Riv]   Rivlin Th. J. *Chebyshev polynomials*, John Wiley & Sons, 1990

[Ro1]   Rogosinski W. W., *Some elementary inequalities for polynomials*, Math. Gaz., v. 39, no. 327 (1955), 7–12.

[Ro2]   Roitman M., *On roots of polynomials and of their derivatives*, J. London Math. Soc. (2) **27** (1983), 248–256.

[Ro3]   Rosen M. I., *Niels Hendrik Abel and equation of the fifth degree*, Amer. Math. Monthly **102** (1995), 495–505.

[Ro4]   Roth K. F., *Rational approximations to algebraic numbers*, Mathematika **2** (1955), 1–20 (Corrigendum p. 168).

[Ru]   Rubinstein Z., *Remarks on a paper by A. Aziz*, Proc. Amer. Math. Soc. **94** (1985), 236–238.

[Sa]   Salmon G., *Lessons introductory to the modern higher algebra*, 5th ed., NY., Chelsea,1964.

[Sc1]   Scharlau W., *Quadratic and hermitian forms*, Springer, Berlin, 1985.

[Sc2]   Schinzel A., *Solution d'un problème de K. Zarankiewicz sur les suites Des puissances consécutives de nombres irrationnels*, Colloc. Math. **9** (1962), 291–296.

[Sc3]   Schinzel A., *Reducibility of polynomials*, Actes Congrès intern. math. 1970. Tome 1, 491–496.

[ScZ]   Schinzel A., Zassenhaus H., *A refinement of two theorems of Kronecker*, Michiga Math. J. **12** (1965), 81–85.

[Sc4]   Schmeisser G., *On Ilieff's conjecture*, Math. Z. **156** (1977), 165–173.

[Sc5]   Schoenberg I. J., *Mathematical time exposures*, MAA, 1982.

[Sc6]   Schur I., *Über die Existenz unendlich vieler Primzahlen in einige speziellen arithmetischen Progressionen*, S.-B. Berlin. Math. Ges. **11** (1912), 40–50.

[Sc7]   Schwarz H., *Über dienige Fälle, in denen Gaussische Reihe $F(\alpha, \beta, \gamma, x)$ eine algebraische Function ihres vierten Elementen ist*, Borchardt's J. **75** (1872).

[Se1]   Serre J.-P., *A course in arithmetic*, N.Y.–Heidelberg–Berlin, Springer–Verlag, 1973.

[Se2]   Serre J.-P., *Lectures on the Mordell — Weil theorem*, Vieweg, 1989.

[SS]   Shapiro B., Shapiro M., *This strange and mysterious Rolle's theorem.* math.CA/0302215, Submitted to Amer. Math. Monthly

[So]   Sonnenschein H., *A representation for polynomials in several variables*, Amer. Math. Monthly **78** (1971), 45–47.

[St1]   Stengle G., *Integral solution of Hilbert's seventeenth problem*, Math. Ann. **246** (1979), 33–39.

[St2]   Stillwell J., *Galois theory for beginners*, Amer. Math. Monthly **101** (1994), 22–27.

[St3]   Strelitz Sh., *On the Routh–Hurwitz problem*, Amer. Math. Monthly **84** (1977), 542–544.

[St4]   Sturmfels B., *Gröbner bases and convex polytopes*, AMS, 1996.

296   References

[Su1]   Sudbery A., *The number of distinct roots of a polynomial and its derivatives*, Bull. London. Math. Soc. **5** (1973), 13–17

[Su2]   Sury B., *The value of Bernoulli polynomials at rational points*, Bull. London Math. Soc. **25** (1993), 327–329.

[Su3]   Suzuki J., *On coefficients of cyclotomic polynomials*, Proc. Japan Acad. **A63** (1987), 279–280.

[Sw]   Swan R. G., *Factorization of polinomials over finite fields*, Pacific J. Math. **12** (1962), 1099–1106.

[Th]   Thom R., *L'équivalence d'une fonction différentiable et d'un polynome*, Topology **3**, Suppl. 2 (1965), 297–307.

[Ta]   Tabachnikov S., *Polynomials*, Fazis, Moscow, 1996 (in Russian).

[Ti]   Tikhomirov V. M., Uspensky V. V., *Ten proofs of the Fundamental Theorem of Algebra*, Matematicheskoe Prosveshchenie (3rd series) **1** (1997), 50–70 (in Russian).

[Tv]   Tverberg H., *On the irreducibility of polynomials taking small values*, Math. Scand. **32** (1973), 5–21.

[vdW]   van der Waerden B. L. *Algebra*. Vol. I, II. Based in part on lectures by E. Artin and E. Noether. Translated from the seventh German edition by Fred Blum and John R. Schulenberger. Springer-Verlag, New York, 1991. xiv+265 pp.; xii+284 pp.

[W]   Wahab J. H., *New cases of irreducibility for Legendre polynomials*, Duke Math. J. **19** (1952), 165–176.

[We]   Weber H., *Lehrbuch der Algebra, Bd. 1–3*, Braunschweig, 1908.

[Wh]   Whiteley J. N., *Some inequalities concerning symmetric functions*, Mathematika **5** (1958), 49–57.

[Wi1]   Wilf H. S., *Perron–Frobenius theory and the zeros of polynomials*, Proc. Amer. Math. Soc. **12** (1961), 247–250.

[Wi2]   Witt E., *Über die Kommutativität endlicher Schiefkörper*, Abh. Math. Sem. Hamburg **8** (1931), 413.

[Ya]   Yang Chung-Chun, *A problem on polynomials*, Rev. Roumaine Math. Pures Appl. **22** (1977), 595–598.

Index